AMERICAN MACHINIST GEAR BOOK

Simplified Tables and Formulas for Designing, and Practical Points in Cutting All Commercial Types of Gears

BY

CHARLES H. LOGUE

FORMERLY ASSOCIATE EDITOR AMERICAN MACHINIST
FORMERLY MECHANICAL ENGINEER R. D. NUTTALL CO.

THOROUGHLY REVISED BY

REGINALD TRAUTSCHOLD, M. E.

THIRD EDITION
SECOND IMPRESSION

McGRAW-HILL BOOK COMPANY, INC.
NEW YORK: 370 SEVENTH AVENUE
LONDON: 6 & 8 BOUVERIE ST., E. C. 4
1922

PREFACE TO THIRD EDITION

The continued demand for a work on gearing along lines as originally planned by Charles H. Logue in 1910 has been so persistent that it has been deemed advisable to enlarge and revise the previous edition of the *American Machinist Gear Book* rather than to publish at the present time a new book on practical gear design and methods of production. In doing so, parts of the original work which are no longer of particular value to the practical man have been omitted and other parts revised to bring the work up to date. The chief additions have consisted of sections devoted to varieties of bevel gears which have demonstrated their commercial value during the past few years—spiral type bevel gears and bevels of the Williams "master form" variety; to an exposition of the Williams System of Internal Gearing; and to the first presentation in book form of the successful development of the rolling process of gear production.

The section devoted to "Costs" in the Revised Edition has been omitted, not so much because it has grown obsolete, but because the developments in production methods have been so marked and radical in recent years that the question of costs has been complicated by a variety of production processes which do not lend themselves to direct comparison. The section on "Practical Points in Gear Cutting" has also been omitted for the reason that gear cutting has now been very largely superseded by gear generation and the helpful hints on gear cutting are of little value in modern processes of gear generation.

REGINALD TRAUTSCHOLD.

NEW YORK, N. Y.
September, 1922.

PREFACE TO FIRST EDITION

This book has been written to fill a pressing want; to give practical data for cutting, molding, and designing all commercial types, and to present these subjects in the plainest possible manner by the use of simple rules, diagrams, and tables arranged for ready reference. In other words, to make it a book for "the man behind the machine," who, when he desires information on a subject, wants it accurate and wants it quick, without dropping his work to make a general study of the subject. At the same time a general outline of the underlying principles is given for the student, who desires to know not only how it is made, but what is made. Controversies and doubtful theories are avoided. Tables and formulas commonly accepted are given without comment. A great deal of this matter has previously been published in the columns of the AMERICAN MACHINIST, but is revised to make the subject more complete. Credit is given in all cases when the author is known: there may be cases however, where record of the original source of information has been lost, as is often the case when data are in daily use and the authority is obscure. Obviously in such cases the author's name cannot be given.

CHARLES H. LOGUE.

May 1, 1910.

PREFACE TO SECOND EDITION

The first edition of this work, published some five years ago, met with gratifying success as it filled a widespread demand for reliable information on the subject of gearing. The last few years have added much to our knowledge concerning certain types of gears however, so that the present is a fitting time to revise and bring this work up to date. In doing so, the original plan of the work has been followed and much of its text retained intact, while considerable new matter has been added. The sections on bevel, worm, helical, and skew bevel gears have been rewritten. The section on pattern work and molding has been omitted and one touching upon the cost of standard gears substituted therefor.

REGINALD TRAUTSCHOLD.

September 1, 1915.

CONTENTS

AMERICAN MACHINIST GEAR BOOK

SECTION I

Tooth gearing furnishes an efficient and simple means of transmitting power at a constant speed ratio, making it possible to time the movements of machine parts positively.

Owing to refinements in the tooth form, the introduction of generating machines and facilities to cut gears of the largest size accurately, loads may now be transmitted at speeds which a comparatively short time ago were considered prohibitive. The need for gears that would answer the exacting requirements of automobile construction has done much to bring this about. Designing automobile gears, however, is a case of fitting the gears to the machine; it is a question of securing material that will stand the strain; the gear dimensions are practically self determined; however, this is not the only kind of gearing that has been designed after this fashion.

We have an excellent formula for the strength of gear teeth, but it contains a variable factor—the allowance to be made on account of impact—concerning which very little is known. The most important question of all, that of wear, has heretofore been left practically untouched. The best data obtainable has been given. Few records have been kept of actual performances, and nothing whatever has been found relative to the abrasion of different materials in tooth contact.

The various ways in which gears are mounted is responsible for the apparent contradiction of what few data are at hand, as a gear driver which is entirely satisfactory on one machine will be worthless on another at the same load and the same speed.

The circumferential speed that may be allowed for gears of different types is another neglected factor, and last, but not least, what do we know of gear efficiency? In fact the most important information relative to gear transmissions has been entirely a matter of guesswork. It is hardly to be assumed that this will ever be reduced to an exact equation, but there should be some basis from which to form our conclusions.

Gears may be roughly divided into three general classes: Gears connecting parallel shafts; gears connecting shafts at any angle in the same plane; gears connecting shafts at any angle not in the same plane.

In the first class are included spur, helical, herringbone, and internal gears. The second class covers bevel gears only. The third class includes worm, spiral and skew bevel gears.

Gears connecting parallel shafts are the most efficient, and from a point of efficiency may be graded into herringbone, internal, spur, and, lastly, helical gears.

The efficiency of the second class, bevel gears, varies with the shaft angle, increasing as the angle approaches zero.

As a general thing the third class should be avoided wherever possible, although worm gears have their peculiar uses; for instance, where a quiet, self-locking drive is required without reference to the loss of power.

Spiral gears are employed where the load is light and the gear ratio is low, say under 10 to 1; worm gears are often employed for low ratios down to 1 to 1, but are extremely difficult to cut and therefore expensive. When the worm is made much coarser than quadruple thread there is generally trouble.

Skew bevel gears are used where the distance between the shafts is not great enough to employ worm or spiral gears. Skew bevel gears are simpler, and easier to cut than has been generally supposed, but are still things to avoid.

These three general classifications are oommercially subdivided as follows:

KIND	RELATION OF AXES	PITCH SURFACES	NOTES
Spur	Parallel	Cylinders	
Bevel	Intersecting at any angle in the same plane.........	Cones	
Helical	Parallel	Cylinders	
Herringbone	Parallel	Cylinders	Double Helical
Spiral	At any angle not in same plane................	Cylinders	For small ratios
Worm	At any angle not in same plane................	Cylinders	For large ratios
Skew Bevel	At any angle not in same plane................	Hyperboloids	Where shaft centers are close
Internal	Parallel or at any angle in same plane...........	Cylinders or cones	With teeth cut on inner surface
Elliptical	Parallel or at any angle in same plane...........	Elliptical cylinders or elliptical cones	
Irregular	Parallel or at right angles in same plane...........	Any	Irregular pitch lines
Intermittent	Parallel or at right angles in same plane...........	Cylinders	To give driven gear a period, or periods of rest during one revolution of driver
Friction	Parallel or at any angle in same plane...........	Cylinders or cones	Contact surfaces representing the pitch surfaces of a toothed gear

COMMERCIAL CLASSIFICATION OF GEARS

TOOTH PARTS

Fixed axes are connected by imaginary pitch surfaces, which roll upon each other and transmit uniform motion without slipping. The object in toothed gearing is to provide these imaginary surfaces with teeth, the action of which will make the uniform motion of the pitch surfaces positive; not depending upon friction produced by lateral pressure as in friction gears, which are an excellent representation of pitch surfaces.

If the teeth are not so formed that this condition is fulfilled the movement of the driven gear will be made up of accelerations and retardations which will not only absorb a large percentage of the power but disintegrate the material of which the gear is constructed and seriously affect the operation of the machine. Tool marks on planer and boring mill work corresponding to the teeth in the driven gear may be traced directly to this.

There is but one form of tooth in common use—the involute; the cycloidal form has practically disappeared. For a thorough understanding of tooth contact, however, it must be included.

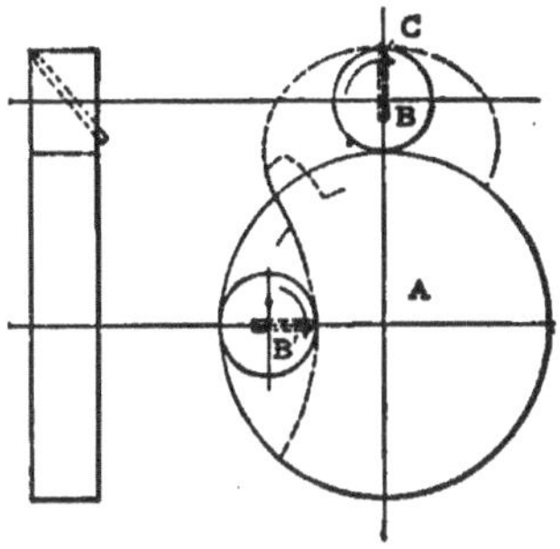

FIG. 1. GENERATING THE CYCLOIDAL TOOTH.

CYCLOIDAL

Generated by rolling a circle above and below the pitch circle of gear; a point on its circumference describing the tooth outline. See Fig. 1.

INVOLUTE

Generated by rolling a straight line on the base circle of gear, any point on this line describing the involute curve. See Fig. 2. The same result is obtained by unwinding a string from the base circle. See Fig. 3.

OCTOID

Conjugated by a tool representing a flat sided crown gear tooth; a modification of the involute. Used only on bevel gear generating machines. See Fig. 33.

THE CYCLOID

An illustration of the manner in which the cycloidal tooth is generated is illustrated by Fig. 1; the wheel *A* being the pitch circle and *B* and *B'* the describing circles which are of the same diameter. The point *C* will describe the face of the tooth as the circle *B* is rolled on the pitch circle, and the flank of the tooth as the circle *B'* is rolled inside the pitch circle. In other words, the exterior cycloid is formed by rolling the describing circle on the outside of the pitch circle, this exterior cycloid engaging the interior cycloid, which is formed by rolling the describing circle on the inside of the pitch circle.

The describing circle is commonly made equal to the pitch radius of a 15-tooth pinion of the same pitch as the gear being drawn.

According to J. Howard Cromwell: "Roomer, a celebrated Danish astronomer, is said to have been the first to demonstrate the value of these curves for tooth profiles." But De la Hire is credited with demonstrating that it was possible to form both the face and flanks of any number of gears with the same describing circle.

The pressure angle of the teeth is not constant in one direction, but varies from zero at the pitch point to about 22 degrees at the end of the contact with a rack tooth. The contact points of all the teeth engaged intersect the line of action, which is a segment of the describing circle drawn from the line of centers. See Fig. 34.

Wilfred Lewis has said: "The practical consideration of cost demands the formation of gear teeth upon some interchangeable system.

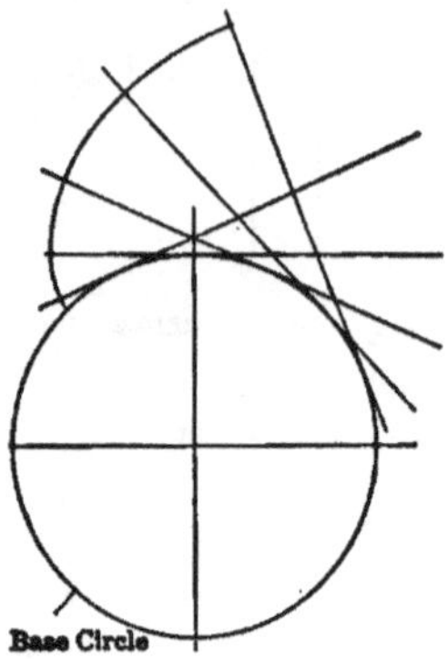

FIG. 2. THE INVOLUTE GENERATED BY A STRAIGHT LINE.

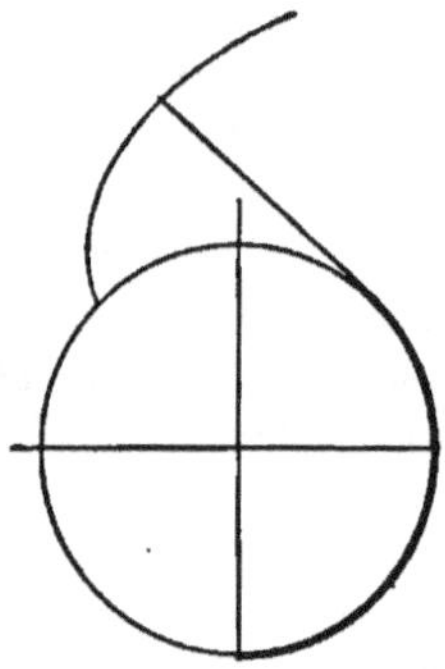

FIG. 3. THE INVOLUTE GENERATED BY A STRING.

"The cycloidal system cannot compete with the involute, because its cutters are formed with greater difficulty and less accuracy, and a further expense is entailed by the necessity for more accurate center distances. Cycloidal teeth must not only be accurately spaced and shaped but their wheel centers must be fixed with equal care to obtain satisfactory results. Cut gears are not only more expensive in this system, but also when patterns are made for castings the double curved faces require far more time and care in chiseling. An involute tooth can be shaped with a straight-edged tool, such as a chisel or a plane, while the flanks of cycloidal teeth require special tools, approximating in curvature the outline desired. It is, therefore, hardly necessary to argue any further against the cycloidal gear teeth, which have been declining in popularity for many years, and the question now to be considered is the angle of obliquity most desirable for interchangeable involute teeth."

In this same connection George B. Grant, of the Philadelphia Gear Works, wrote: "There is no more need of two different kinds of tooth curves for gears of the same pitch than there is need of two different threads for standard screws, or of two different coins of the same value, and the cycloidal tooth would never be missed if it were dropped altogether. But it was first in the field, is simple in theory, is easily drawn, has the recommendation of many well-meaning teachers and holds its position by means of 'human inertia,' or the natural reluctance of the average human mind to adopt a change, particularly a change for the better."

THE INVOLUTE

The pressure on the teeth of involute gears is constantly in the direction of the line of action. The line of action is drawn through the pitch point at an angle from the horizontal equal to the angle of obliquity. All contact between the teeth is along this line. The base circle is drawn inside the pitch circle and tangent to the line of action.

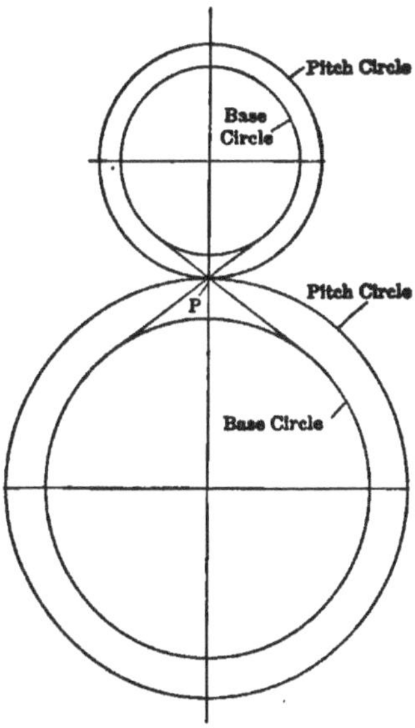

FIG. 4. THE ACTION OF INVOLUTE TEETH ILLUSTRATED BY A CROSSED BELT CONNECTING THE BASE CIRCLES.

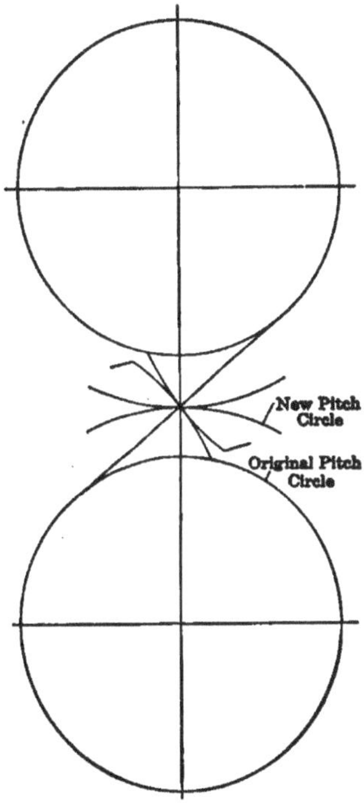

FIG. 5. SEPARATING THE PITCH CIRCLES TO ALLOW THE EXTERIOR CYCLOIDS TO ENGAGE.

The action of a pair of involute gears is the same as if their base circles were connected by a cross belt; the point at which the belt crosses being the pitch point P; the straight portion of the belt not touching the base circles representing the lines of action. See Fig. 4. At the pitch point the velocities of both gears are equal. To show that the involute is but a limiting case of the cycloidal system, consider the describing line as a curve of infinite radius, which is rolled upon the pitch circle. As this describing line cannot be rolled inside the pitch circle to form the interior cycloid that will engage the exterior cycloid formed by rolling the describing line outside the pitch circle of the mating gear, the pitch circles upon which the cycloids are formed must be separated so as to allow the exterior cycloids to engage each other. The original pitch circles becoming the base circles. See Fig. 5.

The distance between the pitch circle and the base circle, and therefore, the angle of obliquity, depends upon the proportionate length of tooth to be used and the smallest number of teeth in the system. To obtain contact for the full length of the tooth, the base circle must fall below the lowest point reached by the teeth of the mating gear. Below the base line there can be no contact of any value.

There is such a difference between the largest possible gear and the rack that it is at first a little difficult to see the application of the methods used to describe the involute to the rack tooth. As the diameter of the gear is increased, the radii used to draw the involute curve are lengthened, and the teeth have less curvature. Until finally, when the radius of the pitch circle is of infinite length, the tooth radii are also infinite, and the involute is a straight line, drawn at right angles to the line of action.

The theoretical rack tooth, therefore, has perfectly flat sides, each side being inclined toward the center of the tooth to an angle equaling the angle of obliquity. See Fig. 6.

ORIGIN OF THE INVOLUTE TOOTH

The origin of the involute curve as applied to the teeth of gears is credited to De la Hire, a French scientist, a complete description and explanation of its use being published about 1694 in Paris. The first English translation of this work was published in London in 1696 by Mandy.* Professor Robinson, of Edinburgh, later describes this theory, references being made to his work in "An Essay on Teeth of Wheels," by Robertson Buchannan, edited by Peter Nicholson and published in 1808. In this essay the involute as applied to the teeth of gears is fully described, Fig. 7 being a copy of a cut used therein for illustration. That the principal advantage of the involute system was then well understood will be shown in the following paragraph, referring to Fig. 7:

"It is obvious that these teeth will work both before and after passing the

* However, the origin of the involute gear tooth is surrounded by mystery, no two authorities agreeing upon the subject. According to Robert Willis, in his "Principles of Mechanism," the involute was first suggested for this purpose by Euler, in his second paper on the Teeth of Wheels. N. C. Petr XI. 209.

line of centers, they will work with equal truth, whether pitched deep or shallow, a quality peculiar to them and of very great importance."

The theory of the involute gear tooth is also described by Sir David Brewster, Dr. Thomas Young, Mr. Thomas Reid and others.

Professor Robert Willis, gives a very complete description of this form of tooth in his "Principles of Mechanism," 1841. Up to this period the involute tooth was not seriously considered, the cycloidal being the favorite. The involute tooth was objected to on account of the great thrust supposed to be put on the bearings by the oblique action of the teeth.

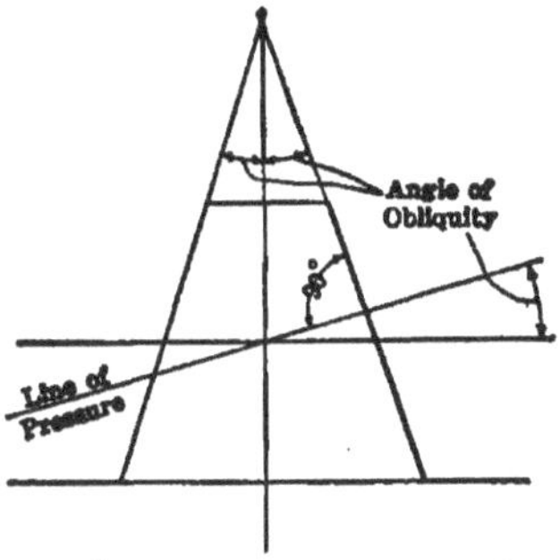

FIG. 6. THE INVOLUTE RACK TOOTH.

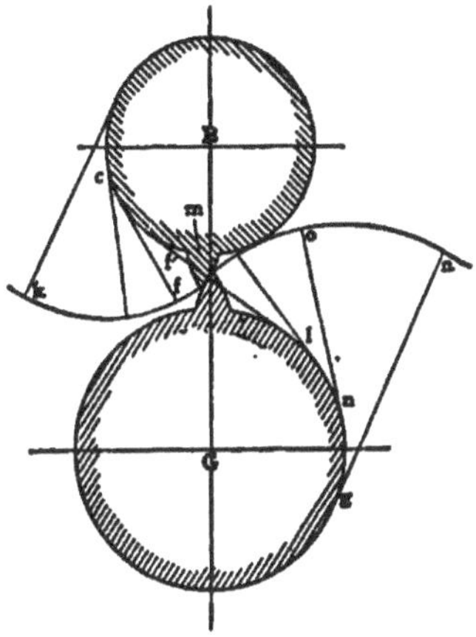

FIG. 7. ACTION OF THE INVOLUTE TOOTH.

In an 1842 edition of M. Camus' work, "A Treatise on the Teeth of Wheels," edited by John I. Hawkins, a series of experiments with wooden models was made to demonstrate the actual thrust occasioned by different angles of obliquity. The result of these experiments is given as follows:

"These experiments, tried with the most scrupulous attention to every circumstance that might affect their result, elicit this important fact—that the teeth of wheels in which the tangent of the surfaces in contact makes a less angle than 20 degrees with the line of centers, possess no tendency to cause a separation of their axes: Consequently, there can be no strain thrown upon the bearings by such an obliquity of tooth."

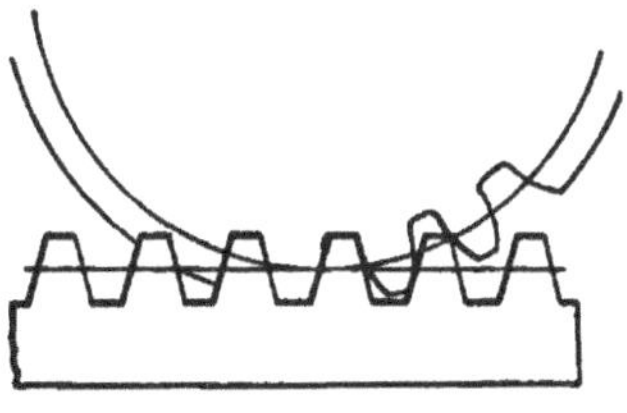

FIG. 8. THE MOLDING PROCESS.

J. Howard Cromwell, in his treatise on Tooth Gearing, 1901, says: "Such an obliquity as 20 degrees must, unless counteracted by an opposing force, tend to separate the axes; and, as suggested by Mr. Hawkins, this opposing force is most probably the friction between the teeth, which would tend to drag the axes together with as much force as that tending to separate them."

That the involute system is closely connected to the cycloidal system is shown by Dr. Brewster in his reference to De la Hire's work.

"De la Hire considered the involute of a circle as the last of the exterior epicycloids; which it may be proved to be, if we consider the generating straight line (see Fig. 2) as a curve of infinite radius."

The 14½ degree angle of obliquity, as proposed by Professor Robert Willis in his "Principles of Mechanism," was adopted by the Brown & Sharpe Company some forty years ago. Since that time this system has come into general use.

THE MOLDING PROCESS

If a gear blank made of some pliable material is forced into contact with a rack, as shown in Fig. 8, the rack tooth would conjugate teeth in the blank.

FIG. 9. ACTION OF THE TOOL IN GENERATING A TOOTH.

It does not matter what form is given the conjugating tooth, as long as it has a regular line of action; all gears formed by it will interchange.

The Bilgram spur and spiral gear generating machine operates upon this principle. See Fig. 9. The cutter *A*, which is a reciprocating or planing tool having the profile of a correct rack tooth—namely, a truncated,

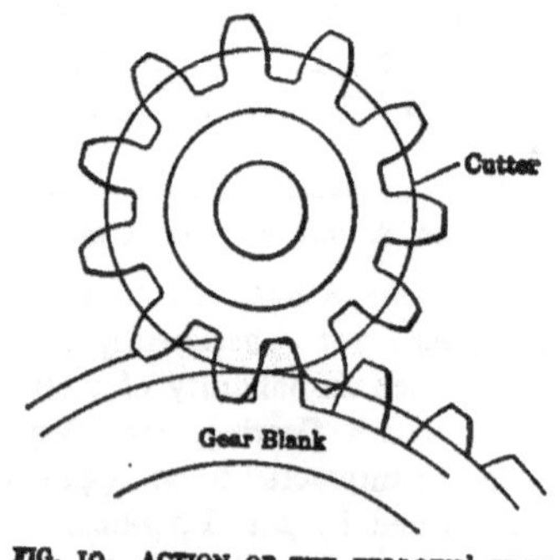

FIG. 10. ACTION OF THE FELLOWS' GEAR CUTTER.

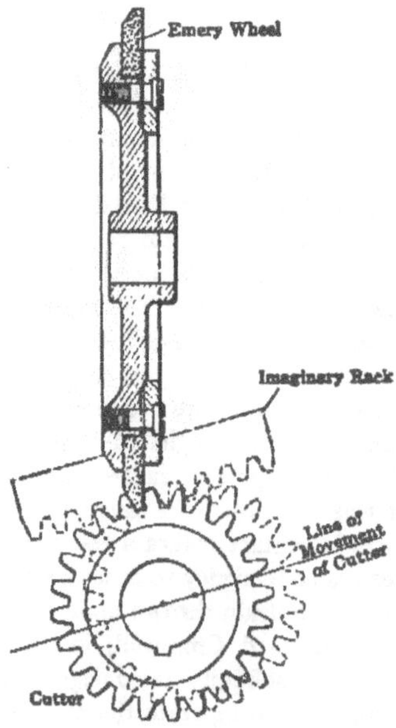

FIG. 11. GENERATION OF THE FELLOWS' GEAR-CUTTER TEETH.

straight-sided wedge. While this tool reciprocates, it also travels slowly to the right, the blank meanwhile turning under it, the motion being that which would exist were the tool a rack tooth and the blank a gear. During this combined movement the tool cuts the tooth space in the manner indicated. In the Bilgram bevel-gear machine the tool does not move sidewise, the blank being rolled upon it as a complete gear might be rolled on a stationary rack, but in the spur-gear machine this action is reversed—the blank turning on a fixed center, while the tool moves over it, as it would be turned by a moving rack.

The Fellows' gear shaper is designed on the same principle, but instead of a rack tooth as a planing tool, a gear of from 12 to 60 teeth is used, the motion of cutter and blank being the same as between gears in mesh. See Fig. 10. These cutters are ground to shape after being hardened as shown in Fig. 11, in which the emery wheel is shaped as the planing tool in Fig. 9. The cutter being ground taking the place of the gear.

TO DRAW THE INVOLUTE CURVE

The involute curve is constructed on the base circle as follows: Draw the pitch circle and through pitch point P, Fig. 12, draw the line of action at the required angle of obliquity. Tangent to this line draw the base circle.

Divide the base circle into any number of equal spaces, 1′, 2′, 3′, 4′, 5′, 6′, as shown in Fig. 13. From each of these points draw lines intersecting at

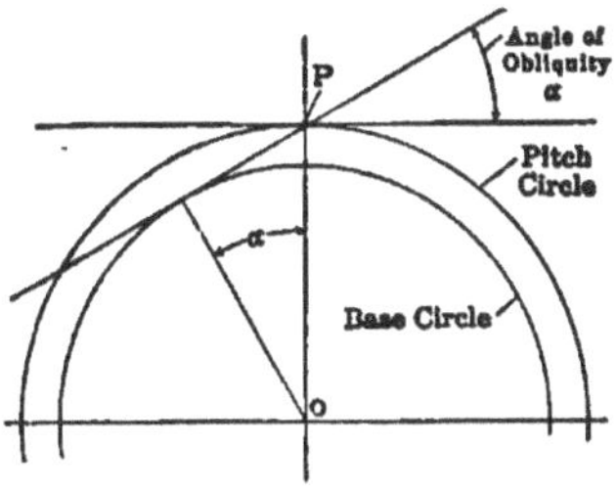

FIG. 12. LOCATING THE BASE CIRCLE.

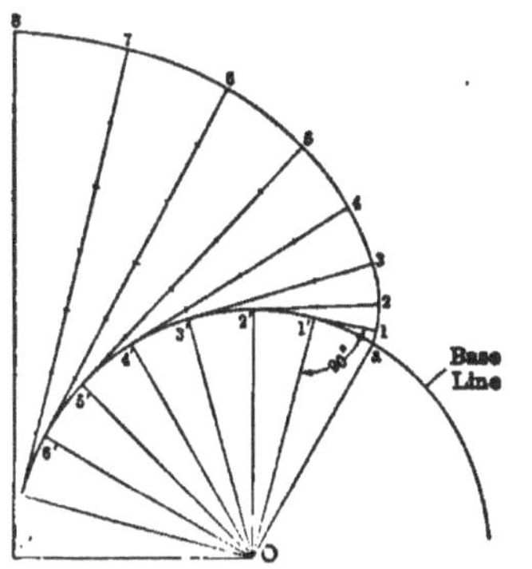

FIG. 13. DRAWING THE INVOLUTE.

center O. Draw lines 1′-1, 2′-2, 3′-3, etc., tangent to base circle and at right angles with lines extending to center. Make the length of line 1′-1 equal to one of the divisions of base circle: Line 2′-2 equal to two divisions, line 3′-3 equal to three divisions, and so on. Then through points 1, 2, 3, 4, 5, 6, etc., trace the involute curve. Find a convenient radius, not necessarily on base circle, from which to draw the balance of the teeth, several radii sometimes being necessary to get the proper curve, especially for a small number of teeth. The involute curve does not extend below the base circle, for within the base circle it is simply a matter of obtaining sufficient clearance to avoid interference with the teeth of the mating gear.

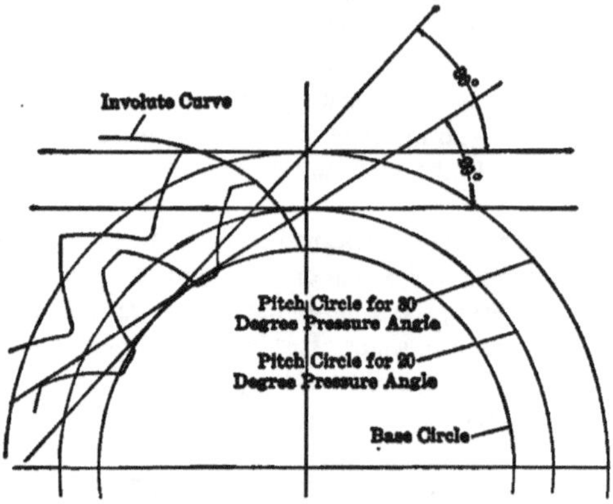

FIG. 14. DIAGRAM SHOWING HOW PROPER ACTION IS MAINTAINED AS THE GEAR AXES ARE SEPARATED.

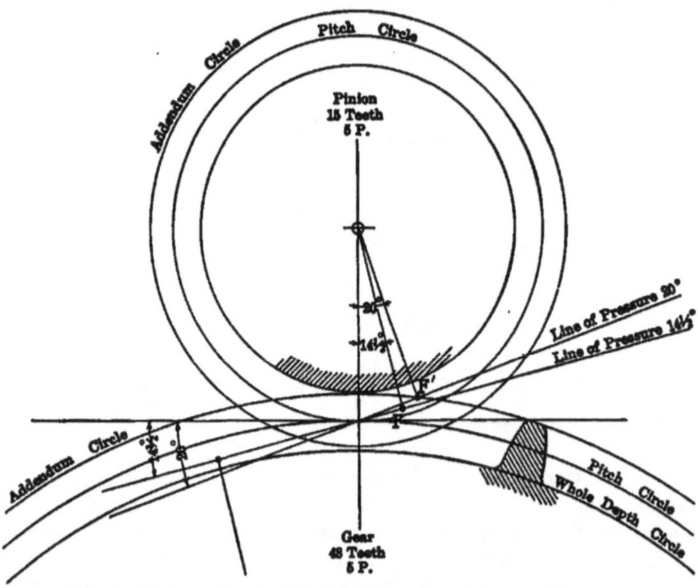

FIG. 15. GRAPHICAL DEMONSTRATION FOR INTERFERENCE OF SPUR GEARS.

The involute curve is always the same for a given base circle diameter, but the angle of obliquity, or the path of tooth contact, is dependent upon the pitch diameter, so the effect of a change in center distance of engaging gears serves simply to alter the obliquity of the path of tooth contact. This is shown clearly in the diagram Fig. 14, in which the angles of obliquity for gear teeth of widely different pitch diameter, but the same base circle diameter, are depicted. It is this peculiarity of the involute system of gearing which permits a certain center adjustment of engaging gears of true involute form without destroying their correct tooth action. The greater the distance between pitch and base circles, the greater becomes the angle of obliquity.

INTERFERENCE IN INVOLUTE GEARS

The limitations and inaccuracies of the involute system are well explained in the following paragraphs by C. C. Stutz:

While the general principles governing the interference of involute gears are well known, the following graphical demonstrations, formulas, and plotted diagrams may place this general information in more efficient form for the use of many.

Fig. 16 shows a graphical demonstration of the interference of a 5-pitch, 15-tooth true involute form spur pinion and a 5-pitch, 48-tooth mating gear. The point F is the right-angled intersection of a line drawn from the center of the pinion, and at an angle of 14½ degrees with the common center line of the pinion and gear, with the line of pressure which is drawn through the point of tangency of the two pitch circles and at an angle of 14½ degrees to the common tangent at that point. If this point falls within the addendum circle of the meshing gear, the tooth of the meshing gear will interfere from this point up to its addendum circle. Therefore the tooth from this point on the curve must be corrected to overcome it.

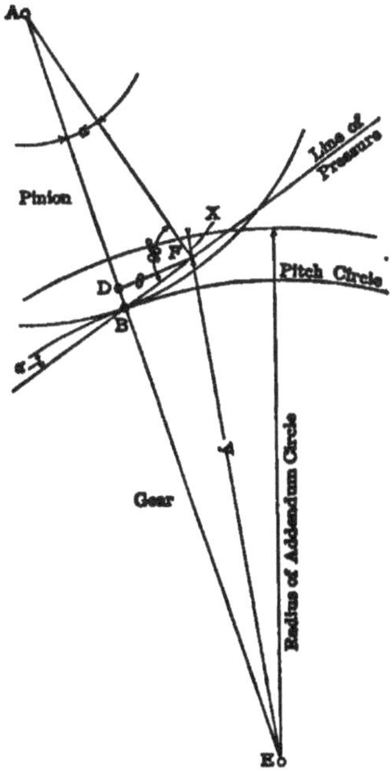

FIG. 16. INTERFERENCE OF SPUR GEARS.

If the point F falls on or outside of the addendum circle of the meshing gear no interference will result. The point F' for an angle of obliquity of 20 degrees falls on the addendum circle and thus the gear and pinion indicated in the illustration would mesh without interference for this angle.

FORMULA FOR LOCATING THE POINT OF INTERFERENCE OF SPUR GEARS

Referring to Fig. 16:

Let $AF = c$. $AB = r_2$. $AD = d$. $BE = r_1$. $DE = f$.

α = the angle of tooth pressure.

y = the distance from the center of the gear to the point at which interference begins.

x = the distance from the point at which interference begins to the addendum circle of the gear measured along a radius.

O = the perpendicular distance from the point at which interference begins to the center line of the pinion and gear.

Then

r_1 = the pitch radius of the gear.

r_2 = the pitch radius of the pinion.

D' = the pitch diameter of the gear.

D = the outside diameter of the gear.

Then

$$c = r_2 \cos \alpha, \text{ and}$$
$$d = c \cos \alpha = r_2 \cos^2 \alpha.$$

Now

$$f = r_1 + r_2 - d.$$
$$= r_1 + r_2 (1 - \cos^2 \alpha),$$

and

$$O = c \sin \alpha = r_2 \sin \alpha \cos \alpha.$$

Now

$$y^2 = f^2 + O^2 \text{ and } y = \sqrt{f^2 + O^2}.$$

Then by substituting

$$y = \sqrt{[r_1 + r_2(1 = \cos^2 \alpha)]^2 + (r_2 \sin \alpha \cos \alpha)^2}$$

For a pressure angle of 14½ degrees

$$y = \sqrt{(r_1 + 0.0627r_2)^2 + (0.2424r_2)^2},$$

and

$$x = \frac{D}{2} - y.$$

For a pressure angle of 20 degrees

$$y = \sqrt{(r_1 + 0.1169\, r_2)^2 + (0.3214\, r_2)^2},$$

and

$$x = \frac{D}{2} - y.$$

Solving for x and y will give the point of interference for any particular case.

DIAGRAM FOR LOCATION OF INTERFERENCE

Fig. 17 shows a diagram giving the location of the point of the beginning of interference for one diametral pitch involute gears from 10 to 135 teeth meshing with a 12-tooth pinion. The ordinates are the distances from the point where interference commences to the addendum circle of the gear measured

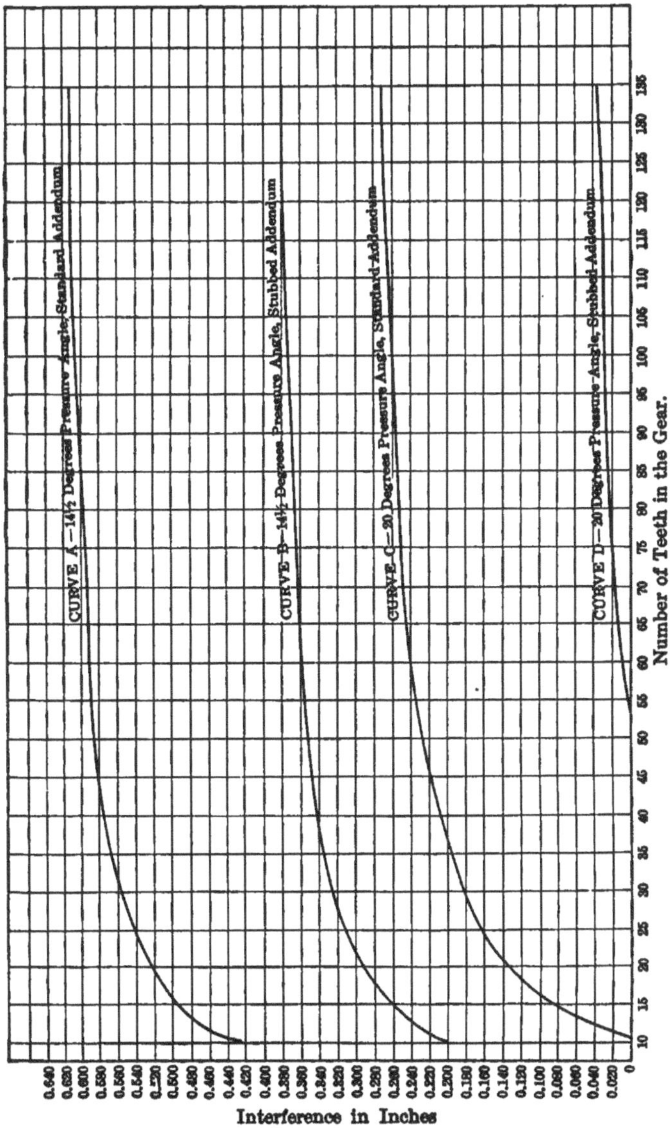

FIG. 17. DIAGRAM SHOWING INTERFERENCE BETWEEN AN INVOLUTE 12-TOOTH, ONE PITCH PINION AND MATING GEARS FROM 10 TO 135 TEETH.

along the radius. They correspond to the quantity x in the preceding equations. From this point to the addendum circle the tooth outline must be corrected.

The upper curve A is for a pressure angle of 14½ degrees and an addendum of 0.3183 × circular pitch. The second, B, is for the same pressure angle and a shorter addendum, 0.25 × circular pitch.

This addendum factor is for what is known as the stubbed tooth standard, as proposed by the author on page 23.

The third curve, C, is for a pressure angle of 20 degrees and an addendum of 0.3183 × circular pitch, while the lowest one, D, is for the 20-degree angle and the stubbed tooth addendum.

The diagram as plotted is for one diametral pitch. To find the corresponding ordinate for any other pitch divide the value given in the diagram by the required pitch. The quotient will be the distance desired.

INTERFERENCE OF RACK AND PINION

Interference will occur between the teeth of a rack and pinion when the point B, Fig. 18, which is the intersection of a perpendicular from the point O to the line of pressure AL falls inside of the rack addendum line EE. In the

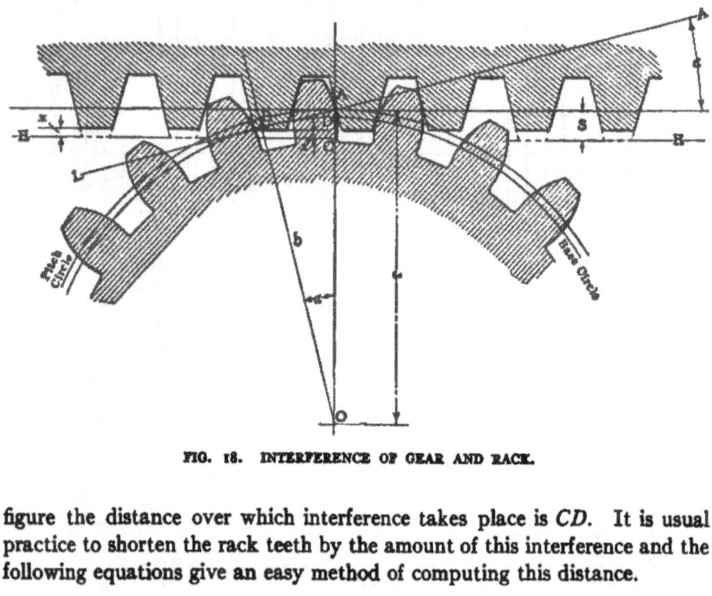

FIG. 18. INTERFERENCE OF GEAR AND RACK.

figure the distance over which interference takes place is CD. It is usual practice to shorten the rack teeth by the amount of this interference and the following equations give an easy method of computing this distance.

Let N = the number of teeth in the pinion.
p = the diametral pitch.
r = the pitch radius.
b = the radius of the base circle.

Let

α = the pressure angle.
x = the distance necessary to shorten the addendum of the rack tooth

and

s = the normal addendum of the rack tooth.

Then

$$s = \frac{1}{p},$$
$$r = \frac{\frac{1}{2}N}{p},$$
$$b = r \cos \alpha,$$
$$OD = b \cos \alpha,$$
$$OD = r \cos^2 \alpha,$$
$$OC = r - s,$$
$$x = OD - OC, \text{ and substituting}$$
$$= r \cos^2 \alpha - (r - s)$$
$$= \frac{\frac{1}{2}N}{p}(\cos^2 \alpha - 1) + \frac{1}{p}:$$

Whence

$$x = \frac{1 - \frac{1}{2}N(1 - \cos^2\alpha)}{p}$$

For a pressure angle of 14½ degrees

$$x = \frac{1 - 0.03135N}{p}.$$

For a pressure angle of 20 degrees

$$x = \frac{1 - 0.05849N}{p}.$$

Solving these equations we find that for the true involute form of tooth and a pressure angle of 14½ degrees interference between the teeth of rack and pinion begins with a pinion of 31 teeth. Similarly for a 20-degree pressure angle the interference begins with a pinion of 17 teeth.

MODERN TOOTH FORM DEVELOPMENTS

The introduction of generating machines, gear shapers and hobbers, for the rapid production of gears with teeth of the involute form has had a very marked effect upon gear development and has brought about radical innovations in the design and form of gear teeth. The limitations and inaccuracies developed in the practical application of the involute system necessitated considerable modification in the profile curvature of the teeth to avoid or to overcome involute interference. Such modifications in the quantity production methods evolved to meet the demand for large numbers of high grade gears of specific sizes, such as is encountered in the automobile, machine tool and other fields, entailed no sacrifice in gear efficiency, as the peculiarity of true involute gearing which permits a certain center adjustment of gears without sacrifice of correct tooth action is no particular advantage. In such mechanisms, the gear centers are definitely fixed and the opportunity of center adjustment does not occur.

The straight line of pressure peculiarity of involute gears, though an advantage in laying out (drawing) such gears, is also sacrificed in modern processes of gear production by generation—the profiles of the gear teeth being modified usually to avoid involute interference—so there remains no particular object in retaining the true involute form of tooth for a gear standard. In fact, there have been developed other highly desirable tooth forms which possess distinctive advantages in the way of ease, economy and accuracy in the actual commercial production of gearing.

Before considering these later developments, however, it may be well to review the theory of gearing which led to the adoption of the involute system as a standard and the theory which justifies a radical departure from the involute form of gear tooth. The objective in all systems of toothed gearing is to secure as nearly as possible that positive uniform motion of the pitch surfaces of the gears upon which the efficiency of the mechanism is so largely dependent. This is basic and underlies the theory upon which the established tooth forms, such as the involute, were developed and also the modern tooth forms, such as that employed in the Williams system of internal gearing and the tooth form employed in the Anderson system of gear rolling.

The chief difference in the older and more modern forms of gear teeth is not so much in the actual form of the teeth, although there is a very noticeable difference in form, as it is in the methods employed to attain the objective of positive uniform pitch surface motion. Before the advent of the gear generating machine and of the modern gear shaper, gear cutting entailed the machining of the individual teeth with the aid of formed cutters, the cutting of each tooth space presenting a separate machining process—a series of similar but nevertheless independent operations. In other words, an attempt was made to form the gear teeth so that they would transmit to the pitch surfaces of engaging gears a positive uniform motion. The gears being machined independently, this necessitated a standardization of gear tooth profiles in order that gears of various sizes might be run together satisfactorily. For various reasons, which need not be enlarged upon, the tooth form having profiles of involute curvature was found to possess distinctive advantages and the involute system of gearing became the generally recognized standard.

The development of gear generating machines brought about by the demand for more rapid and cheaper gear production contributed much to the art of gear cutting and at the same time threw considerable light upon the requirements of tooth form needed for securing positive and uniform motion of pitch surfaces. These machines, whether they employ reciprocating cutters or hobs for machining the gear teeth, are so designed that the motion between the pitch surfaces of the cutting tool and the gear blank is positive and uniform during the machining operations and the teeth formed on the gear blanks are conjugately related. The result is that gears of various size engage one another in a manner to reproduce closely this positive uniform motion of pitch surfaces. In short, the potent difference between gears with teeth cut individually by formed cutters and gears with generated teeth is that in the former case an attempt is made to proportion the teeth so that they will transmit to the gears in operation that desired positive uniform motion of pitch surfaces and in the latter case the positive uniform motion of pitch surfaces is employed in the formation of the teeth—in one case, development toward the objective and in the other, development backward from the objective.

ADAPTATION OF GENERATION TO INVOLUTE GEARING

The cutting of gear teeth of the involute form was naturally the first adaptation of gear generation, the cutting tooth section retaining the true involute rack form in the case of machines employing the principle of the Bilgram machine and subsequently in the case of the more modern gear hobbers and being of developed involute profile in the case of the Fellows' machine. With rack tooth machines and hobbers the generated teeth are obviously free from the involute interference of rack and pinion and that of externally meshing gears which is not so serious. In the case of pinion cutter machines the same result is secured as the pinion cutter teeth are really generated by an involute rack tool, or grinding wheel. See Fig. 11. The effect of this avoidance of tooth interference is that the profiles of the generated teeth depart from the true involute curve over a greater or less distance, depending upon the number of gear teeth and their proportions—particularly in respect to the relative lengths of addendum and dedendum.

In the case of internal gearing in which involute interference is more accentuated than in external gearing or in the rack and pinion, the modification of the profile curvature of the cutter teeth is of necessity considerably more pronounced. In fact, *machining* generation of internal gears is feasible only with pinion type cutters having cutting profiles departing quite noticeably from the involute in curvature. The reason for this greater prevalence of involute interference in internal gearing is that the tooth profiles bounding the spaces between teeth in internal gears are of concave curvature, accentuating the interference occurring between rack and pinion just as the involute interference of externally meshing gears is relatively reduced by reason of the convex curvature of their tooth profiles.

The concave profile curvature of internal gear teeth is well illustrated in

Fig. 19, which depicts a method of interference correction for gears cut with formed tools, and is obvious as well in the equations derived to cover cases in which the pinion of an internal gear combination has less than 55 teeth. These equations establish the length of the various radii for modifying the tooth profiles in order to correct for interference.

INTERFERENCE OF INTERNAL GEAR AND PINION

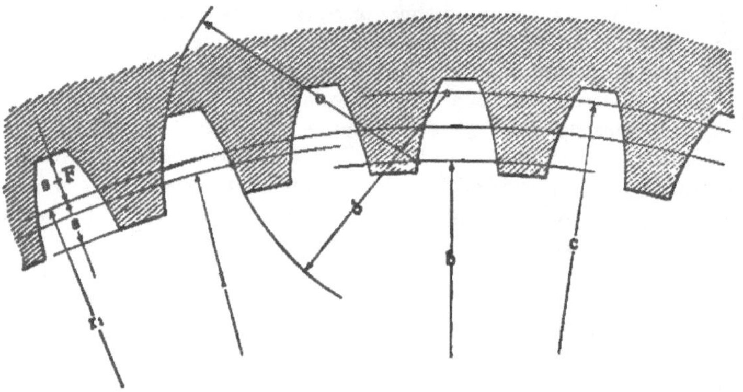

FIG. 19. INTERFERENCE OF INTERNAL GEAR AND PINION.

Referring to Fig. 19:
Let

r_1 = the pitch radius of gear.
r_2 = the pitch radius of pinion.
b = the radius of the base circle.
c = the radius of the correction circle.
d = the radius of the rounding off circle.
l = the radius of the interference circle.
o = the radius of the tooth cutting.
α = the pressure angle.

Then

$$b = r_1 \cos \alpha,$$

$$l = \tfrac{1}{2}(b + r_1),$$

$$c = \frac{r_2}{\cos \alpha} + r_1 r -_2$$

$$o = \tfrac{1}{4} r_1, \text{ and}$$

$$d = \sqrt{\left(\frac{r_1}{4}\right)^2 + \left(\frac{r_2}{4}\right)^2}.$$

These examples of involute interference and the modifications of involute tooth profiles in commercial production of gearing by generation are presented to demonstrate the justification in departure from the former standard involute tooth form in the development of certain modern tooth forms which offer marked advantages in the way of reducing the cost of gear fabrication in large commercial production. Two of these newer forms of gear teeth are deserving of particular notice.

WILLIAMS SYSTEM OF INTERNAL GEARING

The gear teeth in the Williams development of internal gearing are of the involute rack tooth shape, the bounding profiles being straight lines. See Fig. 20. The teeth of the engaging pinions have profiles of conjugate curvature and are considerably more rugged and substantial than those of pinions for internal gears with teeth of involute form. The effect of this modification is a curved line of pressure, or path of tooth contact, which materially lengthens the duration of contact between pairs of pinion and gear teeth.

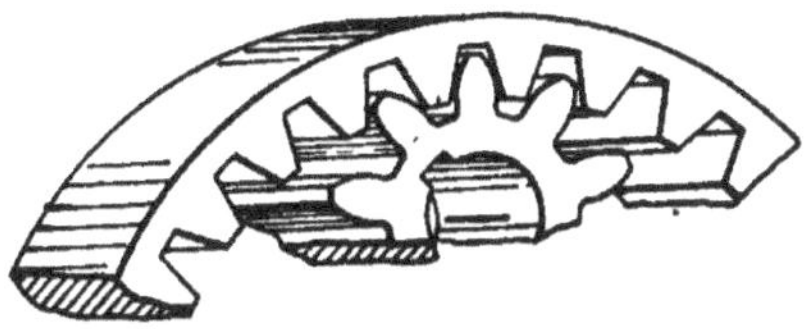

FIG. 20.

ANDERSON ROLLED GEARS

The latest attainment in gear producing methods, that of rolling gear teeth upon heated metal blanks by hardened die rolls under heavy pressure,

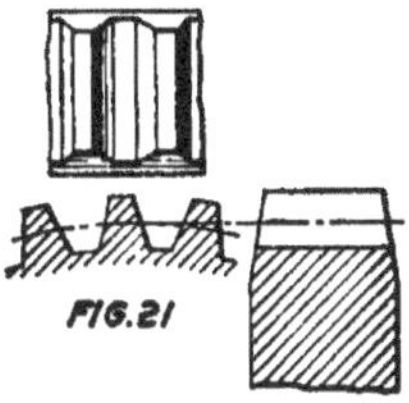

FIG. 21

FIG. 22

FIG. 23

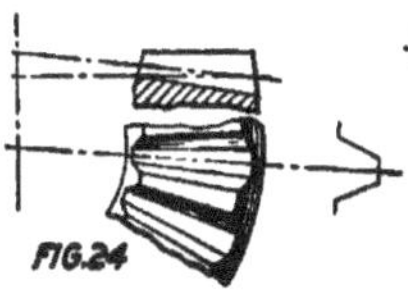

FIG. 24

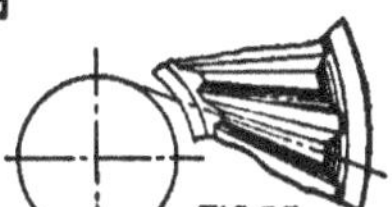

FIG. 25

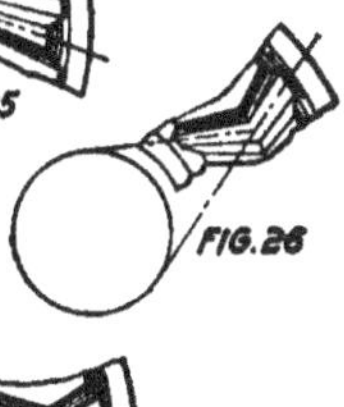

FIG. 26

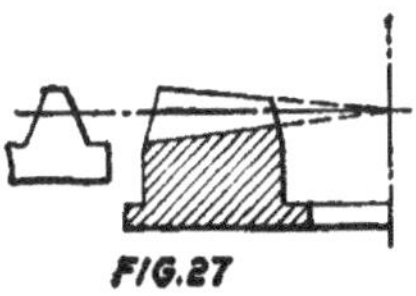

FIG. 27

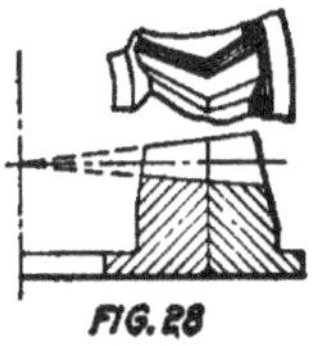

FIG. 28

has also necessitated the adoption of a distinctive form for the functioning die roll teeth. As in the case of the Williams internal gear tooth, the simplest and most easily reproduced form—that of the involute rack tooth has been chosen. This form of tooth (Fig. 21) is in fact a "master form," being basic for the involute system of gearing as well as for those of the Williams internal gears and Anderson rolled gears.

In the production of gears by the hot rolling method, the "master form" is employed for the teeth of the die roll used in forging the teeth on the heated gear blanks and is not, consequently, the exact section of the teeth formed by the moulding generation process. The latter are developed forms which vary in profile curvature somewhat with the class of die roll employed, but they are all developments from the one basic "master form." In the cases of spur, helical and herringbone gears with their pitch surfaces concentric with their axes, the tooth form is very similar to the involute, being modified only by a slight increase in curvature of both addendum and dedendum sections of the tooth profiles. The modification is not apparent to the eye, as in the actual fabrication of gears by this method the number of die roll teeth is considerably greater than the number of teeth rolled on the gear blank, so that the modification in the rolled tooth profile is negligible for all practical purposes.

In the rolling of bevel gears of any type, the same negligible modification occurs when the die roll is of the flat bevel gear form, but no modification from the octoid form of tooth takes place when the die roll is of the crown gear form—see Figs. 21–28 for diagrammatic depictions of Anderson die roll teeth.

EXISTING TOOTH STANDARDS—BROWN & SHARPE'S

The Brown & Sharpe system is perhaps the best known; the angle of obliquity being 14½ degrees.

$$\text{Addendum,} = 0.3183p^1 \text{ or } \frac{1}{p}$$

$$\text{Dedendum,} = 0.3683p^1 \text{ or } \frac{1.157}{p}$$

$$\text{Working depth,} = 0.6366p^1 \text{ or } \frac{2}{p}$$

$$\text{Whole depth,} = 0.6866p^1 \text{ or } \frac{2.157}{p}$$

$$\text{Clearance,} = 0.05p^1 \text{ or } \frac{0.157}{p}$$

In which p^1 = circular pitch, and p = diametral pitch.

GRANT'S

The Grant system has an angle of obliquity of 15 degrees, otherwise it is the same as Brown & Sharpe's. This system is used on the Bilgram generator.

SELLERS'

Wm. Sellers & Co. adopted a form of tooth some 32 years ago in which the angle of obliquity was 20 degrees, with an addendum of 0.3 and a clearance of 0.05 of the circular pitch.

HUNT'S

The C. W. Hunt Co. have a standard in which the angle of obliquity is 14½ degrees; the tooth parts being as follows:

$$\text{Addendum, } = 0.25p^1, \text{ or } \frac{0.7854}{p}$$

$$\text{Dedendum, } = 0.30p^1, \text{ or } \frac{0.9424}{p}$$

$$\text{Working depth, } = 0.50p^1, \text{ or } \frac{1.5708}{p}$$

$$\text{Whole depth, } = 0.55p^1, \text{ or } \frac{1.71278}{p}$$

$$\text{Clearance, } = 0.05p^1, \text{ or } \frac{0.157}{p}$$

THE AUTHOR'S

This system, presented in connection with a discussion of an interchangeable involute gear-tooth system at the December, 1908, meeting of the American Society of Mechanical Engineers, was originally published in AMERICAN MACHINIST, June 6, 1907. Angle of obliquity is 20 degrees. Balance of the tooth parts being the same as the Hunt system described above.

FELLOWS'

The stubbed tooth adopted by the Fellows Gear Shaper Company has an angle of obliquity of 20 degrees. The tooth parts, however, do not bear a definite relation to the pitch; the addendum being made to correspond to a diametral pitch one or two sizes finer, as:

$$\frac{\text{Actual pitch}}{\text{Pitch depth}} = \frac{2}{2\frac{1}{2}}\ \frac{2\frac{1}{2}}{3}\ \frac{3}{4}\ \frac{4}{5}\ \frac{5}{7}\ \frac{7}{9}\ \frac{8}{10}\ \frac{10}{12}\ \frac{12}{14}\ \frac{14}{18}$$

The upper figures indicate the diametral pitch for tooth spacing and the lower figures indicate the diametral pitch from which the depth is taken.

In this system the addendum varies from 0.264 to 0.226 of the circular pitch; 0.25, which is the addendum for the Hunt and the author's standard, is a rough mean.

The author's standard tooth is shown in Fig. 30 for comparison with the 14½-degree standard in Fig. 29.

Wilfred Lewis discussed tooth standards before the American Society of Mechanical Engineers, 1900, as follows:

"About 30 years ago, when I first began to study the subject, the only system of gearing that stood in much favor with machine-tool builders was the cycloidal.

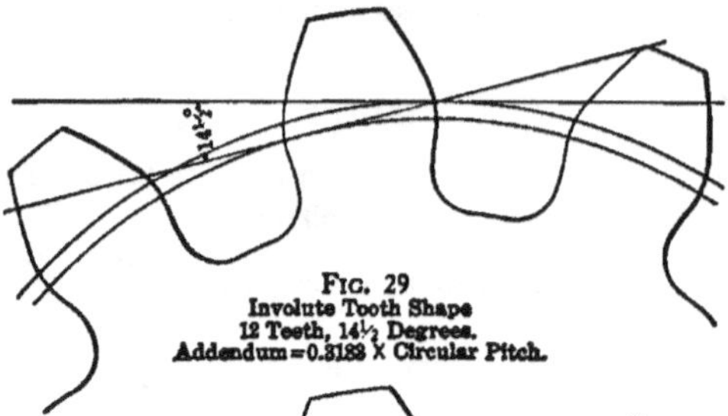

FIG. 29
Involute Tooth Shape
12 Teeth, 14½ Degrees.
Addendum = 0.3183 X Circular Pitch.

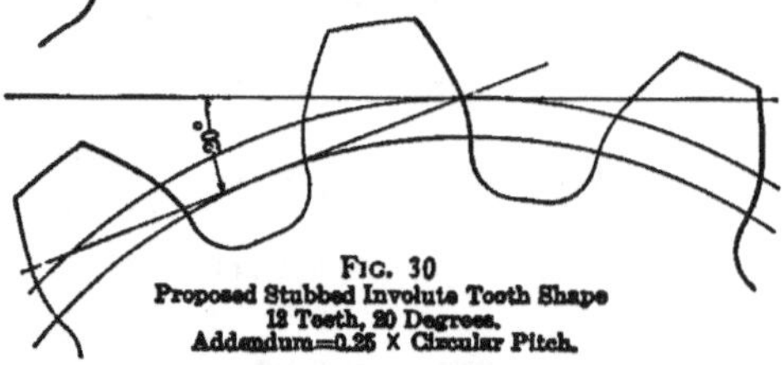

FIG. 30
Proposed Stubbed Involute Tooth Shape
12 Teeth, 20 Degrees.
Addendum = 0.25 X Circular Pitch.

COMPARATIVE FORMS OF 14½-DEGREE AND 20-DEGREE STANDARDS.

"For some time thereafter William Sellers & Co., with whom I was connected, continued to use cylindrical gearing made by cutters of the true cycloidal shape, but the well-known objection to this form of tooth began to be felt, and possibly 25 years ago my attention was turned to the advantages of an involute system. The involute systems in use at that time were the ones here described as standard, having 14½ degrees' obliquity, and another recommended by Willis having an obliquity of 15 degrees. Neither of these satisfied the requirements of an interchangeable system, and with some hesitation I recommended a 20-degree system, which was adopted by William Sellers & Co., and has worked to their satisfaction ever since. I did not at that time have quite the courage of my convictions that the obliquity should be 22½ degrees or one-fourth of a right angle. Possibly, however, the obliquity of 20 degrees may still be justified by reducing the addendum from a value of one to some fraction thereof, but I would not undertake at this time to say which of the two methods I would prefer.

"I brought up the same question nine years ago before the Engineers' Club of Philadelphia, and said at that time that a committee ought to be appointed to investigate and report on an interchangeable system of gearing. We have an interchangeable system of screw threads, of which everybody knows the advantage, and there is no reason why we should not have a standard system of gearing, so that any gear of a given pitch will run with any other gear of the same pitch."

A UNIVERSAL STANDARD

At that time, it was felt that to do away with the then existing multiplicity of standards, especially in connection with the involute form of tooth, and adopt an acceptable standard gear-tooth system would be a much-needed reform. A special committee was appointed by the American Society of Mechanical Engineers, under the chairmanship of Wilfred Lewis, to bring this about, but the committee accomplished little and finally gave up the work. More recently other committees and other organizations have attempted the same reform, but to little avail. The modifications of tooth section found necessary in automobile work and the development of new and efficient forms of gear teeth has had much to do with this and with the attainment of a practical and highly efficient process of rolling gear teeth on heated blanks—the Anderson process, which will be discussed in a subsequent section—the acceptable standard today will of necessity have to disregard the form of the tooth and permit considerable latitude in overall tooth dimensions.

It is desirable, however, to have a gear-tooth system which will permit some uniformity in formulas for computations, even though profile curvatures, etc., may differ. Such an elastic system is even now generally recognized and the subsequent equations and calculations will be based on present day practice as developed for the involute system. The illustrations will also depict involute gearing for the most part, the new gear systems being discussed specifically and individually.

MODIFIED TEETH

A common method of modifying the involute tooth to avoid either interference, undercutting, or the necessity of departing from the true outline is by shortening the dedendum and lengthening the addendum of the pinion tooth. The opposite treatment is given the gear tooth, the dedendum being made deeper to accommodate the added addendum of the pinion and the addendum of the gear correspondingly short. This method is employed on all bevel-gear generating machines for angles less than 20 degrees to avoid interference, the amount of correction depending, of course, upon both the number of teeth being cut and the number of teeth in the engaging gear, or, in other words, depending upon the position of the base line.

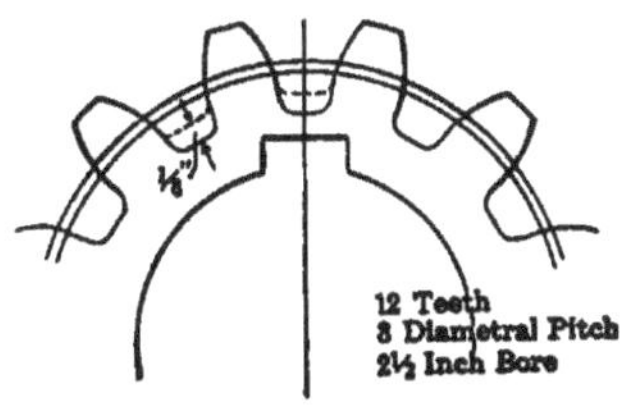

FIG. 31. SHORTENING THE DEDENDUM TO STRENGTHEN KEYWAY.

On bevel-gear generating machines it is the practice to make no modification in the angle for a 20-degree tooth when cutting a depth equal to $0.6866p'$. For this depth of tooth and a pressure angle of 20 degrees interference beginning at 17 teeth, enough roll, however, can be given the blank to allow the generating tool to undercut the flank of the tooth, and avoid interference without any correction of the tooth parts. This is not the case, however, when cutting the standard 14½ or 15 degrees on account of limitation in the movements of the machine. This modification in the tooth parts for bevel gears is accomplished by shifting the face angles and outside diameters, the pinion being enlarged and the gear reduced.

The dedendum of the pinion is sometimes shortened for another reason: Often the bore is so large as to leave insufficient stock between the bottom of the teeth and the keyseat. See Fig. 31. When the pinion cannot be enlarged or the bore reduced the only possible recourse is to shorten the dedendum, taking the amount shortened from the point of the gear tooth. This practice is not to be recommended although extensively used; it would be much better to apply the short tooth of increased obliquity to such cases.

THE OCTOID

All bevel-gear generating machines operate on the octoid system, and not the involute, as is generally supposed.

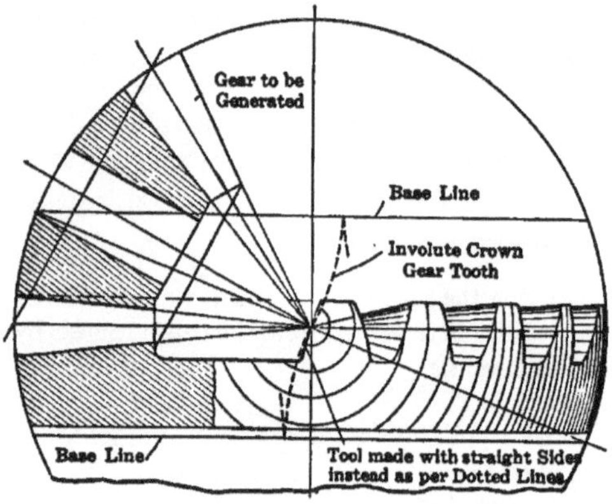

FIG. 32. GENERATING THE OCTOID TOOTH.

FIG. 33.

An involute spur gear may be generated by the action of a tool representing the rack tooth, as illustrated by Fig. 9. In generating a bevel gear, however, the tool representing the engaging rack tooth must always travel toward the apex of the gear being cut, swinging in a partial circle instead of travelling on a straight line in the direction of the rotation of the gear, as is the case when cutting a spur gear. The base of the bevel-gear tooth is, therefore, a crown gear instead of a rack.

An involute crown gear theoretically correct will have curved instead of straight sides as shown in Fig. 32. As it is not practical to make the generating tools this peculiar shape, they are made straight sided and the octoid tooth is the result.

THE LINE OF ACTION

There is a definite relation between the circle or line which will describe the tooth outline and the line of action. Thus, if the line of action is in the form of a circle, as shown in Fig. 34, that circle of which this line is a segment will describe the tooth outline if rolled upon the pitch circle. The difficulties encountered in the general application of this law are well illustrated by George B. Grant in section 32 of his "Treatise on Gear Wheels," as follows: "This accidental and occasionally useful feature of the rolled curve has generally been made to serve as a basis for the general theory of the tooth curve, and it is responsible for the usually clumsy and limited treatment of that theory. The general law is simple enough to define, but it is so difficult to apply, that but one tooth curve, the cycloidal, which happens to have the circle for a roller, can be intelligently handled by it, and the natural result is that that curve has received the bulk of the attention.

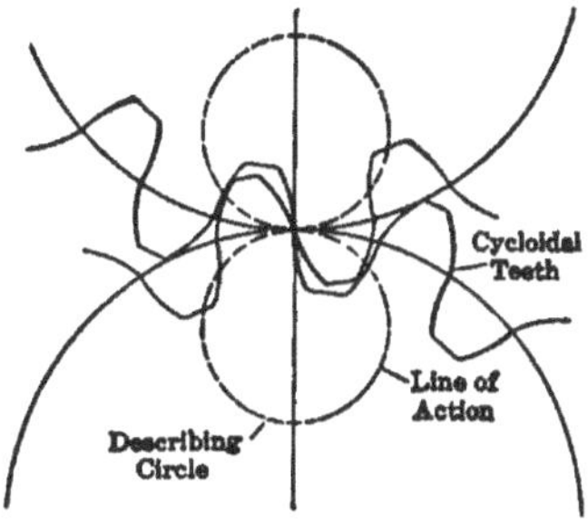

FIG. 34. RELATION OF THE LINE OF ACTION TO THE DESCRIBING CIRCLE.

For example, the simplest and best of all the odontoids (pure form of tooth curve), the involute, is entirely beyond its reach, because its roller is the logarithmic spiral, a transcendental curve that can be reached only by the higher mathematics.

No tooth curve, which, like the involute, crosses the pitch line at any angle but a right angle, can be traced by a point in a simple curve. The tracing point must be the pole of a spiral, and therefore a mechanical impossibility. A practicable rolled odontoid must cross the pitch line at right angles.

To use the rolled curve theory as a base of operations will confine the discussion to the cycloidal tooth, for the involute can only be reached by abandoning its true logarithmic roller, and taking advantage of one of its peculiar properties, and the segmental, sinusoidal, parabolic, and pin tooth, as well as most other available odontoids, cannot be discussed at all."

THE LAW OF TOOTH CONTACT

To transmit uniform motion, any form of tooth curve is subject to this general law: "The common normal to the tooth must pass through the pitch point." That is, a line drawn from the pitch point P through the contact point of any pair of teeth, as at b, must be at right angles or normal to the engaging portions of both teeth. See Fig. 35.

In the involute system the line of action always passes through the pitch point P, and the engaging teeth take their base from the points f and y, where the line of action intersects the base circles. Conversely, a line drawn from the instantaneous radii of any two teeth engaged will pass through their point of contact if the teeth are correctly formed. For example: In Fig. 35, the point of contact between the teeth *C* and *D* is at *b*, on the line of action,

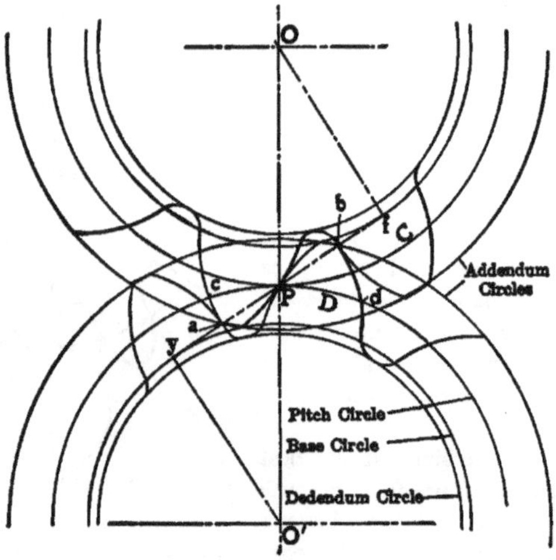

FIG. 35. THE ARC OF ACTION.

the radius of the engaged portion of the tooth *C* is at *f*, and the radius of the tooth *D* is at *y*, fulfilling the required conditions.

THE ARC OF ACTION

The tooth action between two gears is between the points *a* and *b*, at which points the line of action intersects the addendum circles of the two gears. The actual length of contact is along the pitch lines occupied by the teeth whose addendum circles intersect the line of action, or between the points *c* and *d*. See Fig. 35.

The distance *P*—*d* passed over while the point of contact approaches the pitch point is the arc of approach, the distance *P*—*c* passed over as the point of contact leaves the pitch point is the arc of recess.

By increasing the addendum of the driving gear the arc of approach is reduced and the arc of recess is increased. The friction of the arc of approach is greater than in the arc of recess.

THE BUTTRESSED TOOTH

The buttressed tooth shown in Fig. 36 is described by Professor Robert Willis in a paper published in the Transactions of the Institute of Civil Engineers, London, 1838. It is apparent that the object is to obtain a strong tooth for a pair of gears operating continuously in one direction. This is accomplished by increasing the angle of obliquity of the back of the tooth, the face of the tooth being any angle desired. If the back of the tooth is correctly

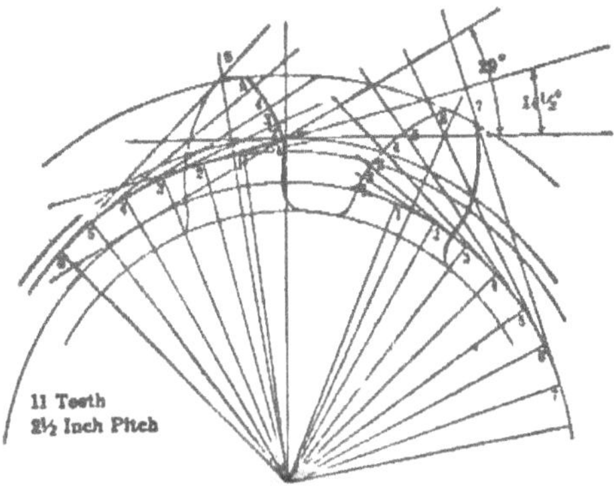

FIG. 36. THE BUTTRESSED TOOTH.

formed the gears will operate satisfactorily in either direction although with an increased pressure on their bearings when using the back face of the teeth owing to the increased obliquity of action. For many purposes there is no

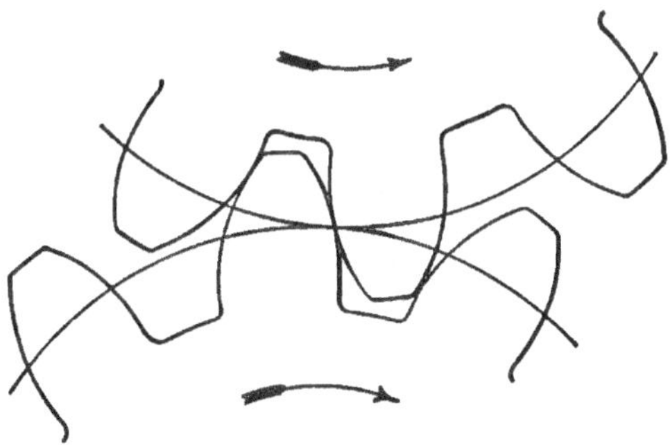

FIG. 37. BUTTRESSED TEETH IN CONTACT.

objection to this, and it is a great wonder that this tooth is not more extensively used.

Of course, there must be a limit to the angle of the back of the tooth. For practical purposes the curve at the top of tooth at the back should not extend farther than the center line of the tooth; for an addendum of $\frac{I}{p}$ or $0.6866p_1$, this will occur at an angle of about 32 degrees. A greater angle than this will subject the tooth to breakage at the point. In Fig. 37 is shown a pair of buttressed tooth gears in contact.

STEPPED GEARS

A stepped spur gear consists of two or more gears keyed to the same shaft, the teeth on each gear being slightly advanced. If mated with a simillar gear the tooth contact will be increased, which increases the smoothness of action. A common form of this type of gear is that of two gears cast in one piece with a separating shroud. For a cut gear there must be a groove turned between the faces of sufficient width to allow the planing tool or cutter to clear. A tooth is placed opposite a space, when the gear is made in two sections.

HUNTING TOOTH

It has been customary to make a pair of cast tooth gears with a hunting tooth, in order that each tooth would engage all of the teeth in the mating gear, the idea being that they would eventually be worn into some indefinite but true shape. Some designers have even gone so far as to specify a pair of "hunting-tooth miter gears." That is, one "miter" gear would have, say, 24 teeth and its mate 25 teeth.

There never was any call for the introduction of the hunting tooth even in cast gears, but in properly cut gears any excuse for its use has certainly ceased to exist.

TEMPLET MAKING

In making the templets for gear teeth there are several points of importance to be kept in mind, namely:

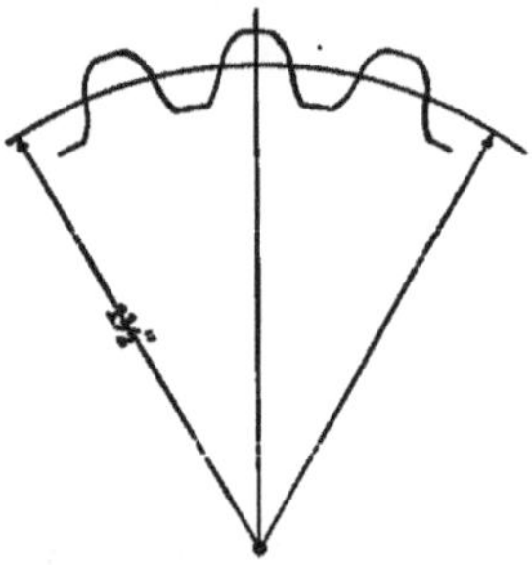

FIG. 38. TOOTH OUTLINE AS PHOTOGRAPHED FROM LARGE SCALE DRAWING.

Templets should be made of light sheet steel instead of zinc which is often employed; the surface of steel should be coppered by the application of blue vitriol.

For spiral or worm gears, templets should always be made for the normal pitch.

For spacing and tooth thickness, always use chordal measurements. Check the chordal distance over the end teeth of templet. This is of the utmost importance.

Put enough teeth in the templet to show the entire tooth action, and try the templets on centers before making up the cutters or formers.

It is a good idea to make a standard templet of each pitch as they are required, to try out other templets that must be made later on.

When a templet is required for a fine pitch gear it is good practice to lay out the teeth on white paper 10 or even 20 times the actual size and reduce by photography. On this drawing the center should be plainly marked and inclosed in a heavy circle, also a short section of the pitch line should be made heavy with a connecting radial line indicating the radius of pitch circle.

If the pitch radius required is 1½ inches, it should be made, say, 15 inches on the drawing. The drawing is then photographed, the camera being set until the radial line, which was drawn 15 inches, measures 1½ inches on the ground glass. See Fig. 38.

DEFINITION OF PITCHES

Diametral pitch is the number of teeth to each inch of the pitch diameter.

Circular pitch is the distance from the edge of one tooth to the corresponding edge of the next tooth measured along the pitch circle.

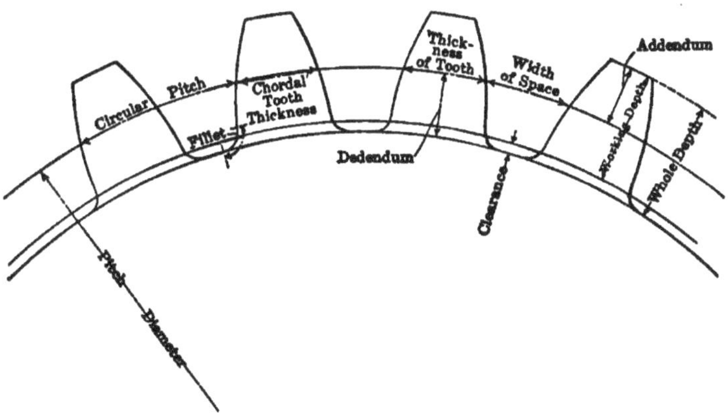

FIG. 39. TOOTH PARTS.

DIAMETRAL PITCH	CIRCULAR PITCH	THICKNESS OF TOOTH ON PITCH LINE	WHOLE DEPTH	DEDENDUM	ADDENDUM
1/2	6.2832″	3.1416″	4.3142″	2.3142″	2.0000″
3/4	4.1888	2.0944	2.8761	1.5728	1.3333
1	3.1416	1.5708	2.1571	1.1571	1.0000
1 1/4	2.5133	1.2566	1.7257	0.9257	0.8000
1 1/2	2.0944	1.0472	1.4381	0.7714	0.6666
1 3/4	1.7952	0.8976	1.2326	0.6612	0.5714
2	1.5708	0.7854	1.0785	0.5785	0.5000
2 1/4	1.3963	0.6981	0.9587	0.5143	0.4444
2 1/2	1.2566	0.6283	0.8628	0.4628	0.4000
2 3/4	1.1424	0.5712	0.7844	0.4208	0.3636
3	1.0472	0.5236	0.7190	0.3857	0.3333
3 1/2	0.8976	0.4488	0.6163	0.3306	0.2857
4	0.7854	0.3927	0.5393	0.2893	0.2500
5	0.6283	0.3142	0.4314	0.2314	0.2000
6	0.5236	0.2618	0.3595	0.1928	0.1666
7	0.4488	0.2244	0.3081	0.1653	0.1429
8	0.3927	0.1963	0.2696	0.1446	0.1250
9	0.3491	0.1745	0.2397	0.1286	0.1111
10	0.3142	0.1571	0.2157	0.1157	0.1000
11	0.2856	0.1428	0.1961	0.1052	0.0909
12	0.2618	0.1309	0.1798	0.0964	0.0833
13	0.2417	0.1208	0.1659	0.0890	0.0769
14	0.2244	0.1122	0.1541	0.0826	0.0714
15	0.2094	0.1047	0.1438	0.0771	0.0666
16	0.1963	0.0982	0.1348	0.0723	0.0625
17	0.1848	0.0924	0.1269	0.0681	0.0588
18	0.1745	0.0873	0.1198	0.0643	0.0555
19	0.1653	0.0827	0.1135	0.0609	0.0526
20	0.1571	0.0785	0.1079	0.0579	0.0500
22	0.1428	0.0714	0.0980	0.0526	0.0455
24	0.1309	0.0654	0.0898	0.0482	0.0417
26	0.1208	0.0604	0.0829	0.0445	0.0385
28	0.1122	0.0561	0.0770	0.0413	0.0357
30	0.1047	0.0524	0.0719	0.0386	0.0333
32	0.0982	0.0491	0.0674	0.0362	0.0312
34	0.0924	0.0462	0.0634	0.0340	0.0294
36	0.0873	0.0436	0.0599	0.0321	0.0278
38	0.0827	0.0413	0.0568	0.0304	0.0263
40	0.0785	0.0393	0.0539	0.0289	0.0250
42	0.0748	0.0374	0.0514	0.0275	0.0238
44	0.0714	0.0357	0.0490	0.0263	0.0227
46	0.0683	0.0341	0.0469	0.0252	0.0217
48	0.0654	0.0327	0.0449	0.0241	0.0208
50	0.0628	0.0314	0.0431	0.0231	0.0200
56	0.0561	0.0280	0.9385	0.0207	0.0178
60	0.0524	0.0262	0.0360	0.0193	0.0166

TABLE 1—DIAMETRAL PITCH

Relation between Diametral Pitch and Circular Pitch, with corresponding Tooth Dimensions

CIRCULAR PITCH	DIAMETRAL PITCH	THICKNESS OF TOOTH OF PITCH LINE	WHOLE DEPTH	DEDENDUM	ADDENDUM
6 "	0.5236	3.0000"	4.1196"	2.2098"	1.9098"
5 "	0.6283	2.5000	3.4330	1.8415	1.5915
4 "	0.7854	2.0000	2.7464	1.4732	1.2732
3½"	0.8976	1.7500	2.4031	1.2890	1.1140
3 "	1.0472	1.5000	2.0598	1.1049	0.9550
2¾"	1.1424	1.3750	1.8882	1.0028	0.8754
2½"	1.2566	1.2500	1.7165	0.9207	0.7958
2¼"	1.3963	1.1250	1.5449	0.8287	0.7162
2 "	1.5708	1.0000	1.3732	0.7366	0.6366
1⅞"	1.6755	0.9375	1.2874	0.6906	0.5968
1¾"	1.7952	0.8750	1.2016	0.6445	0.5570
1⅝"	1.9333	0.8125	1.1158	0.5985	0.5173
1½"	2.0944	0.7500	1.0299	0.5525	0.4775
1 7/16"	2.1855	0.7187	0.9870	0.5294	0.4576
1⅜"	2.2848	0.6875	0.9441	0.5064	0.4377
1 5/16"	2.3936	0.6562	0.9012	0.4837	0.4178
1¼"	2.5133	0.6250	0.8583	0.4604	0.3979
1 3/16"	2.6465	0.5937	0.8156	0.4374	0.3780
1⅛"	2.7925	0.5625	0.7724	0.4143	0.3581
1 1/16"	2.9568	0.5312	0.7295	0.3913	0.3382
1 "	3.1416	0.5000	0.6866	0.3683	0.3183
15/16"	3.3510	0.4687	0.6437	0.3453	0.2984
⅞"	3.5904	0.4375	0.6007	0.3223	0.2785
13/16"	3.8666	0.4062	0.5579	0.2993	0.2586
¾"	4.1888	0.3750	0.5150	0.2762	0.2387
11/16"	4.5696	0.3437	0.4720	0.2532	0.2189
⅝"	5.0265	0.3125	0.4291	0.2301	0.1989
9/16"	5.5851	0.2812	0.3862	0.2071	0.1790
½"	6.2832	0.2500	0.3433	0.1842	0.1592
7/16"	7.1808	0.2187	0.3003	0.1611	0.1393
⅖"	7.8540	0.2000	0.2746	0.1473	0.1273
⅜"	8.3776	0.1875	0.2575	0.1381	0.1194
⅓"	9.4248	0.1666	0.2287	0.1228	0.1061
5/16"	10.0531	0.1562	0.2146	0.1151	0.0995
2/7"	10.9956	0.1429	0.1962	0.1052	0.0909
¼"	12.5664	0.1250	0.1716	0.0921	0.0796
2/9"	14.1372	0.1111	0.1526	0.0818	0.0707
⅕"	15.7080	0.1000	0.1373	0.0737	0.0637
3/16"	16.7552	0.0937	0.1287	0.0690	0.0592
⅙"	18.8496	0.0833	0.1144	0.0614	0.0531
1/7"	21.9911	0.0714	0.0981	0.0526	0.0455
⅛"	25.1327	0.0625	0.0858	0.0460	0.0398
1/9"	28.2743	0.0555	0.0763	0.0409	0.0354
1/10"	31.4159	0.0500	0.0687	0.0368	0.0318
1/16"	50.2655	0.0312	0.0429	0.0230	0.0199

TABLE 2—CIRCULAR PITCH

Relation between Circular Pitch and Diametral Pitch, with corresponding Tooth Dimensions

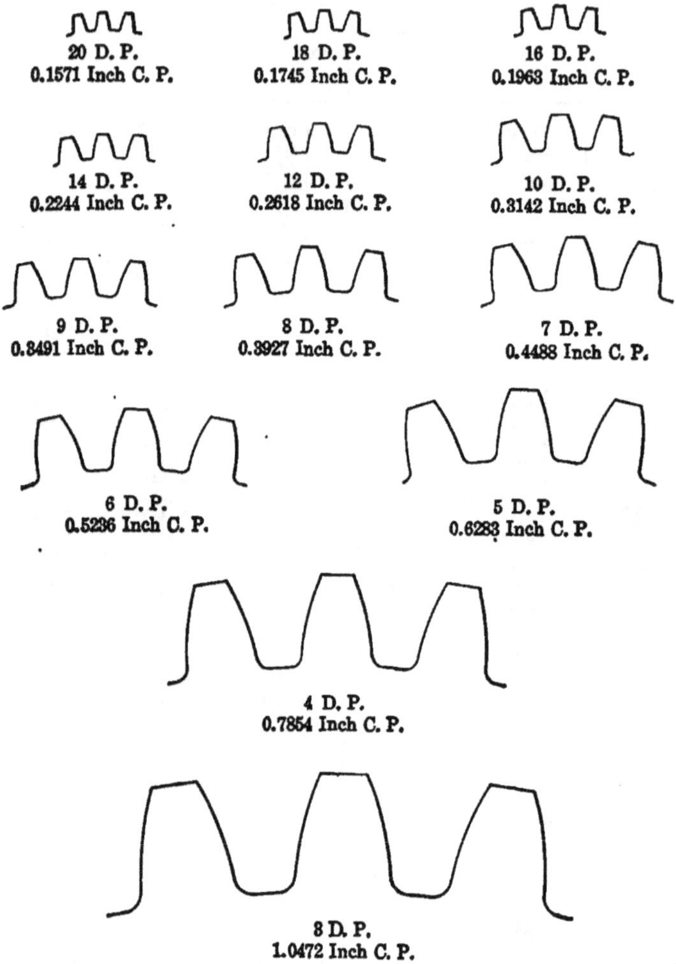

COMPARATIVE SIZES OF GEAR TEETH—INVOLUTE FORM.

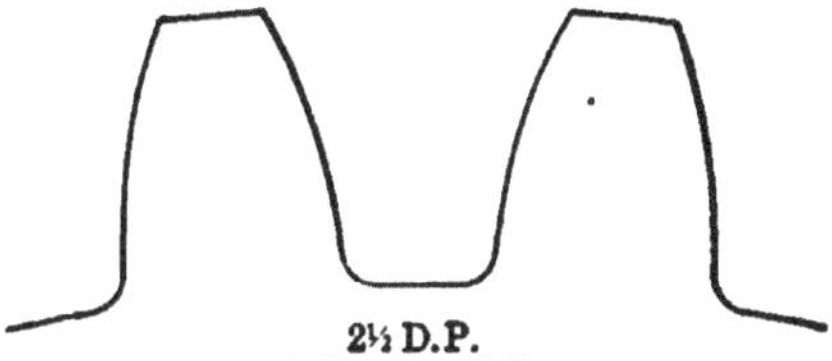

2½ D.P.
1.2566 In. C.P.

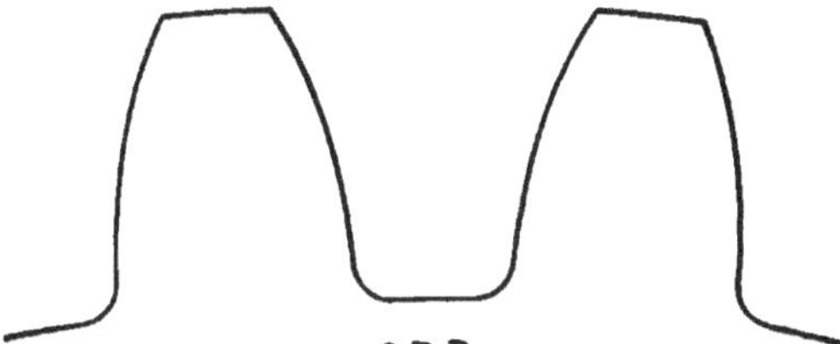

2 D.P.
1.5708 In. C.P.

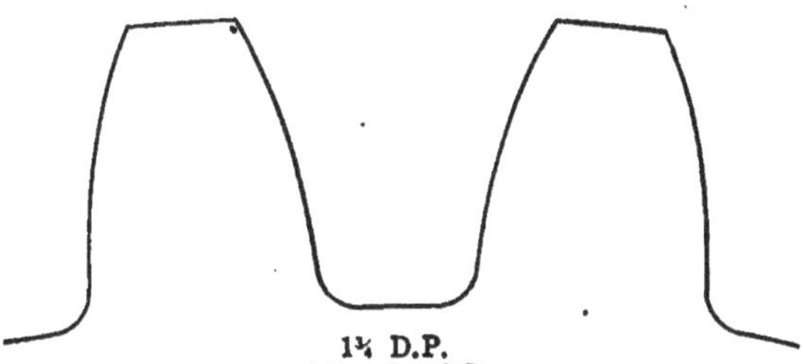

1¾ D.P.
1.7952 In. C.P.

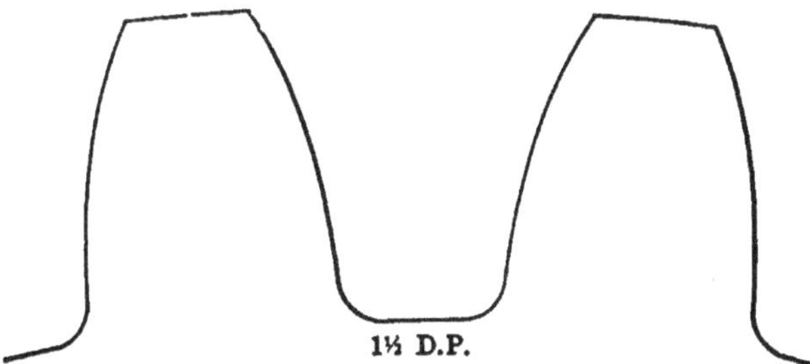

1½ D.P.
2.0944 In. C.P.

COMPARATIVE SIZES OF GEAR TEETH—INVOLUTE FORMS.

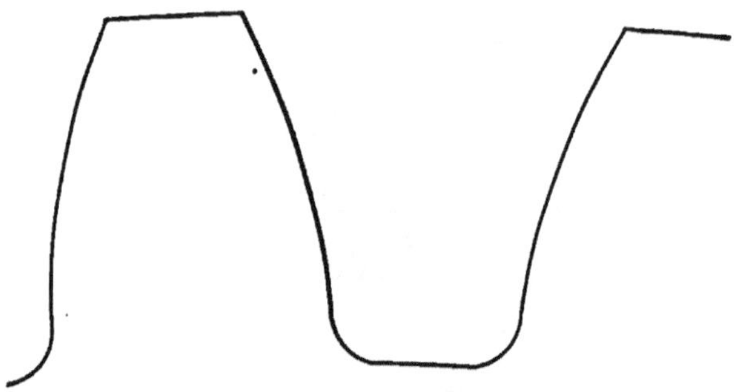

1¼ D. P.
2.5133 Inch C. P.

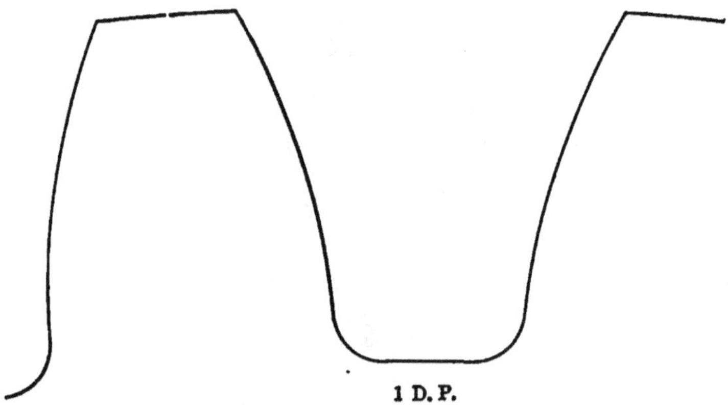

1 D. P.
3.1416 Inch C. P.

COMPARATIVE SIZES OF GEAR TEETH—INVOLUTE FORM.

NO. TEETH	PITCH DIAMETER	NO. TEETH	PITCH DIAMETER	NO. TEETH	PITCH DIAMETER	NO. TEETH	PITCH DIAMETER
8	2.550	43	13.687	78	24.828	113	35.968
9	2.870	44	14.006	79	25.146	114	36.286
10	3.183	45	14.324	80	25.465	115	36.605
11	3.501	46	14.642	81	25.783	116	36.923
12	3.820	47	14.961	82	26.101	117	37.241
13	4.138	48	15.279	83	26.420	118	37.560
14	4.456	49	15.597	84	26.738	119	37.878
15	4.775	50	15.915	85	27.056	120	38.196
16	5.093	51	16.234	86	27.375	121	38.514
17	5.411	52	16.552	87	27.693	122	38.833
18	5.730	53	16.870	88	28.011	123	39.151
19	6.048	54	17.189	89	28.330	124	39.469
20	6.366	55	17.507	90	28.648	125	39.788
21	6.684	56	17.825	91	28.966	126	40.106
22	7.003	57	18.144	92	29.284	127	40.424
23	7.321	58	18.462	93	29.603	128	40.743
24	7.639	59	18.780	94	29.921	129	41.061
25	7.958	60	19.099	95	30.239	130	41.379
26	8.276	61	19.417	96	30.558	131	41.697
27	8.594	62	19.735	97	30.876	132	42.016
28	8.913	63	20.053	98	31.194	133	42.334
29	9.231	64	20.372	99	31.513	134	42.652
30	9.549	65	20.690	100	31.831	135	42.971
31	9.868	66	21.008	101	32.148	136	43.289
32	10.186	67	21.327	102	32.468	137	43.607
33	10.504	68	21.645	103	32.785	138	43.926
34	10.822	69	21.963	104	33.103	139	44.243
35	11.141	70	22.282	105	33.421	140	44.562
36	11.459	71	22.600	106	33.740	141	44.881
37	11.777	72	22.918	107	34.058	142	45.199
38	12.096	73	23.237	108	34.376	143	45.517
39	12.414	74	23.555	109	34.695	144	45.835
40	12.732	75	23.873	110	35.013	145	46.154
41	13.051	76	24.192	111	35.331	146	46.472
42	13.369	77	24.510	112	35.650	147	46.790

TABLE 3—PITCH DIAMETERS FOR ONE-INCH CIRCULAR PITCH

Teeth from 8 to 147

FOR ANY OTHER PITCH—MULTIPLY BY THAT PITCH

METRIC PITCH

The module is the addendum, or the pitch diameter in millimeters divided by the number of teeth in the gear.

The pitch diameter in millimeters is the module multiplied by the number of teeth in the gear. All calculations are in millimeters.

M = module (addendum)
D' = pitch diameter
D = outside diameter
N = number of teeth
W = working depth of tooth
W' = whole depth of tooth
t = thickness of tooth at pitch line
f = clearness
r = root

$$M = \frac{D'}{N} \text{ or } \frac{D}{N+2}, \qquad D' = NM, \qquad D = (N+2)\,M$$

$$N = \frac{D'}{M} \text{ or } \frac{D}{M} - 2, \qquad W = 2M \qquad W' = W + f$$

$$t = M\,1.5708, \qquad f = \frac{M\,1.5708}{10}, \text{ or } M\,0.157, \; r = M + f$$

MODULE	ENGLISH DIAMETRAL PITCH	MODULE	ENGLISH DIAMETRAL PITCH	MODULE	ENGLISH DIAMETRAL PITCH
0.5	50.800			7	3.628
1	25.400	3	8.466	8	3.175
1.25	20.320	3.5	7.257	9	2.822
1.5	16.933	4	6.350	10	2.540
1.75	14.514	4.5	5.644	11	2.309
2	12.700	5	5.080	12	2.117
2.25	11.288	5.5	4.618	14	1.814
2.5	10.160	6	4.233	16	1.587
2.75	9.236				

Module in Millimeters

TABLE 4—PITCHES COMMONLY USED

CHORDAL PITCH

The chordal pitch is the shortest distance between two teeth measured on the pitch line, in other words, the distance to which the dividers would be set to space around the gear on the pitch line. This pitch is not used except for laying out large gears and segments that cannot be cut on a gear cutter. For such cases, also for laying out templets, it is absolutely necessary to use the chordal pitch, as the chordal pitch of the pinion is different from the chordal pitch of the gear, the circular pitch of each being equal.

N = number of teeth,
C' = chordal pitch.
D' = pitch diameter,
e = sine of one half of angle subtended by side at center.

$$e = \text{sine } \frac{180^\circ}{N}.$$

$$D' = \frac{C'}{e}.$$

$$C' = D' e, \text{ or } D' \text{ sine } \frac{180^\circ}{N}.$$

Table 5, diameters for chordal pitch, will be found useful for sprocket gears.

NO. TEETH	PITCH DIAMETER	NO. TEETH	PITCH DIAMETER	NO. TEETH	PITCH DIAMETER	NO. TEETH	PITCH DIAMETER
4	1.414	39	12.427	74	23.562	109	34.701
5	1.701	40	12.746	75	23.880	110	35.019
6	2.000	41	13.064	76	24.198	111	35.337
7	2.305	42	13.382	77	24.517	112	35.655
8	2.613	43	13.699	78	24.835	113	35.974
9	2.924	44	14.018	79	25.153	114	36.292
10	3.236	45	14.335	80	25.471	115	36.610
11	3.549	46	14.653	81	25.790	116	36.929
12	3.864	47	14.972	82	26.108	117	37.247
13	4.179	48	15.291	83	26.426	118	37.565
14	4.494	49	15.608	84	26.744	119	37.883
15	4.810	50	15.927	85	27.063	120	38.202
16	5.126	51	16.244	86	27.381	121	38.520
17	5.442	52	16.562	87	27.699	122	38.838
18	5.759	53	16.880	88	28.017	123	39.156
19	6.076	54	17.200	89	28.335	124	39.475
20	6.392	55	17.516	90	28.654	125	39.793
21	6.710	56	17.835	91	28.972	126	40.111
22	7.027	57	18.152	92	29.290	127	40.429
23	7.344	58	18.471	93	29.608	128	40.748
24	7.661	59	18.789	94	29.927	129	41.066
25	7.979	60	19.107	95	30.245	130	41.384
26	8.297	61	19.425	96	30.563	131	41.703
27	8.614	62	19.744	97	30.881	132	42.021
28	8.931	63	20.062	98	31.200	133	42.339
29	9.249	64	20.380	99	31.518	134	42.657
30	9.567	65	20.698	100	31.836	135	42.976
31	9.884	66	21.016	101	32.154	136	43.294
32	10.202	67	21.335	102	32.473	137	43.612
33	10.520	68	21.653	103	32.791	138	43.931
34	10.838	69	21.971	104	33.109	139	44.249
35	11.156	70	22.289	105	33.428	140	44.567
36	11.474	71	22.607	106	33.740	141	44.890
37	11.791	72	22.926	107	34.058	142	45.204
38	12.110	73	23.244	108	34.376	143	45.522

TABLE 5—PITCH DIAMETERS FOR ONE-INCH CHORDAL PITCH

Teeth from 4 to 143

FOR ANY OTHER PITCH MULTIPLY BY THAT PITCH

CHORDAL THICKNESS OF TEETH

In order to correctly measure the teeth, the chordal thickness must be used, as illustrated by Fig. 40. Also as the location of the pitch line on the sides of the teeth falls below the pitch line at the center of tooth. The measurement for the addendum must also be corrected, if any degree of

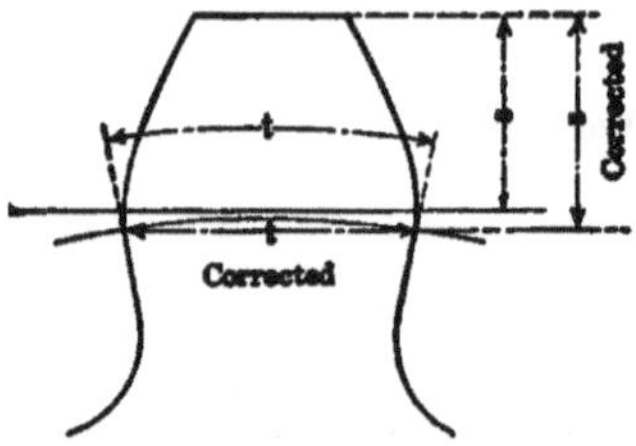

FIG. 40. CHORDAL TOOTH THICKNESS.

accuracy is expected. Table 6, gives these corrected dimensions for various standard pitches.

Number of Teeth.	1 D. P.		1½ D. P.		2 D. P.		2½ D. P.		Number of Teeth.
	Thickness.	Addendum.	Thickness.	Addendum.	Thickness.	Addendum.	Thickness.	Addendum.	
8	1.5607	1.0769	1.0405	0.7179	0.7804	0.5385	0.6243	0.4308	8
9	1.5628	1.0684	1.0419	0.7123	0.7814	0.5342	0.6251	0.4273	9
10	1.5643	1.0616	1.0429	0.7077	0.7821	0.5308	0.6257	0.4246	10
11	1.5654	1.0559	1.0436	0.7039	0.7827	0.5279	0.6261	0.4224	11
12	1.5663	1.0514	1.0442	0.7009	0.7831	0.5257	0.6265	0.4206	12
14	1.5675	1.0440	1.0450	0.6960	0.7837	0.5220	0.6270	0.4176	14
17	1.5686	1.0362	1.0457	0.6908	0.7843	0.5181	0.6274	0.4145	17
21	1.5694	1.0294	1.0463	0.6863	0.7847	0.5147	0.6277	0.4118	21
26	1.5698	1.0237	1.0465	0.6825	0.7849	0.5118	0.6279	0.4095	26
35	1.5702	1.0176	1.0468	0.6784	0.7851	0.5088	0.6281	0.4070	35
55	1.5706	1.0112	1.0471	0.6741	0.7853	0.5056	0.6282	0.4045	55
135	1.5707	1.0047	1.0471	0.6698	0.7853	0.5023	0.6283	0.4019	135

Number of Teeth.	3 D. P.		3½ D. P.		4 D. P.		5 D. P.		Number of Teeth.
	Thickness.	Addendum.	Thickness.	Addendum.	Thickness.	Addendum.	Thickness.	Addendum.	
8	0.5202	0.3589	0.4459	0.3077	0.3902	0.2692	0.3121	0.2154	8
9	0.5209	0.3561	0.4465	0.3052	0.3907	0.2671	0.3126	0.2137	9
10	0.5214	0.3538	0.4469	0.3033	0.3911	0.2654	0.3129	0.2123	10
11	0.5218	0.3519	0.4473	0.3017	0.3913	0.2640	0.3131	0.2112	11
12	0.5221	0.3505	0.4475	0.3004	0.3916	0.2628	0.3133	0.2103	12
14	0.5225	0.3480	0.4479	0.2983	0.3919	0.2610	0.3135	0.2088	14
17	0.5228	0.3454	0.4482	0.2961	0.3921	0.2590	0.3137	0.2072	17
21	0.5231	0.3431	0.4485	0.2941	0.3923	0.2573	0.3139	0.2059	21
26	0.5233	0.3412	0.4485	0.2925	0.3925	0.2559	0.3140	0.2047	26
35	0.5234	0.3392	0.4486	0.2907	0.3926	0.2544	0.3140	0.2035	35
55	0.5235	0.3371	0.4487	0.2889	0.3927	0.2528	0.3141	0.2022	55
135	0.5236	0.3349	0.4488	0.2871	0.3927	0.2512	0.3141	0.2009	135

TABLE 6—CHORDAL THICKNESSES AND ADDENDA OF GEAR TEETH OF DIAMETRAL PITCH

Boston Gear Works

Number of Teeth.	6 D. P. Thickness.	6 D. P. Addendum.	7 D. P. Thickness.	7 D. P. Addendum.	8 D. P. Thickness.	8 D. P. Addendum.	9 D. P. Thickness.	9 D. P. Addendum.	Number of Teeth.
8	0.2601	0.1795	0.2230	0.1538	0.1951	0.1346	0.1734	0.1197	8
9	0.2605	0.1781	0.2233	0.1526	0.1954	0.1336	0.1736	0.1187	9
10	0.2607	0.1769	0.2235	0.1517	0.1955	0.1327	0.1738	0.1180	10
11	0.2609	0.1760	0.2236	0.1508	0.1957	0.1320	0.1739	0.1173	11
12	0.2610	0.1752	0.2238	0.1502	0.1958	0.1314	0.1740	0.1168	12
14	0.2612	0.1740	0.2239	0.1491	0.1959	0.1305	0.1742	0.1160	14
17	0.2614	0.1727	0.2241	0.1480	0.1961	0.1295	0.1743	0.1151	17
21	0.2616	0.1716	0.2242	0.1471	0.1962	0.1287	0.1744	0.1144	21
26	0.2616	0.1706	0.2243	0.1462	0.1962	0.1280	0.1744	0.1137	26
35	0.2617	0.1696	0.2243	0.1454	0.1963	0.1272	0.1745	0.1131	35
55	0.2618	0.1685	0.2244	0.1445	0.1963	0.1264	0.1745	0.1124	55
135	0.2618	0.1675	0.2244	0.1435	0.1963	0.1256	0.1745	0.1116	135

Number of Teeth.	10 D. P. Thickness.	10 D. P. Addendum.	11 D. P. Thickness.	11 D. P. Addendum.	12 D. P. Thickness.	12 D. P. Addendum.	13 D. P. Thickness.	13 D. P. Addendum.	Number of Teeth.
8	0.1561	0.1077	0.1419	0.0979	0.1301	0.0897	0.1201	0.0828	8
9	0.1563	0.1068	0.1421	0.0971	0.1302	0.0890	0.1202	0.0822	9
10	0.1564	0.1061	0.1422	0.0965	0.1304	0.0885	0.1203	0.0816	10
11	0.1565	0.1056	0.1423	0.0960	0.1305	0.0880	0.1204	0.0812	11
12	0.1566	0.1051	0.1424	0.0956	0.1305	0.0876	0.1205	0.0809	12
14	0.1567	0.1044	0.1425	0.0949	0.1306	0.0870	0.1206	0.0803	14
17	0.1569	0.1036	0.1426	0.0942	0.1307	0.0863	0.1207	0.0797	17
21	0.1569	0.1029	0.1427	0.0936	0.1308	0.0858	0.1207	0.0792	21
26	0.1570	0.1024	0.1427	0.0931	0.1308	0.0853	0.1207	0.0787	26
35	0.1570	0.1018	0.1427	0.0925	0.1309	0.0848	0.1208	0.0782	35
55	0.1571	0.1011	0.1428	0.0919	0.1309	0.0843	0.1208	0.0777	55
135	0.1571	0.1005	0.1428	0.0913	0.1309	0.0837	0.1208	0.0772	135

Number of Teeth.	14 D. P. Thickness.	14 D. P. Addendum.	15 D. P. Thickness.	15 D. P. Addendum.	16 D. P. Thickness.	16 D. P. Addendum.	17 D. P. Thickness.	17 D. P. Addendum.	Number of Teeth.
8	0.1115	0.0769	0.1040	0.0718	0.0975	0.0673	0.0918	0.0633	8
9	0.1116	0.0763	0.1042	0.0712	0.0977	0.0669	0.0919	0.0628	9
10	0.1117	0.0758	0.1043	0.0709	0.0978	0.0664	0.0920	0.0624	10
11	0.1118	0.0754	0.1044	0.0704	0.0978	0.0659	0.0921	0.0621	11
12	0.1119	0.0751	0.1044	0.0701	0.0979	0.0657	0.0921	0.0618	12
14	0.1119	0.0746	0.1045	0.0696	0.0980	0.0652	0.0922	0.0614	14
17	0.1120	0.0740	0.1046	0.0691	0.0980	0.0648	0.0923	0.0609	17
21	0.1121	0.0735	0.1046	0'0686	0.0981	0.0643	0.0923	0.0605	21
26	0.1121	0.0731	0.1046	0.0682	0.0981	0.0640	0.0923	0.0602	26
35	0.1122	0.0727	0.1047	0.0678	0.0981	0.0636	0.0924	0.0598	35
55	0.1122	0.0722	0.1047	0.0674	0.0981	0.0632	0.0924	0.0595	55
135	0.1122	0.0718	0.1047	0.0670	0.0981	0.0628	0.0924	0.0591	135

CHORDAL THICKNESSES AND ADDENDA OF GEAR TEETH OF DIAMETRAL PITCH—*Continued*
Boston Gear Works

Number of Teeth.	18 D. P. Thickness.	18 D. P. Addendum.	19 D. P. Thickness.	19 D. P. Addendum.	20 D. P. Thickness.	20 D. P. Addendum.	24 D. P. Thickness.	24 D. P. Addendum.	Number of Teeth.
8	0.0867	0.0598	0.0821	0.0567	0.0780	0.0538	0.0650	0.0448	8
9	0.0868	0.0593	0.0822	0.0562	0.0781	0.0534	0.0651	0.0445	9
10	0.0869	0.0589	0.0823	0.0558	0.0782	0.0530	0.0651	0.0443	10
11	0.0869	0.0586	0.0824	0.0555	0.0783	0.0528	0.0652	0.9439	11
12	0.0870	0.0584	0.0824	0.0553	0.0784	0.0525	0.0653	0.0437	12
14	0.0871	0.0580	0.0825	0.0549	0.0784	0.0522	0.0653	0.0435	14
17	0.0871	0.0575	0.0826	0.0545	0.0784	0.0518	0.0653	0.0432	17
21	0.0872	0.0572	0.0826	0.0542	0.0785	0.0514	0.0654	0.0429	21
26	0.0872	0.0568	0.0826	0.0538	0.0785	0.0511	0.0654	0.0426	26
35	0.0872	0.0565	0.0826	0.0535	0.0785	0.0508	0.0654	0.0424	35
55	0.0873	0.0562	0.0827	0.0532	0.0785	0.0505	0.0654	0.0421	55
135	0.0873	0.0558	0.0827	0.0528	0.0785	0.0502	0.0654	0.0419	135

Chordal Thicknesses and Addenda of Gear Teeth of Diametral Pitch—*Continued*

Number of Teeth.	5/8″ C. P. Thickness.	5/8″ C. P. Addendum.	3/4″ C. P. Thickness.	3/4″ C. P. Addendum.	7/8″ C. P. Thickness.	7/8″ C. P. Addendum.	1″ C. P. Thickness.	1″ C. P. Addendum.	Number of Teeth.
8	0.3105	0.2142	0.3725	0.2570	0.4347	0.2997	0.4968	0.3426	8
9	0.3109	0.2125	0.3730	0.2550	0.4353	0.2976	0.4974	0.3400	9
10	0.3112	0.2112	0.3734	.02534	0.4357	0.2957	0.4978	0.3378	10
11	0.3114	0.2100	0.3737	0.2520	0.4360	0.2941	0.4982	0.3360	11
12	0.3116	0.2091	0.3739	0.2510	0.4363	0.2938	0.4986	0.3346	12
14	0.3118	0.2077	0.3741	0.2492	0.4366	0.2908	0.4988	0.3322	14
17	0.3120	0.2061	0.3744	0.2473	0.4369	0.2886	0.4992	0.3298	17
21	0.3122	0.2048	0.3746	0.2457	0.4371	0.2868	0.4994	0.3276	21
26	0.3123	0.2036	0.3748	0.2443	0.4372	0.2851	0.4997	0.3258	26
35	0.3124	0.2024	0.3748	0.2429	0.4373	0.2833	0.4999	0.3238	35
55	0.3124	0.2011	0.3748	0.2414	0.4374	0.2816	0.4999	0.3218	55
135	0.3124	0.1999	0.3748	0.2398	0.4374	0.2798	0.4999	0.3198	135

Number of Teeth.	1 1/4″ C. P. Thickness.	1 1/4″ C. P. Addendum.	1 1/2″ C. P. Thickness.	1 1/2″ C. P. Addendum.	1 3/4″ C. P. Thickness.	1 3/4″ C. P. Addendum.	2″ C. P. Thickness.	2″ C. P. Addendum.	Number of Teeth.
8	0.6210	0.4284	0.7450	0.5140	0.8694	0.5994	0.9936	0.6852	8
9	0.6218	0.4250	0.7460	0.5100	0.8706	0.5952	0.9948	0.6800	9
10	0.6224	0.4224	0.7468	0.5068	0.8714	0.5914	0.9956	0.6756	10
11	0.6228	0.4200	0.7474	0.5040	0.8720	0.5882	0.9964	0.6720	11
12	0.6232	0.4182	0.7478	0.5020	0.8726	0.5876	0.9972	0.6692	12
14	0.6236	0.4154	0.7482	0.4984	0.8732	0.5816	0.9976	0.6644	14
17	0.6240	0.4122	0.7488	0.4946	0.8738	0.5772	0.9984	0.6596	17
21	0.6244	0.4096	0.7492	0.4914	0.8742	0.5736	0.9988	0.6552	21
26	0.6246	0.4072	0.7496	0.4886	0.8744	0.5702	0.9994	0.6516	26
35	0.6248	0.4048	0.7498	0.4858	0.8746	0.5666	0.9998	0.6476	35
55	0.6250	0.4022	0.7499	0.4828	0.8748	0.5632	0.9999	0.6436	55
135	0.6250	0.3998	0.7499	0.4796	0.8748	0.5596	0.9999	0.6396	135

Table 7—Chordal Thicknesses and Addenda of Gear Teeth of Circular Pitch
Boston Gear Works

INVOLUTE CUTTERS

Until quite recently involute cutters were made in sets of eight, as follows:

Number of Cutter	
1	for 135 teeth to rack
2	for 55 to 134 teeth
3	for 35 to 54 teeth
4	for 26 to 34 teeth
5	for 21 to 26 teeth
6	for 17 to 20 teeth
7	for 14 to 16 teeth
8	for 12 to 13 teeth

Modern conditions, however, require a more accurate tooth than can be produced by this number of cutters. A set of fifteen, utilizing the half numbers is now in common use.

Number of Cutter	
1	for 135 teeth to a rack
1½	for 80 to 134 teeth
2	for 55 to 79 teeth
2½	for 42 to 54 teeth
3	for 35 to 41 teeth
3½	for 30 to 34 teeth
4	for 26 to 29 teeth
4½	for 23 to 25 teeth
5	for 21 to 22 teeth
5½	for 19 to 20 teeth
6	for 17 to 18 teeth
6½	for 15 to 16 teeth
7	for 14 teeth
7½	for 13 teeth
8	for 12 teeth

To produce accurate gears, templets for tooth thickness, made according to Tables 6 and 7, should be used instead of using one templet for each pitch and depending upon the workman's judgment as to how much shake to allow for different numbers of teeth. These templets, made up according to Tables 6 and 7, which are based on the use of eight cutters for each pitch, should be sufficiently accurate for all practical purposes.

SECTION II

SPUR GEAR CALCULATIONS

To find the pitch diameters of two gears, the number of teeth in each and the distance between centers being given: Divide twice the distance between centers by the sum of the number of teeth: Find the pitch diameter of each gear separately by multiplying this quotient by its number of teeth.

Example: Find the pitch diameters of a pair of spur gears, 21 and 60 teeth, for 25-inch centers.

$$\frac{2 \times 25}{21 + 60} = 0.617284,$$

$0.617284 \times 21 = 12.96296$ inches, or the pitch diameter of the pinion
$0.617284 \times 60 = 37.03704$ inches, or the pitch diameter of the gear

The distance between the centers is one-half the sum of the pitch diameters. In the above example the center distance would prove to be:

$$\frac{12.96296 + 37.03704}{2} = 25 \text{ inches}$$

NOTATIONS FOR FORMULAS

p = diametral pitch

$$\left.\begin{array}{l} \left.\begin{array}{l} D' = \text{pitch diameter} \\ D = \text{outside diameter} \\ V = \text{velocity} \end{array}\right\} \text{gear} \\ \\ \left.\begin{array}{l} d' = \text{pitch diameter} \\ d = \text{outside diameter} \\ v = \text{velocity} \end{array}\right\} \text{pinion} \end{array}\right\} \text{These gears run together}$$

a = distance between the centers
b = number of teeth in both

TO FIND	HAVING	RULE	FORMULA	EXAMPLE
b	a and p	The continued product of center distance, pitch and 2	$a\,p\,2$	$15 \times 3 \times 2 = 90$
a	D' and d'	One-half the sum of the pitch diameters	$\frac{D' + d'}{2}$	$\frac{20 + 10}{2} = 15''$
a	b and p	Divide the total number of teeth by twice the pitch	$\frac{b}{2p}$	$\frac{90}{2 \times 3} = 15$
N	$n\,v$ and V	Divide the product of the number of teeth and velocity of pinion by the velocity of gear	$\frac{n\,v}{V}$	$\frac{30 \times 2}{1} = 60$
N	$b\,v$ and V	Divide the product of the total number of teeth and velocity of pinion by the sum of the velocities	$\frac{b\,v}{v + V}$	$\frac{90 \times 2}{2 + 1} = 60$
n	$b\,v$ and V	Divide the product of the total number of teeth and the velocity of gear by the sum of the velocities	$\frac{b\,V}{v + V}$	$\frac{90 \times 1}{2 + 1} = 30$
n	$N\,v$ and V	Divide the product of the number of teeth in gear and its velocity by the velocity of pinion	$\frac{N\,V}{v}$	$\frac{60 \times 1}{2} = 30$
n	$p\,D'\,V$ and v	Divide the continued product of the pitch, pitch diameter and velocity of the gear by the velocity of pinion.	$\frac{p\,D'\,V}{v}$	$\frac{3 \times 20 \times 1}{2} = 30$
v	$N\,V$ and n	Divide the product of the number of teeth and velocity of gear by the number of teeth in pinion	$\frac{N\,V}{n}$	$\frac{60 \times 1}{30} = 2$
v	$p\,D'\,V$ and n	Divide the continued product of the pitch, pitch diameter and velocity of gear by the number of teeth in pinion	$\frac{p\,D'\,V}{n}$	$\frac{3 \times 20 \times 1}{30} = 2$
V	$n\,v$ and N	Divide the product of the number of teeth in pinion and its velocity by the number of teeth in gear	$\frac{n\,v}{N}$	$\frac{30 \times 2}{60} = 1$
D'	$a\,v$ and V	Divide the continued product of the center distance, velocity of pinion and 2, by the sum of the velocities.	$\frac{2\,a\,v}{v + V}$	$\frac{2 \times 15 \times 2}{2 + 1} = 20$
d'	$a\,v$ and V	Divide the continued product of the center distance, velocity of gear and 2, by the sum of the velocities	$\frac{2\,a\,V}{v + V}$	$\frac{2 \times 15 \times 1}{2 + 1} = 10$

TABLE 8—FORMULAS FOR A PAIR OF MATING SPUR GEARS

TO FIND	HAVING	RULE	FORMULA
The Diametral Pitch.	The Circular Pitch..	Divide 3.1416 by the Circular Pitch....................	$p = \frac{3.1416}{p'}$
The Diametral Pitch.	The Pitch Diameter and the Number of Teeth...........	Divide Number of Teeth by Pitch Diameter.............	$p = \frac{N}{D'}$
The Diametral Pitch.	The Outside Diameter and the Number of Teeth.........	Divide Number of Teeth plus 2 by Outside Diameter........	$p = \frac{N+2}{D}$
Pitch Diameter.	The Number of Teeth and the Diametral Pitch............	Divide Number of Teeth by the Diametral Pitch.............	$D' = \frac{N}{p}$
Pitch Diameter.	The Number of Teeth and Outside Diameter..............	Divide the product of Outside Diameter and Number of Teeth by Number of Teeth plus 2...	$D' = \frac{DN}{N+2}$
Pitch Diameter.	The Outside Diameter and the Diametral Pitch.........	Subtract from the Outside Diameter the quotient of 2 divided by the Diametral Pitch.........	$D' = D - \frac{2}{p}$
Pitch Diameter.	Addendum and the Number of Teeth..	Multiply Addendum by the Number of Teeth............	$D' = sN$
Outside Diameter.	The Number of Teeth and the Diametral Pitch............	Divide Number of Teeth plus 2 by the Diametral Pitch.......	$D = \frac{N+2}{p}$
Outside Diameter.	The Pitch Diameter and the Diametral Pitch............	Add to the Pitch Diameter the quotient of 2 divided by the Diametral Pitch............	$D = D' + \frac{2}{p}$
Outside Diameter.	The Pitch Diameter and the Number of Teeth...........	Divide the product of the Pitch Diameter and Number of Teeth plus 2 by the Number of Teeth	$D = \frac{(N+2)D'}{N}$
Outside Diameter.	The Number of Teeth and Addendum....	Multiply the Number of Teeth plus 2 by Addendum.........	$D = (N+2)s$
Number of Teeth.	The Pitch Diameter and the Diametral Pitch............	Multiply Pitch Diameter by the Diametral Pitch.............	$N = D'p$
Number of Teeth.	The Outside Diameter and the Diametral Pitch.........	Multiply Outside Diameter by the Diametral Pitch and subtract 2....................	$N = Dp - 2$
Thickness of Tooth.	The Diametral Pitch.	Divide 1.5708 by the Diametral Pitch.....................	$t = \frac{1.5708}{p}$
Addendum.	The Diametral Pitch.	Divide 1 by the Diametral Pitch, or $s = \frac{D'}{N}$..............	$s = \frac{1}{p}$
Dedendum.	The Diametral Pitch.	Divide 1.157 by the Diametral Pitch....................	$s + f = \frac{1.157}{p}$
Working Depth.	The Diametral Pitch.	Divide 2 by the Diametral Pitch.	$W = \frac{2}{p}$
Whole Depth.	The Diametral Pitch.	Divide 2.157 by the Diametral Pitch....................	$W' + f = \frac{2.157}{p}$
Clearance.	The Diametral Pitch.	Divide 0.157 by the Diametral Pitch....................	$f = \frac{0.157}{p}$
Clearance.	Thickness of Tooth..	Divide Thickness of Tooth at pitch line by 10............	$f = \frac{t}{10}$

TABLE 9—SPUR GEAR CALCULATIONS FOR DIAMETRAL PITCH

14½ Degree Standard

R. D. Nuttall Company

TO FIND	HAVING	RULE	FORMULA
The Circular Pitch.	The Diametral Pitch.	Divide 3.1416 by the Diametral Pitch....................	$p' = \frac{3.1416}{p}$
The Circular Pitch.	The Pitch Diameter and the Number of Teeth...........	Divide Pitch Diameter by the product of 0.3183 and Number of Teeth...................	$p' = \frac{D'}{0.3183\ N}$
The Circular Pitch.	The Outside Diameter and the Number of Teeth.........	Divide Outside Diameter by the product of 0.3183 and Number of Teeth plus 2.............	$p' = \frac{D}{0.3183\ N+2}$
Pitch Diameter.	The Number of Teeth and the Circular Pitch...........	The continued product of the Number of Teeth, the Circular Pitch and 0.3183...........	$D' = N\ p'\ 0.3183$
Pitch Diameter.	The Number of Teeth and the Outside Diameter...........	Divide the product of Number of Teeth and Outside Diameter by Number of Teeth plus 2...	$D' = \frac{N\ D}{N+2}$
Pitch Diameter.	The Outside Diameter and the Circular Pitch...........	Subtract from the Outside Diameter the product of the Circular Pitch and 0.6366...........	$D' = D - (p'\ 0.6366)$
Pitch Diameter.	Addendum and the Number of Teeth..	Multiply the Number of Teeth by the Addendum.............	$D' = N\ s$
Outside Diameter.	The Number of Teeth and the Circular Pitch...........	The continued product of the Number of Teeth plus 2 the Circular Pitch and 0.3183....	$D = (N+2)\ p'\ 0.3183$
Outside Diameter.	The Pitch Diameter and the Circular Pitch...........	Add to the Pitch Diameter the product of the Circular Pitch and 0.6366..................	$D = D + (p'\ 0.6366)$
Outside Diameter.	The Number of Teeth and the Addendum	Multiply Addendum by Number of Teeth plus 2............	$D = s\ (N+2)$
Number of Teeth.	The Pitch Diameter and the Circular Pitch...........	Divide the product of Pitch Diameter and 3.1416 by the Circular Pitch................	$N = \frac{D'\ 3.1416}{p'}$
Thickness of Tooth.	The Circular Pitch...	One-half the Circular Pitch.....	$t = \frac{p'}{2}$
Addendum.	The Circular Pitch...	Multiply the Circular Pitch by 0.3183, or $s = \frac{D'}{N}$........	$s = p'\ 0.3183$
Dedendum.	The Circular Pitch...	Multiply the Circular Pitch by 0.3683....................	$s + f = p'\ 0.3683$
Working Depth.	The Circular Pitch...	Multiply the Circular Pitch by 0.6366....................	$W = p'\ 0.6366$
Whole Depth.	The Circular Pitch...	Multiply the Circular Pitch by 0.6866....................	$W' = p'\ 0.6866$
Clearance.	The Circular Pitch...	Multiply the Circular Pitch by 0.05......................	$f = p\ 0.05$
Clearance.	Thickness of Tooth..	One-tenth the Thickness of Tooth at Pitch Line..............	$f = \frac{t}{10}$

TABLE 10—SPUR GEAR CALCULATIONS FOR CIRCULAR PITCH

14½ Degree Standard

R. D. Nuttall Company

SECTION III

Speeds and Powers

TRANSMISSION OF POWER BY GEARING WITH PARTICULAR REFERENCE TO SPUR AND BEVEL GEARS

SPEED RATIO

The problem of finding the proper diameter or speed of a gear or pulley is simple enough when once thoroughly understood.

The gear may be represented by its number of teeth, pitch diameter, pitch radius, or speed ratio, as the case may be. In the explanation to follow the number of teeth is used. The speed is in revolutions per minute.

Rule: Divide the product of the speed and number of teeth of one gear by the speed *or* number of teeth of its mate to secure the lacking dimension.

That is, if both the speed and number of teeth are known for one gear, multiply the speed by the number of teeth, and divide this product by the known quantity of the mating gear to secure *its* number of teeth or speed, as the case may be.

Or the same result may be obtained by proportion, the values being placed as follows:

$$n : N : R : r \qquad (1)$$

n = number of teeth in pinion
r = revolutions per minute of pinion
N = number of teeth in gear
R = revolutions per minute of gear

Example: A gear having 60 teeth makes 300 revolutions per minute, what will be the speed of an engaging pinion having 15 teeth?

$$n : N : R : r$$
$$15 : 60 : 300 : x$$

Therefore, $x = \frac{60 \times 300}{15}$, or 1,200 revolutions per minute for pinion n

To compute these values for a train of gears, use the continued product of the pinions and the continued product of the gears as a single gear and pinion and proceed as above.

Example: In Fig. 41, the gear N has 100 teeth, N', 70 teeth, N'', 60 teeth, n, 15 teeth; n', 18 teeth; and n'', 24 teeth. The gear N makes 10 revolutions per minute. What will be the speed of the pinion n''?

$$N, N' \text{ and } N'' = 100 \times 70 \times 60 = 420{,}000$$
$$n, n' \text{ and } n'' = 15 \times 18 \times 14 = 6{,}480.$$
$$n : N :: R : r$$
$$6{,}480 : 420{,}000 :: 10 : x$$

Therefore, $x = \dfrac{420{,}000 \times 10}{6{,}480}$, or 648 revolutions per minute for pinion n''.

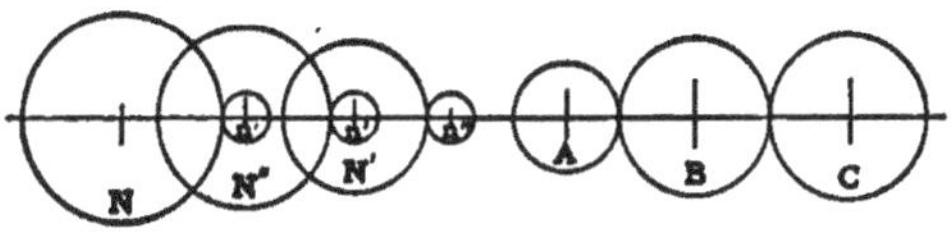

FIG. 41. GEAR TRAIN. FIG. 42. INTERMEDIATE GEAR DOES NOT AFFECT THE SPEED RATIO.

The velocities of a train of gears may also be found as follows: N, N', N'' and n, n', n'', etc., representing the number of teeth in the gears and pinions.

$$r = \frac{RNN'N''}{n\,n'\,n''}. \tag{2}$$

$$R = \frac{r\,n\,n'\,n''}{NN'N''}. \tag{3}$$

The intermediate gear B, as shown in Fig. 42, while it changes the direction of the rotation of the gears, A and C does not alter their speed ratio, the circumferential velocities of all three gears being equal.

ARRANGEMENT OF GEAR TRAINS

For compound reduction there must be four gears, as per Fig. 43, the gears B and C being keyed to an intermediate shaft, the power being transferred to the machine by the shaft-carrying gear D.

When a great reduction is required, say 64 to 1, there may be two intermediate shafts, as in Fig. 44.

This reduction might be accomplished by using a drive, as in Fig. 43, dividing the total reduction between two sets of gears, but a triple reduction is used by way of illustration. The best results are always obtained by dividing the reduction as evenly as possible among the different pairs of gears. For instance: For a double reduction, as in Fig. 43, the ratio of each pair should be made as near the square root of the total reduction as possible. In case of the triple reduction, Fig. 44, the ratio of each pair should be the cube root of the total reduction, or $\sqrt[3]{64} = 4$. That is, there are three sets of gears, each having a speed ratio of 4 to 1. If double reduction had been used the reduction of each gear would have been $\sqrt{64} = 8$, or two sets of gears each having a speed ratio of 8 to 1.

Gear trains proportioned in this way give the highest possible efficiency. For instance: An unsuccessful single gear reduction of 16 to 1 might be made efficient by substituting two pairs of gears, each having a ratio of 4 to 1. Making the compound gears 8 to 1 and 2 to 1 would help, but would not be as

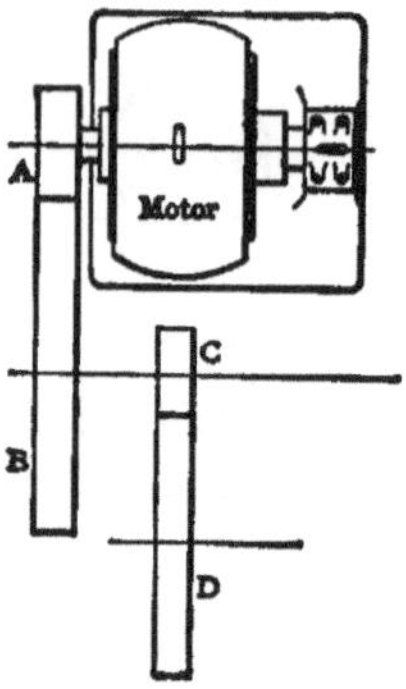

FIG. 43. DOUBLE GEAR REDUCTION.

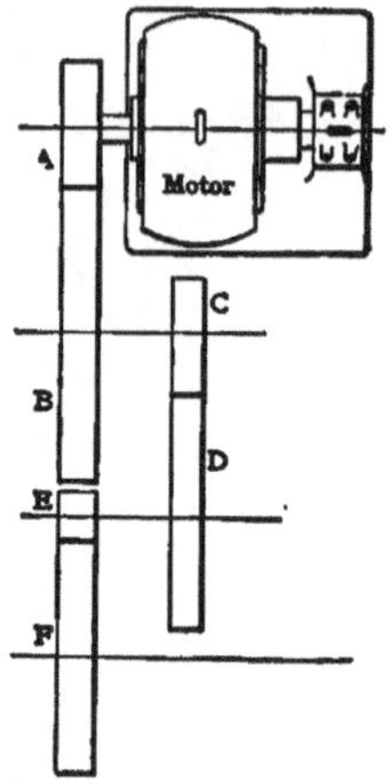

FIG. 44. TRIPLE GEAR REDUCTION.

efficient as the equal reduction. This will be especially noticeable in long leads in the lathe or milling machines.

POWER RATIO

The relative powers of a train of gears are inversely proportional to their circumferential velocities. The circumferential velocity of each pair of gears

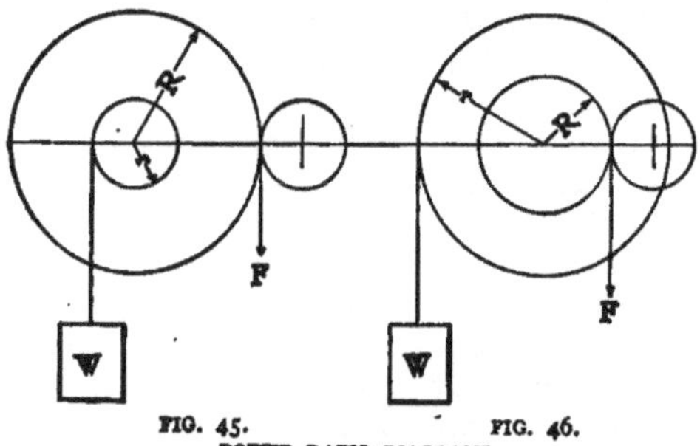

FIG. 45. FIG. 46.
POWER RATIO DIAGRAMS.

in a train being equal, the driving pinion, as shown in Figs. 45 and 46, is ignored in the calculations for a single pair, the circumferential velocity and the load on the teeth being the same as for the mating gear. The problem is to determine the power ratio between the drum r and the gear R.

Ignoring friction, the values of this drive may be found by proportion, arranged as follows:

$$W : R : : F : r \tag{4}$$

Enough must be added to the load W or taken from the effective lifting force F to overcome the frictional resistance of the teeth and bearings. This loss must be estimated and the percentage of loss added to the load W, the ratio of R and r being determined according to this new ratio.

Example: Referring to Fig. 45: If the radius of the gear R is 18 inches, the radius of the drum r three inches, what power will be required at F to raise 300 pounds at W?

$$W : F : : F : r$$

$$300 : 18 : : x : 3$$

Therefore, $x = \dfrac{300 \times 3}{18}, = 50$ pounds required at F.

Suppose the loss in efficiency to be 10 per cent and the radius of the gear R 18 inches. What must be the radius of the drum r to raise 300 pounds at W?

$$300 + 10 \text{ per cent} = 330 \text{ pounds.}$$

$$W : R : : F : T$$

$$300 : 18 : : 50 : x$$

Therefore, $x = \dfrac{18 \times 50}{330}, = 2.7$ inches for the radius of drum r.

For a train of gears, the continued products of the driving and driven gears may be considered as single gears. Or the power ratio may be considered between each pair inversely proportional to their velocity ratios.

Example: Referring to Fig. 47; what force is required at F to raise 2,500 pounds at W, the loss in efficiency being 30 per cent?

$$R\,R'\,R'' = 20 \times 18 \times 10 = 3{,}600$$

$$r\,r'\,r'' = 6 \times 8 \times 5 = 240$$

$$W = 2{,}500 + 30 \text{ per cent} = 3{,}250$$

$W : R : :$

$3{,}250 : 3{,}600 : : x : 240,\ x = \dfrac{3{,}250 \times 240}{3{,}600} F : r$ or 217 pounds at F.

$$\text{Also } F = \frac{W\,r\,r'\,r''}{R\,R'\,R''}, = \frac{3{,}250 \times 6 \times 8 \times 5}{20 \times 18 \times 10} = 217 \text{ pounds.} \tag{5}$$

$$\text{And } W = \frac{F\,R\,R'R''}{r\,r'\,r''}, = \frac{217 \times 20 \times 18 \times 10}{6 \times 8 \times 5}, = 3{,}250 \text{ pounds.} \tag{6}$$

AN EXAMPLE IN HOIST GEARING

Example: What gears will be required to lift a load of 2,400 pounds at a uniform rate of speed, employing a 10 horse-power motor running 1,120 revolutions per minute, driving with a rawhide pinion 4 inches pitch diameter? See Fig. 48.

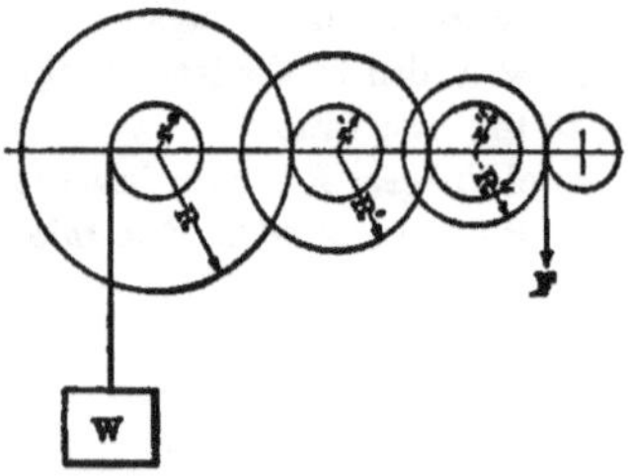

FIG. 47. POWER RATIO OF GEAR TRAINS.

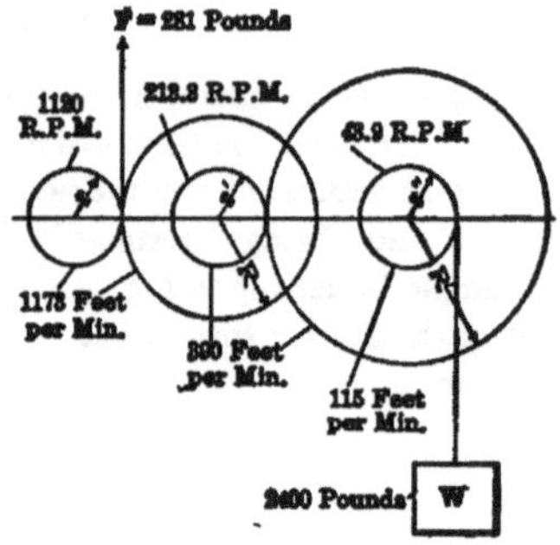

FIG. 48. EXAMPLE OF GEAR DRIVE FOR HOIST.

Velocity of pinion in feet per minute, $V = d' \, 0.2618 \, R.P.M.$ (7)

The safe load, $W + \dfrac{HP \times 33{,}000}{V}$ (8)

Therefore, $V = 4 \times 0.2618 \times 1{,}120 = 1{,}173$ feet per minute.

And $W = \dfrac{10 \times 33{,}000}{1{,}173}$, or 281 pounds, which is the load to be carried by the pinion.

Assuming that 20 per cent is lost by the friction of the gear teeth, bearings, etc., the real load to be raised by the force of 280 pounds at the pitch line of the driving pinion is:

2,400 + 20 per cent. = 2,880 pounds.

The necessary velocity ratio of the gears to equal this ratio of power is, therefore:

$$\frac{2{,}880}{281}, = \frac{10.25}{1}.$$

This reduction must be made between R' and r'', and R and r', the pinion r not being considered as its velocity is the same as that of the gear R, therefore the load on the teeth will be the same.

Since it is always best to make the reduction in even steps, and double reduction is desirable for a ratio of 10.25 to 1 take the square root of the total reduction, 10.25, which is approximately 3.2 to 1 for each reduction. Practically, however, a reduction of $\frac{3.4}{1}$ and $\frac{3}{1}$ will answer.

The ratio between R' and r'' is made $\frac{3.4}{1}$. Assuming the diameter of the drum r'' to be 10 inches, the pitch diameter of the gear R' will be $3.4 \times 10 =$ 34 inches. The ratio between R and r' is $\frac{3}{1}$. Assuming the pitch diameter of the pinion r' to be 7 inches, the pitch diameter of the gear R will be $7 \times 3 =$ 21 inches.

The power or circumferential force of the gear R is, of course, that of the driving pinion r, 281 pounds. Therefore, the power of the pinion r', and consequently that of the gear R', is $281 \times 3 = 843$ pounds.

The problem is now reduced to two simple ones, that is—to design a pair of gears r and R to transmit a force of 281 pounds at a speed of 1,173 feet per minute, and a second pair r' and R' to transmit a force of 843 pounds at a speed of 390 feet per minute.

It is necessary to assume a pitch judged to be suitable for the different drives and to try its value for carrying the required load by the Lewis formula, obtaining the safe load per inch of face, and make the face sufficiently wide to transmit the power.

For the first pair of gears, r and R, assume 4 diametral pitch—0.7854-inch circular pitch—allowing 5000 pounds per square inch as a safe stress for rawhide. Number of teeth in pinion $r = 4 \times 4 = 16$.

$$\text{Safe load per inch of face} = spy \frac{600}{600 + V}. \qquad \text{(See formula 24.)}$$

$$\text{Or } 5{,}000 \times 0.785 \times 0.077 \frac{600}{600 + 1{,}173} = 100 \text{ pounds per inch of face.}$$

Making the face of the gears r and R 3 inches it will safely carry $3 \times 100 = 300$ pounds, which is sufficient.

For the second pair, r' and R', try 3 diametral pitch—1.0472-inch circular pitch—both gears of cast iron. Figure the strength of the pinion, as it is the weaker of the two. Allow 8,000 pounds per square inch as a safe stress.

For a pinion 7 inches pitch diameter, 3 pitch, the number of teeth equals $7 \times 3 = 21$ teeth. Factor y for 21 teeth equals 0.092. $W = 8{,}000 \times 1.047 \times 0.092 \frac{600}{600 + 390}$, or 467 pounds per inch of face.

Making the face 3 inches, the gears will carry a load of $3 \times 467 = 1{,}401$ pounds. These gears will therefore be heavier than necessary, but owing to the nature of the service this should be the case, especially as they are made of cast iron.

From the ratio of this train of gears it will be found that the load will be raised at $\frac{1{,}173}{3 \times 3.4} = 115$ feet per minute, using the full speed of the motor. If the load must be raised at a greater speed than 115 feet, a more powerful motor would be required, and if at a lower speed there must be a greater gear reduction. For instance, if the hoisting speed had been 80 feet per minute the speed ratio would be $\frac{1{,}173}{80} = \frac{14.7}{1}$, instead of $\frac{10.2}{1}$ as in the example.

The above problem is generally put before the designer in a different manner—that is, the load and speed at which the load is to be raised are given, the size of motor and ratio of gearing, etc., to be determined.

Example: A load of 2400 pounds is to be raised at the uniform rate of 115 feet per minute; what size motor and what gears will be required?

Assuming as before a loss of 20 per cent in efficiency in the driven gears, bearing, etc., this load will require:

$$\frac{2{,}880 \times 115}{33{,}000}, = 10 \text{ horse-power } (2{,}400 \text{ pounds} + 20 \text{ per cent} = 2{,}880 \text{ pounds}).$$

Using a rawhide pinion 4-inch pitch diameter on the motor, we consider the problem in the same manner as in previous examples making the ratio of the gears; $\frac{2{,}880}{281} = \frac{10.25}{1}$

The problem of determining the proper gears is the same.

RAILWAY GEARS

Speed in feet per minute at rim of car wheel $V' = 88 \times$ speed of car in miles per hour. (9)

Speed in feet per minute at pitch line of gear $V = 88 \times$ miles per hour $\times R$. (10)

Ratio of gear to wheel $R = \frac{\text{pitch diameter of gear}}{\text{diameter of wheel}}$. (11)

Force at pitch line of gear $F = \frac{HP \times 33{,}000}{V}$, or $\frac{Kw \times 44{,}102}{V}$. (12)

Fiber stress in tooth $S = \frac{F}{p' f y \frac{600}{600 + V}}$. (See formula 18.) (13)

Traction effort at wheel $T = \frac{F}{R}$. (14)

Horse power $HP = \frac{T M}{0.375}$. (15)

Speed of car in miles per hour $M = \frac{\text{Dia. of wheel} \times \text{teeth in pinion} \times \text{revolution per minute of pinion}}{\text{teeth in gear} \times 336}$. (16)

Traction effort $T = \frac{\text{teeth in gear} \times 24 \times \text{gear efficiency} \times \text{torque of motor}}{M \times \text{diameter of wheel}}$. (17)

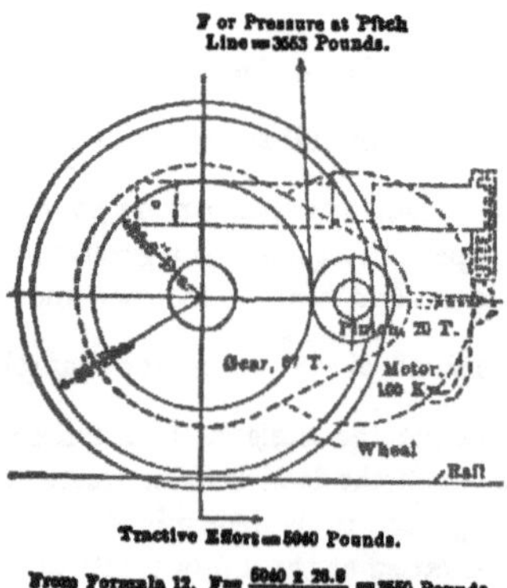

From Formula 12, $F = \frac{5040 \times 28.8}{38} = 3550$ Pounds

FIG. 49. RAILWAY GEARS.

Example: A car weighing 60 tons driven by four motors accelerating at the rate of 1½ miles per hour, per second, reaches the peak of its starting torque when at a speed of 20 miles per hour. The gears are 20 and 67 teeth 2½ diametral pitch (1.26 inches circular pitch) 5¼-inch face. The diameter of the car wheels is 38 inches. It is required to know the maximum fiber stress in pinion tooth. The power exerted by motors at its peak is 400 kilowatts (800 amperers at 500 volts). See Fig. 49.

Kilowatts per motor $Kw = \frac{400}{4} = 100\ Kw$

Pitch diameter of gear $D' = \frac{67}{2\frac{1}{2}} = 26.8$ in.

Ratio of gear to wheel $R = \frac{26.8}{38} = 0.705$

Speed of gear in feet per minute

$$V = 88 \times 20 \times 0.705 = 1{,}241 \text{ feet per minute.}$$

Force at pitch line $F = \frac{100 \times 44{,}102}{1{,}241} = 3{,}553$ pounds.

Fiber stress in pinion tooth

$$S = \frac{3{,}553}{1.26 \times 5.25 \times 0.09 \times \frac{600}{600 + 1{,}241}} = 18{,}300 \text{ pounds per square inch.}$$

STRENGTH OF GEAR TEETH

Lewis

W = load transmitted in pounds (same as value F),
p' = circular pitch,
f = face,
y = factor for different numbers and forms of teeth (Table 11),
S = safe working stress of material,
V = velocity in feet per minute,

$$W = Sp'fy\frac{600}{60 + V}. \qquad (18)$$

Number of Teeth	Value of Factor y Involute 20°	Value of Factor y Involute 15° Cycloidal	Value of Factor y Radial Flanks	Number of Teeth	Value of Factor y Involute 20°	Value of Factor y Involute 15° Cycloidal	Value of Factor y Radial Flanks
12	0.078	0.067	0.052	27	0.111	0.100	0.064
13	0.083	0.070	0.053	30	0.114	0.102	0.065
14	0.088	0.072	0.054	34	0.118	0.104	0.066
15	0.092	0.075	0.055	38	0.122	0.107	0.067
16	0.094	0.077	0.056	43	0.126	0.110	0.068
17	0.096	0.080	0.057	50	0.130	0.112	0.069
18	0.098	0.083	0.058	60	0.134	0.114	0.070
19	0.100	0.087	0.059	75	0.138	0.116	0.071
20	0.102	0.090	0.060	100	0.142	0.118	0.072
21	0.104	0.092	0.061	150	0.146	0.120	0.073
23	0.106	0.094	0.062	300	0.150	0.122	0.074
25	0.108	0.097	0.063	Rack	0.154	0.124	0.075

Table 11—Values of Factor y for Lewis Formula

Safe working stress S for 0.30 carbon steel = 15,000
Safe working stress S for 0.50 carbon steel = 25,000
Safe working stress S for cast iron = 8,000
Safe working stress S for rawhide = 5,000

AVERAGE VALUES FOR S

Mr. Lewis' formula for the strength of gears originally read: $W = Sp'fy$, a table being given in which the allowable stress of the material S was reduced as the speed of the gear was increased as follows:

SPEED OF TEETH IN FEET PER MINUTE	100 OR LESS	200	300	600	900	1200	1800	2400
Cast Iron........	8,000	6,000	4,800	4,000	3,000	2,400	2,000	1,700
Steel............	20,000	15,000	12,000	10,000	7,500	6,000	5,000	4,300

SAFE WORKING STRESSES S IN POUNDS PER SQUARE INCH FOR DIFFERENT SPEEDS

Later Carl G. Barth introduced an equation, $\frac{600}{600 + V}$, which gives practically the same result as the table when added to the formula, the value S being the safe working stress per square inch of the material used, or

$$W = Sp'fy \frac{600}{600 + V}.$$

Mr. Barth's equation is the one commonly accepted. It is evident, however, that this value will vary for different conditions, the design and workmanship being important factors in its proper determination.

The load is reduced as the speed increases on account of impact. It is evident that an accurately spaced and generated gear should have a much higher value than one cut by ordinary methods.

It is also evident that helical and herringbone gears, owing to the nature of their tooth contact, should have a much higher value, as they operate under entirely different conditions, therefore are capable of heavier loads at higher speeds for the same area of tooth contact. Rawhide gears should also have a higher factor, as rawhide will absorb shocks that would affect harder materials.

In the absence of all vibration, and with an indeflectable material, this equation could be eliminated from the formula for strength and wear. These are conditions that can never be attained, but it is evident that this value will stand extended investigation.

STRENGTH OF BEVEL GEARS

In general apply the Lewis formula for spur gears, figuring the safe load from the average pitch diameter, or, stated a little differently, the velocity in feet per minute and the pitch is to be taken at the average pitch diameter, otherwise the gear is to be treated as a spur.

Let N = number of teeth.
p' = circular pitch.
D' = pitch diameter.
b = face width.
p'_a = average circular pitch.
D_a = average pitch diameter.
a = apex distance.
E = center angle.
S = safe working stress.
V_a = velocity (average) in feet per minute.

In order to get the average pitch we must first determine the apex distance a. Now

$$a = \frac{D'}{2 \sin. E}.$$

The average pitch is the pitch at the center of the gear face f. Above this section the tooth strength increases and below this point it decreases. The

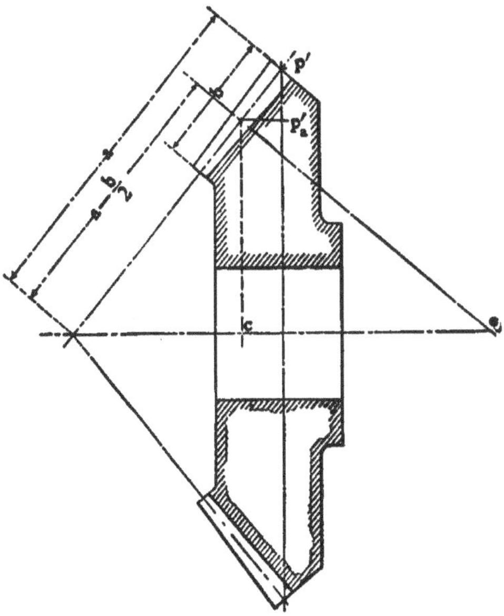

FIG. 50. DIAGRAM FOR STRENGTH OF BEVEL GEARS.

mean strength of the tooth is, therefore, located at this point. Thus it is the proper dimension to use in determining the strength of the tooth. This average pitch is found from the following equation:

$$p'_a = \frac{p'\left(a - \frac{b}{2}\right)}{a}.$$

This formula is derived as follows: By referring to Fig. 51, it is evident that the pitch of the tooth at p' is to the apex distance a as the pitch at p'_a is to the mean apex distance $\left(a - \frac{b}{2}\right)$.

The average pitch diameter D_a is found from formula (3) by substituting the average pitch. The average velocity V_a is found from formula (4) by using the average pitch diameter. Then

$$a = \frac{D}{2 \sin E}. \qquad (1)$$

$$p'_a = \frac{p'\left(a - \frac{b}{2}\right)}{a}. \qquad (2)$$

$$D_a = Np'_a \, 0.3183. \qquad (3)$$

$$V_a = 0.2618 D_a (\text{r. p. m.}). \qquad (4)$$

$$\text{Safe load} = Sp'_a by\left(\frac{600}{600 + V_a}\right). \qquad (5)$$

$$\text{Horse-power} = \frac{\text{Safe load} \times V_a}{33{,}000}. \qquad (6)$$

The values for the factor y are from the Lewis formula.

An illustrative example is as follows: What power may be transmitted by a pair of miter gears of the following dimensions: 30 teeth, 2-inch pitch, 5-inch face, 19.107-inch pitch diameter, at 50 revolutions per minute? The material is cast iron.

$a = \frac{19.107}{2 \times 0.707} = 13.51$ inches, the apex distance;

$p'_a = \frac{2 \times (13.51 - 5/2)}{13.51} = 1.63$ inches, the average pitch;

$D_a = 30 \times 1.63 \times 0.3183 = 15.5$ inches, the average pitch diameter;

$V_a = 0.2618 \times 15.5 \times 50 = 203$, the average velocity in feet per minute;

The safe load $= 8{,}000 \times 1.63 \times 5 \times 0.102 \times \left(\frac{600}{600 + 203}\right) = 4{,}970$ pounds.

The horse-power $= \frac{4{,}970 \times 203}{33{,}000} = 30.5$.

The teeth in bevel gears are more strongly shaped than the teeth of spur gears of the same pitch and number. This increase is represented by the radius $e - p'_a$ in Fig. 51, compared with the radius at the point p'. The corresponding number of teeth for this larger radius is found by the expression

$$\frac{N}{\cos E}.$$

When selecting the constant y, however, it is well to disregard this increase, as it will tend to compensate for the loss in efficiency due to the use of bevel gears.

NOTE ON FORMULAS FOR THE STRENGTH OF GEAR TEETH

The Lewis formula is not directly applicable to all types of gears, unless they are of the straight tooth variety with teeth of standard proportions. The absence of shock—sudden transference of load between successive teeth—in certain types of gears and the axial thrust introduced by an oblique arrangement of teeth bring in conditions which necessitate the modification of the original formula as corrected by Barth and as adapted to bevel gearing. W. C. Bates found that the transference of load from tooth to tooth without shock in herringbone gearing modified the speed factor ratio added to the Lewis formula by Barth by changing the constant 600 to 1,200.

In the case of spiral gears, and also helical gears, the same modification of the formula is necessary, as these types are similar so far as elimination of sudden application of load is concerned. A second modification is necessary for gears with obliquely arranged teeth, as the total load on any type of spiral gear tooth is the resultant of the transmitted load and the thrust and is equal to the transmitted load multiplied by the secant of the spiral angle. The safe load is then the load which could be carried were it not for the thrust multiplied by the cosine of the spiral angle.

As the factor y for the Lewis formula, given in Table 11, is based upon the condition that the proportions of the teeth conform to standard still another modification in this factor is necessary when employing the Lewis factor to gears with long or short addenda. In the case of modified teeth (see page 26), the breaking moment arm is altered and the thickness of the teeth at the top of the root fillets is affected.

These various modifications of the original Lewis formula convert that equation into a series of formulas, as follows, the strength factors for which are given in Table 12.

MODIFIED LEWIS FORMULAS

STANDARD ADDENDUM,

Straight Tooth Spur Gears $\quad \text{Safe Load} = Sp'fy \dfrac{600}{600 + V}$

Herringbone Gear $\quad \text{Safe Load} = Sp'fy \dfrac{1,200}{1,200 + V}$

Helical and Spiral Gears $\quad \text{Safe Load} = Sp'fy \cos A \dfrac{1,200}{1,200 + V}$

Straight Tooth Bevel Gear $\quad \text{Safe Load} = Sp''fy \dfrac{600}{600 + V'}$

Herringbone Bevel Gear $\quad \text{Safe Load} = Sp''fy \dfrac{1,200}{1,200 + V'}$

Helical and Spiral Types Bevel Gears $\quad \text{Safe Load} = Sp''fy \cos A \dfrac{1,200}{1,200 + V'}$

SHORT ADDENDUM (0.3 × working depth of tooth)

Straight Tooth Spur Gears Safe Load $= Sp'fy' \frac{600}{600 + V}$

Herringbone Gear Safe Load $= Sp'fy' \frac{1,200}{1,200 + V}$

Helical and Spiral Gears Safe Load $= Sp'fy' \cos A \frac{1,200}{1,200 + V}$

LONG ADDENDUM (0.7 × working depth of tooth)

Straight Tooth Spur Gears Safe Load $= Sp'fy'' \frac{600}{600 + V}$

Herringbone Gear Safe Load $= Sp'fy'' \frac{1,200}{1,200 + V}$

Helical and Spiral Gears Safe Load $= Sp'fy'' \cos A \frac{1,200}{1,200 + V}$

SHORT ADDENDUM (0.3 × working depth of tooth)

Straight Tooth Bevel Gear Safe Load $= Sp''fy' \frac{600}{600 + V'}$

Herringbone Bevel Gear Safe Load $= Sp''fy' \frac{1,200}{1,200 + V'}$

Helical and Spiral Types Bevel Gears Safe Load $= Sp''fy'' \cos A \frac{1,200}{1,200 + V'}$

LONG ADDENDUM (0.7 × working depth of tooth)

Straight Tooth Bevel Gear Safe Load $= Sp''fy'' \frac{600}{600 + V'}$

Herringbone Bevel Gear Safe Load $= Sp''fy'' \frac{1,200}{1,200 + V'}$

Helical and Spiral Types Bevel Gears Safe Load $= Sp''fy'' \cos \frac{1,200}{1,200 + V'}$

Notation:

Safe working stress (pounds per square inch) . . . . S
Circular pitch . . . . p'
Average circular pitch (bevel gears) . . . . p''
Face of gear . . . . f
Pitch circle velocity (feet per minute) V
Average pitch circle velocity—bevel gears (ft. per minute.) . . . . V'
Strength factor—standard addendum . . . . y
balanced thrust, short addendum . . . . y'
balanced thrust, long addendum . . . . y''
unbalanced thrust, short addendum . . . . Y'
unbalanced thrust, long addendum . . . . Y''
Spiral angle—helical and spiral gears, unbalanced thrust . . . . A

TABLE 12.—STRENGTH FACTORS FOR MODIFIED LEWIS FORMULAS

NUMBER OF TEETH	VALUES OF FACTORS					NUMBER OF TEETH
	y	y'	y''	Y'	Y''	
8	0.052	0.051	0.062	0.053	0.111	8
9	0.055	0.054	0.066	0.058	0.127	9
10	0.059	0.056	0.070	0.064	0.142	10
11	0.063	0.059	0.074	0.069	0.157	11
12	0.067	0.062	0.078	0.076	0.172	12
13	0.070	0.063	0.081	0.079	0.185	13
14	0.072	0.065	0.083	0.082	0.195	14
15	0.075	0.067	0.086	0.086	0.208	15
16	0.077	0.069	0.089	0.089	0.216	16
17	0.080	0.071	0.092	0.093	0.227	17
18	0.083	0.073	0.095	0.097	0.238	18
19	0.087	0.076	0.099	0.101	0.251	19
20	0.090	0.079	0.103	0.105	0.263	20
21	0.092	0.080	0.105	0.108	0.272	21
23	0.094	0.082	0.108	0.111	0.282	23
25	0.097	0.084	0.111	0.115	0.294	25
27	0.100	0.086	0.114	.....	.....	27
30	0.102	0.088	0.117	.....	.....	30
34	0.104	0.089	0.120	.....	.....	34
38	0.107	0.091	0.122	.....	.....	38
43	0.110	0.093	0.126	.....	.....	43
50	0.112	0.095	0.128	.....	.....	50
60	0.114	0.096	0.130	.....	.....	60
75	0.116	0.097	0.132	.....	.....	75
100	0.118	0.099	0.134	.....	.....	100

FACTOR OF SAFETY FOR GEARS

The load on the teeth of gears made from forgings may be such as to strina the material close to its elastic limit (based upon the worn thickness of tooth), if it is free from flaws. For castings this is not a safe rule, as there are always hidden defects to a greater or less extent. As long as the strain is kept below this point, excessive wear will not take place, but if this point is exceeded but slightly, rapid wear, or fracture of the teeth, is sure to result. For reasonable service, a factor of safety of 1.5 should be used if the load is uniform. Thus, for a forged steel gear having an elastic limit of 20,000 pounds per square inch, the safe load would be $\frac{20,000}{1.5} = 13,330$ pounds per square inch. For cast steel, free from apparent defects, a factor of 2 is recommended; thus, for this same strength in steel in a casting the safe load would be $\frac{20,000}{2} = 10,000$ pounds per square inch.

The elastic limit meant in this connection is the real elastic limit of the material as taken by an accurate extensometer and not by the drop of the beam, or by caliper measurement, as has been commercial practice. This instrument detects the first indications of permanent set in the test piece, showing that the safe load for that material has been exceeded; the drop of the beam is not apparent for some time after this. For untreated mild steels this point is sometimes at one-half the drop of beam; for the higher grades, however, the two points are closer together.

Such an instrument is described by T. O. Lynch in a paper on "The Use of the Extensometer for Commercial Work" read before the American Society for Testing Materials; Philadelphia, published in the Proceedings for 1908, volume 8.

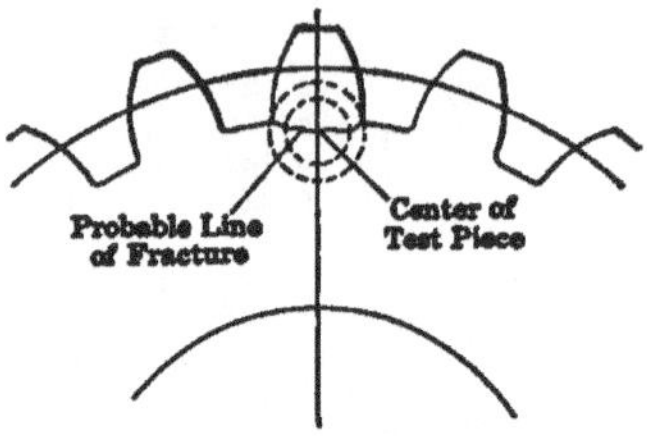

FIG. 51. LOCATION OF TEST PIECE.

It must be pointed out that gear steels should have an ample reduction of area to guard against sudden fracture. Test pieces should be cut with the center of tooth a little below the bottom line, say 0.07 of the circular pitch, as illustrated by Fig. 51, as it is through this point that the tooth generally breaks out.

The strength of the material in gears will be found to vary as much as 30 per cent. in different parts of the tooth and rim, therefore a settled point for cutting out test pieces is necessary if uniform, safe, or accurate results are to be expected. It does not greatly matter if the threaded portion of the test piece projects into the tooth space, as it will on all gears 2½ diametral pitch and finer, so long as the 0.505-inch portion of the piece is clear. When a 0.505-inch test piece (⅕ of a square inch in area) cannot be obtained in this manner, make one 0.2525 inch in diameter (1/20 of a square inch in area), leaving the threaded portion ⅝ inch instead of ¾ inch, which is standard.

Note that elastic limits given in table for wear of gear teeth is by drop of beam.

THE STRENGTH OF SHROUDED GEAR TEETH

In regard to strength of shrouded gear teeth, Wilfred Lewis submits the following, originally published in AMERICAN MACHINIST of Jan. 30, 1902.

"I do not know of any careful analyses of or experiments on the strength of shrouded gear teeth, but I have some recollection of a tradition in vogue about twenty-five years ago that from one-fourth to one-half might be added to the strength by shrouding.

"There are, however, a number of cases to be considered; the shrouding may extend to the pitch line only or to the ends of the teeth, and it may be single or double. Formerly the practice of shrouding pinion was more common than it is to-day, because the advent of steel as a cheap construction material makes it possible to obtain unshrouded pinions of greater strength than the cast-iron gears with which they engage, and now steel pinions have generally supplanted the old cast-iron shrouded ones which were naturally more roughly shaped, because harder to fit, and more difficult to assemble by reason of the shrouding. In my investigation of the strength of gear teeth I therefore assumed that the time for shrouded gears has passed, at least as far as machine tools were concerned, but they are possibly used as freely as ever on roll trains and some other classes of machinery, so that the problem may still be worthy of consideration from a practical standpoint. Rankine, in his 'Applied Mechanics,' rather summarily disposes of the strength of gear teeth by assuming the load that may be carried on one corner to be all that any tooth is good for. When so loaded, it is shown that the corner will break off at an angle of 45 degrees, and the strength of a tooth of any width is then no greater than that of a tooth whose width is twice its hight. So, if the hight of a tooth is 0.65 pitch, the strength, according to Rankine, should be taken for a width of only 1.3 pitch. Faces wider than this would be no stronger, and shrouding at one end would make no difference.

"But his assumption is untenable, because no maker of machinery who values his reputation will put gears into service bearing only at one end, and should they be so started, an even distribution of pressure is sooner or later effected by the natural process of wear.

"A comparison of strength between shrouded and unshrouded gears should therefore be made on the assumption of uniform distribution of pressure across the faces of their teeth, and for this purpose it will be expedient to neglect the influence of tooth forms, which would complicate and prolong the investigation, and treat all teeth simply as rectangular prisms, which may or may not be supported by shrouding. Rankine and Unwin have both been contented to estimate the actual strength of teeth as though they were rectangular prisms, and, although this is far from the truth, it is certainly more admissible as a basis of comparison for another variable than as an approximation for a direct result. The effect of this assumption will be to exaggerate the value of shrouding, and for the present it will be sufficient to indicate roughly the maximum benefit to be anticipated.

"In Fig. 52 a gear tooth is shrouded at one end, and the problem is to determine its strength as compared with the same tooth not shrouded. For convenience, the thickness, or half the pitch may be taken as unity, and the hight as 1.3. The load W is assumed to be applied uniformly along the end of the tooth over the face b, making the full load bW. If there were no shrouding, the strength of this tooth would be measured by the transverse resistance at its root t, and if broken at the root as shown in Fig. 53, we may consider how much strength could be given to it by shrouding alone.

"A tooth thus broken would have some strength as a cantilever imbedded in the shrouding, but more as a shaft subjected to torsion, and for the shape here assumed the torsional strength alone will probably exceed the combined strength due to torsion, and flexure for any actual shape.

"The rectangular tooth whose sides are h and t cannot be treated as an ordinary shaft because its neutral axis is at one side instead of, as usual, at the center of gravity. It must therefore be treated as one half of a shaft whose sides are $t \times 2h$ or 1×2.6, for which the moment of resistance is about $0.6S$, where S is the shearing stress at the end of a tooth. For the unit load W we have $0.6S = 1.3W$, or $S = 2.17W$, and for the width b we have $S_1 = bS = 2.17\,bW$.

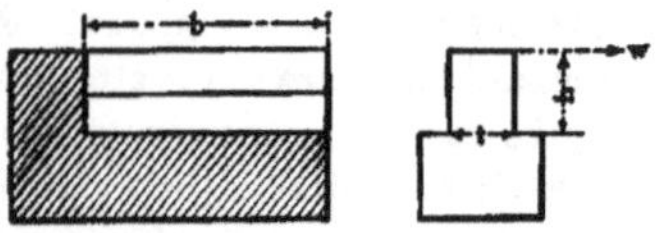

FIG. 52.

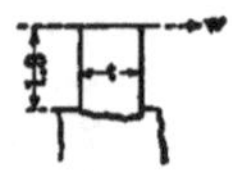

FIG. 53.

DIAGRAMS ILLUSTRATING THE STRENGTH OF SHROUDED GEAR TEETH.

"Thus the maximum intensity of shearing S_1 is found to be a little more than twice the full load bW.

"On the other hand, for an unshrouded tooth the transverse stress f at the root of a tooth depends only upon W and the relation is expressed by the equation $f = 7.8W$. In these terms,

$$W = \frac{f}{7.8},$$
$$= \frac{S_1}{2.17b},$$

whence

$$f = \frac{3.5S_1}{b},$$

and assuming that the shearing stress S_1 may be $0.8f$, we have $b = 2.8$. This means that a tooth 2.8 wide has as much strength as may possibly be added by shrouding at one end, but the question remains to be considered, under what conditions and to what extent can this possible strength be made effective? The development of stress is always accompanied by strain, and in the case of a shrouded tooth the unit load W must be divided between torsion and flexure. Obviously, if the tooth is very long, its stiffness under torsion will be so little as compared with its stiffness under flexure, that the benefit from shrouding will be inappreciable, and on the other hand if very short, the torsional stiffness will be preponderate. When a uniform distribution of load has been attained, as it must be by the action of wear, that part of a tooth farthest from the shrouding will sustain the greatest transverse stress and the load W will be divided at all points along the face of the tooth between torsion and bending directly as the stiffness encountered in these two directions or inversely as the relative strains. Each strain is relieved by the other, but the limit of strength is reached when either attains its maximum.

"Considering a cantilever loaded at the end, we have for the deflection y, under the stress f,

$$y = \frac{fh^2}{1.5\,E},$$

where $h = 1.3$ and E is the modulus of elasticity for flexure. Substituting this value of h we have

$$y = \frac{1.1f}{E}. \qquad (1)$$

"Considering the tooth as a rectangular shaft in torsion, it will be seen that the shearing stress for a distributed load decreases from the maximum S_1 at the shrouding to nothing at the other end of the tooth. For the unit load W the shearing stress is S, for an element dx, the stres is Sdx and for this stress at the distance x from the shrouding, the torsional deflection

$$dz = \frac{Sxdx}{G},$$

where G is the modulus for shearing. The total deflection for the load distributed over a length x therefore is

$$z = \frac{Sx^2}{2G},$$

or for the face b we have

$$z = \frac{Sb^2}{2G}.$$

"But since $G = 0.4E$, we may write

$$z = \frac{Sb^2}{0.8E}. \qquad (2)$$

"The value of S has been found to be $S = 2.17\,W$, and we have also found $= 7.8W$, therefore

$$S = \frac{2.17}{7.8}f,$$
$$= 0.28f.$$

and

$$z = \frac{Sb^2}{8E}.$$

may be written

$$z = \frac{0.35f\,b^2}{E}. \qquad (3)$$

"The deflection of an unshrouded tooth under the load W has been shown by equation (1), and, dividing this into equation (3), we have for the relation between y and z

$$\frac{z}{y} = 0.36^2.$$

For a very narrow tooth, letting $b = 1$, we have $z = 0.3y$; but since y and y are necessarily equal, when the tooth under consideration is attached at its root and also to the shrouding, the load W will be supported at both points, and it will necessarily be divided between them in the proportion of y to z, or as 1 to 3. The shrouding will carry $\frac{W}{1.3}$, or $0.77W$, and the root of the tooth $0.23W$.

"Similarly making, $b = 2$, or one pitch, we have $z = 1.2y$, and at this point the shrouding will carry $0.45W$ and the root of the tooth $0.55W$. Also, when $b = 3$, we have $z = 2.7y$, reducing the load on shrouding, to $0.27y$ and increasing the load at the root to $0.73W$. At $b = 4$, or 2 pitch, $z = 4.8y$ the load on shrouding drops to $0.17W$ and the load at the root rises to $0.83W$.

"The average width of gear faces is probably about 2.5 pitch, and for $b = 5$ we have about $0.12W$ carried by the shrouding and $0.88W$ carried by the tooth acting as a cantilever. We may therefore conclude that the strength of an ordinary pinion shrouded at one end only is not increased more than 12–88 or about 13 per cent., by the shrouding. Indeed, this is only the result of a first approximation, and for the successive proportions of W thus credited to the shrouding new values of z might be estimated to be used as the basis of a second approximation. But we will not continue the process—it is sufficient to know that our valuation of the effect of shrouding is high. A double-shrouded pinion running with a gear whose face is $2.5p'$ will be about 3 pitch between shroudings and its strength will be about the same as that just found for $b = 3$. An ordinary pinion will not therefore be increased in strength by double shrouding more than 37 per cent., and it is probably safe to say that a more elaborate investigation will reduce the additional strength to 10 per cent. for single shrouding and 30 per cent. for double shrouding.

"When the shrouding extends to the pitch line only, the shearing strength of its attachment to a tooth is reduced, but the elastic relations upon which the strength at the root depends remain practically the same. In this case the shearing strength instead of the transverse strength limits the strength of a tooth, and the strength is apparently less than for full shrouding.

"The effect of shrouding is clearly to prevent the adjacent part of a tooth from exercising its strength as a cantilever. The shrouding carries what the tooth itself might carry almost as well. A heavy link in a light chain adds nothing to the strength of the chain, and teeth which are not strong all over need not be strengthened in spots. A little more face covering the space occupied by shrouding is more to the purpose for durability as well as for strength, and when this fact is appreciated I believe the practice of shrouding will disappear in rolling mills, as it has done in machine shops.

"In regard to the working stress allowable for cast iron and steel, I may say that 8,000 pounds was given as safe for cast-iron teeth, either cut or cast, and that 20,000 pounds was intended for ordinary steel whether cast or forged. These were the unit stresses recommended for static loads, and as the speed increased they were reduced by an arbitrary factor, depending upon the speed.

"The iron should be of good quality capable of sustaining about a ton on a test bar 1 inch square between supports 12 inches apart, and of course the steel should be solid and of good quality. The value given for steel was intended to include the lower grades, but when the quality is known to be high, correspondingly higher values may be assigned.

"In conclusion I may say that the crude investigation here given seems to justify the traditions referrred to that from $\frac{1}{4}$ to $\frac{1}{2}$ may be added to the strength of teeth by shrouding. If the teeth are very narrow, $\frac{1}{2}$ may be added, but generally, I believe, $\frac{1}{4}$ is enough and since writing the above I find that D. K. Clark almost splits the difference by adding $\frac{1}{3}$ for double shrouding. But the development of the full strength of gear teeth depends nearly as much upon the strength and stiffness of the gear journals as upon the teeth themselves, and no rules can be given for indiscriminate use."

NOTES ON SHROUDED GEARS

The use of shrouded gears, except in the case of cast units, has been discouraged, heretofore, by the difficulties imposed in making the shrouds integral parts of the gear, tying the gear teeth securely to the rim under the teeth. In machining generation of gears, the formation of such shrouds cannot be accomplished, so that if shrouding is resorted to the shrouds have to be made independent rings and subsequently attached in some manner to the unshrouded gear. This is a difficult and expensive task at best and it is quite problematic whether the slight increase in gear strength realized—the increase in strength being only a fraction of that attained could the shrouds be made integral parts of the gear—is justified.

The recent commercial development of the rolling method of gear production—see Section XVI—entirely alters this situation, for in the rolling process shrouds are formed on the forged gear blanks. These webs, tying the forged teeth to the gear rim, are cut away in the production of unshrouded rolled gearing, but they may be retained in their entirety or in any proportion for the purpose of increasing the strength of the gears with rolled teeth. This distinctive peculiarity of rolled gearing may exert considerable influence on the design of gearing in the future. The addition of the shrouding not only increases the safe load the gears can carry but also increases substantially the speed at which they can be operated. Its chief advantage is, however, the fact that by properly proportioned shrouding the individual power capacities of the gears constituting a train may be brought up to that of the strongest gear—the weakest made the equal of the strongest.

WEAR OF GEAR TEETH

The Lewis formula is the only accurate method of figuring the power of gears so far as the strength of the teeth is concerned, but takes no account whatever of wear, and the value of the tooth surfaces to resist crushing of the material. Trouble from this source is a common experience, although not properly understood, as it is sometimes difficult to account for the "mysterious" failure of gears that were apparently of ample strength. It is noticed that gears generally fail through wear and not by fracture of the teeth, also that the teeth often break at a load far below that which is considered safe. More attention, therefore, should be given to this point.

A certain combination of diameters will carry but a certain load per unit of face, irrespective of the pitch of the gears, so that there is no gain in an increased pitch above that just sufficient to resist fracture. This pitch may be found as usual by the Lewis formula, but the actual strength of the material should be used. The material in a 1-inch pitch tooth is stronger proportionately than the same material in a 2-inch pitch tooth. This is caused by the fact that nearer the exterior the material is stronger and of a closer grain, due to rapid cooling. A tooth is stronger at the top than at its root. It would seem as if tests of material should be made for flexure and not for tensile strength, as the tooth breaks through bending. The average ultimate strength of cast iron for flexure is 38,000 pounds per square inch, while the tensile strength is 24,000 pounds per square inch.

Aside from this feature the surface hardness should also be considered irrespective of the strength of material. For instance, a pressure of 5,000 pounds per unit of contact would be allowed for a case-hardened steel surface, while but 1,500 would be allowed for the same material in its unhardened condition.

The relative hardness of material, in conjunction with the co-efficient of friction for different grades and hardness of material engaging, will supply the safe load A per unit of area.

It is true that the arc of rolling contact in gears is very small; the balance is sliding contact, which increases proportionately over the rolling contact as the pitch points separate, or as the tooth disengages, and decreases as the tooth enters contact until the pitch points again engage where it is rolling.

The wearing qualities of the teeth depend greatly upon their condition when put into service. If a little care is used to obtain a smooth surface at the start and allow the teeth to find their proper bearing, the gears will wear indefinitely longer than if put under full load when new, no matter how accurately the teeth are cut. Also a gear once started to cut can often be saved by the timely application of a fine file, finally smoothing the teeth with and oil stone.

A series of experiments to determine the proper load per unit of contact (A) for different grades and hardness of material used would certainly lead to a fuller knowledge of the capacity of gears for the transmission of power and leave less to supposition on the part of the designer. On account of the peculiar nature of the tooth contact it is quite likely that the best manner to reach accurate results would be with gears made from th ematerials under consideration. The values given in Chart 2 are the best obtainable at the present writing.

The idea of limiting the load to the proportion of the gear diameters irrespective of the pitch (for a unit of face) may at first appear startling, but when we consider that the radii from which the tooth is drawn are always proportional to the *pitch diameter* of the gear and not to the *pitch*, and that the teeth in contact are actually two cylinders rolling and slipping upon each other, it appears more reasonable. See Fig. 54. It should be understood, however, that the diameter and position of these rollers change constantly throughout the contact, and that a gear made in strict accordance with Fig. 54 would not give a uniform movement. It illustrates the principle however.

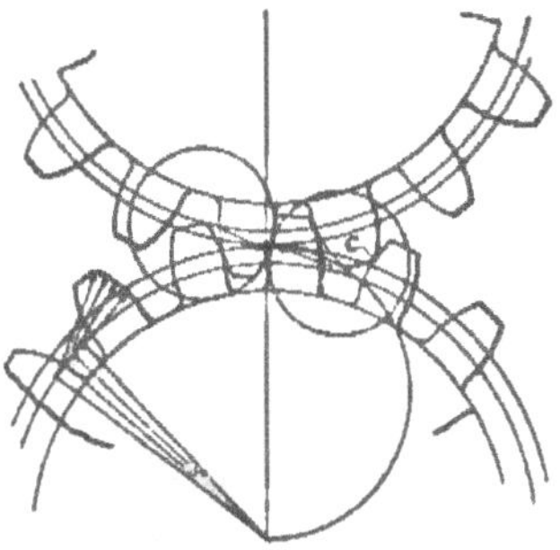

FIG. 54. GEAR TOOTH ACTION.

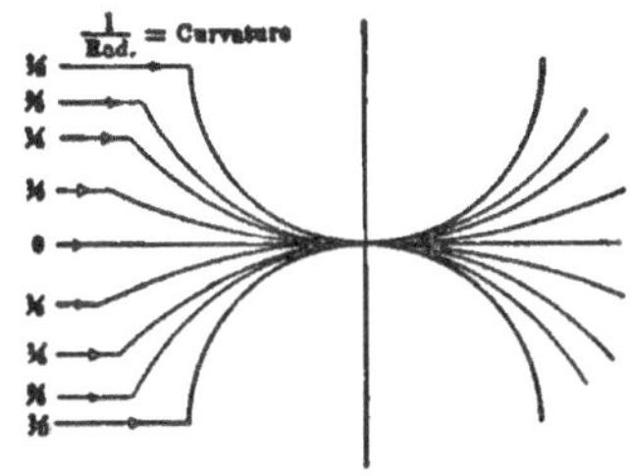

FIG. 55. CURVATURES.

To secure safe results the flank or shortest radius is used in these formulas.

The following is the gist of an article by Harvey D. Williams, in the American Machinist, with its application to gear transmissions:

"The curvature of a plane curve is defined by mathematicians as the change of direction per unit of length, and is equal to the reciprocal of the radius of curvature at the point considered. Thus in going once around a circle of radius R the distance traversed is the circumference $2\pi R$, while the change of direction is in circular measure or radius 2π.

"The change in direction per unit of length is therefore

$$\frac{2\pi}{2\pi R} = \frac{1}{R}.$$

"Accordingly the curvature of a 2-inch circle is 1, that of a 1-inch circle is 2, that of a ½-inch circle is 4, and that of a straight line is $\frac{1}{\infty} = 0$, etc. . . .

"The curvature of a straight line being zero, that of an arc may be said to be its curvature in relation to the straight line, or its *relative curvature* to the straight line. Similarly, in comparing the curvature of two arcs, it will be convenient to use the term 'relative curvature' instead of the difference of curvature, meaning thereby the algebraic difference of the curvature as dimensioned in Fig. 55.

"It will be seen that when two plane curves are tangent to each other externally the relative curvature is to be found by adding the respective curvatures, and when they are tangent internally the relative curvature is to be found by subtracting.

"The amount of contact between plane curved profiles is measured by the reciprocal of the relative curvature." See formulas 21, 22, and 23.

In each of the cases shown by Figs. 56, 57, and 58 the contact is 4, and if these profiles were made of the same material and the same width of face, they would be equally efficient as regards their ability to withstand pressure.

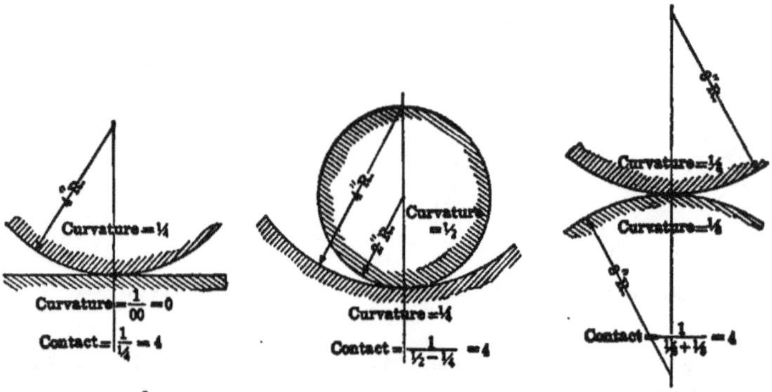

FIG. 56. FIG. 57. FIG. 58.

CONTACTS OF THE SAME CURVATURE.

DEVELOPMENT OF FORMULAS

Let

C = contact.
R^1 = flank radius of the gear.
r^1 = flank radius of the pinion.
f = face width.
V = velocity in feet per minute.
A = safe crushing load per unit of contact.
D^1 = pitch diameter of the gear.
d^1 = pitch diameter of the pinion.
a = angle of obliquity.
W^c = safe load on the tooth to resist crushing and wear.
W = safe load on the tooth to resist fracture.

Then

$$R^1 = \frac{D^1}{2} \sin a, \tag{19}$$

and

$$r^1 = \frac{d^1}{2} \sin a. \tag{20}$$

For spur gears

$$C = \frac{1}{\frac{1}{r^1} + \frac{1}{R^1}}. \tag{21}$$

For internal gears

$$C = \frac{1}{\frac{1}{r^1} - \frac{1}{R^1}}. \tag{22}$$

For racks

$$C = \frac{1}{\frac{1}{r^1} \pm 0} = r_1 = \frac{r^1}{1}. \tag{23}$$

$$W^c = CfA\left(\frac{600}{600 + V}\right). \tag{24}$$

It is desirable that the gear and pinion should wear equally to avoid the necessity of engaging a new pinion with a partly worn gear, thereby decreasing the life of both. It is assumed that wear is proportional to the hardness of the material; obviously the pinion should be harder than the gear in proportion to the ratio of the drive. Therefore, to secure equal wear in a pair of gears having, say, a ratio of 4 to 1, the pinion should be made four times as hard as the gear.

I have thought that a hard pinion would tend to preserve a softer gear, but as no data are found to sustain this theory, and as the value calculated for Chart 1 tends toward a much softer gear than was originally thought proper to make the best wearing combination for certain ratios, this has not been taken into account. It is assumed, therefore, that a hard pinion will neither preserve nor influence the wear of a softer gear. Therefore, it may be assumed safely that a hard gear will not influence the wear of a softer pinion. Chart 1 is made on this basis. The wear of gear and pinion is determined independently if the proper combinations of hardness are not used.

The wear is based entirely on the pinion hardness, the gear performing the same amount of work, but having less wear on account of the greater number of teeth in use in proportion to the ratio of the gears. For instance, in a gear drive having a ratio of 4 to 1, the gear may be 75 per cent. softer than the pinion for equal wear.

Thus the wear of the gear is found according to the pinion hardness that is proper for the ratio of the gears irrespective of the material actually used for the pinion.

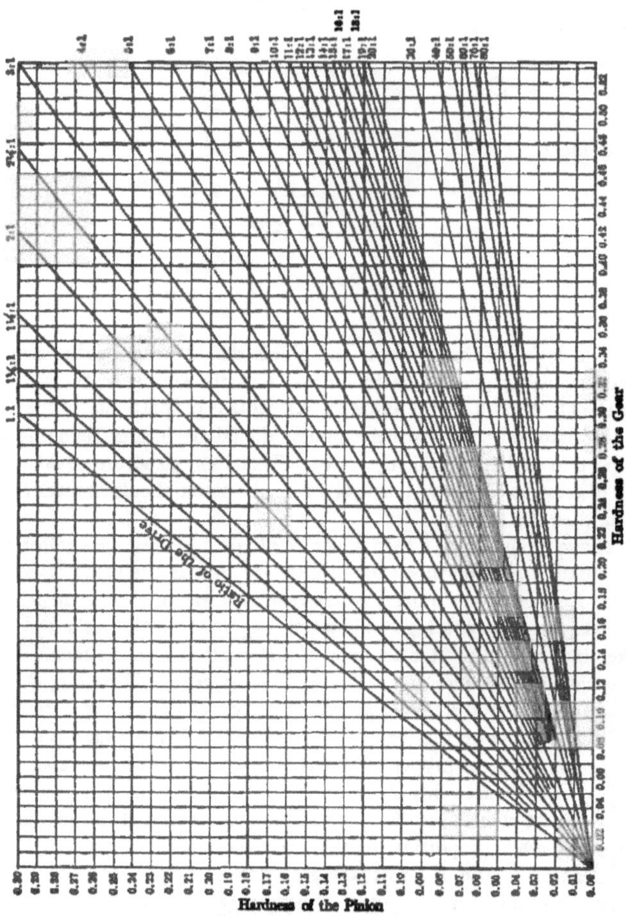

CHART 1. RELATIONS BETWEEN GEAR AND PINION HARDNESS FOR VARIOUS RATIOS OF DRIVES.

For example: If a pinion of a hardness represented by 0.15 (see Chart 1) engages a gear of 0.35 hardness, the ratio being 4 to 1, the wear of the gear will be in accordance with the pinion hardness found opposite the line of ratio and over the gear hardness (0.35), which in this case is 0.17½. In this event the gear will wear out first. On the other hand, if the pinion had been of 0.20 hardness, the pinion would wear out first. The value for the wear of the gear would remain 0.17½.

It is thought that the elastic limit of a material follows the hardness, therefore the wear may be determined from the elastic limit. The points of hardness, however, are used in the accompanying charts for convenience; Chart 2

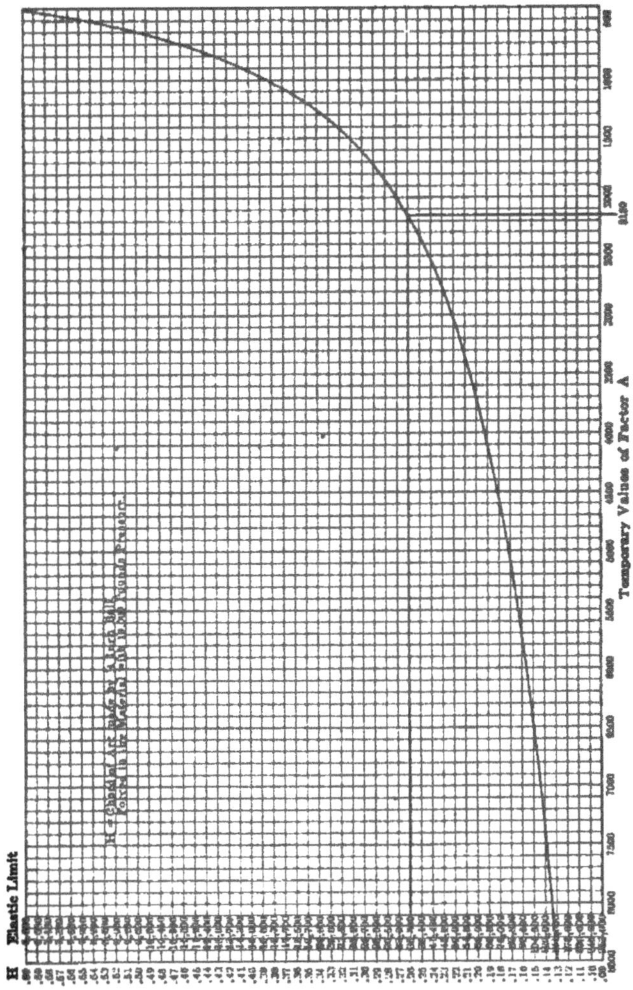

CHART 2. RELATIONS BETWEEN THE ELASTIC LIMIT AND VALUES OF THE FACTOR A.

gives the corresponding values. The hardness values given in these tables were obtained by pressing a 7/8-inch hardened steel ball into the surface of the material with a pressure of 10,000 pounds. The dimensions given are diameters of the indentations thus made. See Fig. 59. The comparisons in hardness were made inversely to the square of these diameters. The elastic limit was determined for one of these values; the comparison for others may be found by the same inverse proportion.

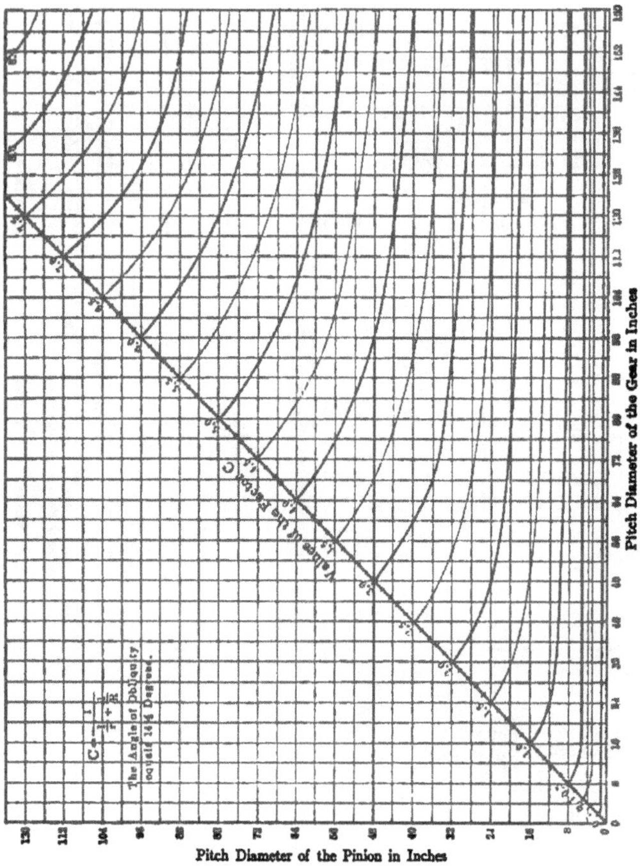

CHART 3. RELATIONS BETWEEN THE PITCH DIAMETERS OF MATING SPUR GEARS AND THE FACTOR C.

Thus, the square of 0.20 = 0.04 and the square of 0.30 = 0.09. Therefore, 0.20 would be $\frac{0.09}{0.04} = 2\frac{1}{4}$ times harder than 0.30. This is a proper combination for a gear ratio of $2\frac{1}{4}$ to 1. The elastic limit for 0.20 is 60,000 pounds per square inch. The elastic limit of $0.30 = \frac{60,000}{2\frac{1}{4}} = 26,700$ pounds per square inch.

For comparison with the Brinell scale; hardness value 0.22 inch in Chart 2 measures 4.6 millimeters; the hardness numeral for this impression being 167, the impression is made with a 10-millimeter ball at a pressure of 3,000 kilograms.

As the elastic limit of cast iron is very close to its ultimate tensile strength the ultimate strength may be used to determine the hardness. This was at first very confusing before it was found that the hardness followed the elastic limit, as cast iron under the ball test referred to would show a hardness equal to machine steel of twice its ultimate tensile strength.

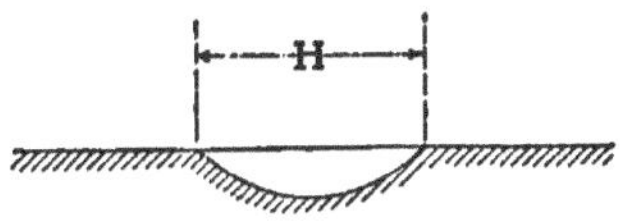

FIG. 59. CHORD MEASUREMENT OF HARDNESS TEST.

The values of *A* according to the hardness or elastic limit of the material (Chart 2) have been assumed as correct for gears operating 10 hours per day for a period of two years. If found in error their multiplier given in Table 15 for the time and conditions of service may be shifted without changing the original values of Charts 1 and 2. This table may be elaborated to any desired extent to cover various conditions. It is evident that a pair of gears will not last as long fastened to the ceiling or to insecure timbers as if mounted upon a proper concrete foundation. There are all manner of machine constructions to be considered as well as unknown overloads, the influence of flywheels and other things that are usually neglected. All this, however, is simple indeed when the Stygian darkness in which we are now wandering is considered.

TIME OF SERVICE	UNIFORM LOAD		
	CONTINUOUS	10 HOURS DAILY	5 HOURS DAILY
For 3 months	3.00	8.00	18.00
For 6 months	1.50	4.00	9.00
For 9 months	1.00	2.67	6.00
For 1 year	0.75	2.00	4.50
For 2 years	0.38	1.00	2.25
For 3 years	0.25	0.67	1.50
For 4 years	0.19	0.50	1.13
For 5 years	0.15	0.40	0.90
For 6 years	0.13	0.33	0.75
For 7 years	0.11	0.29	0.65
For 8 years	0.10	0.25	0.56
For 9 years	0.09	0.22	0.50
For 10 years	0.08	0.20	0.45

TABLE 13—MULTIPLIERS FOR FACTOR 'A'

According to the Conditions of Service and Desired Life of Gears

For gears subjected to 25 per cent. overload, multiply result by 0.80; for gears subjected to 50 per cent. overload, multiply result by 0.70; for gears subjected to 75 per cent. overload, multiply result by 0.60; for gears subjected to 100 per cent. overload, multiply result by 0.50; for gears operating in dust-proof oil case, multiply result by 1.50.

EXAMPLES

What is the safe load for a pair of spur gears properly mounted on concrete foundations to operate continuously for a period of five years before replacing? The gears are to run in oil in a dust-proof case driving an electric generator making 300 revolutions per minute from a turbine revolving at 1,200 revolutions per minute. The overload at no time will exceed 25 per cent. The gear has 84 teeth, 3 diametral pitch, 12-inch face and 28-inch pitch diameter. The pinion has 21 teeth, 3 diametral pitch, 12-inch face and 7-inch pitch diameter. The ratio is 4 to 1. The circumferential speed at the pitch line is 2,200 feet per minute. The pinion should be four times as hard as the gear for equal wear and, according to Chart 1, if a cast-steel gear of 30,500 pounds' elastic limit which is assumed to have a hardness value of 0.28, the pinion hardness should be represented by 0.14, which represents an elastic limit of 120,000 pounds per square inch. This may be obtained by hardening chrome nickel or other high-grade steel. High-carbon steel should be avoided for this purpose on account of its tendency to crystallize.

Referring to Chart 2 it is found that the temporary value of A is 7,500 pounds. The multiplier for time of service is 0.15; the multiplier for overload is 0.80; the multiplier for the oil case is 1.50. Thus the final value of A = 7,500 × 0.15 × 0.80 × 1.50 = 1,350 pounds.

In the formula

$$W^{e} = CfA\,\frac{600}{6{,}00 + V},$$

where

$$\begin{aligned} W^{e} &= \text{the safe load in pounds (to be determined)},\\ C &= 0.70 \text{ (from Chart 3)},\\ f &= \text{the face, 12 inches},\\ A &= 1{,}350 \text{ and } \frac{600}{600 + 2{,}200} \text{ (from Chart 4)} = 0.21, \end{aligned}$$

then

$$W^{e} = 0.70 \times 12 \times 1{,}350 \times 0.21 = 2{,}380 \text{ pounds},$$

or

$$\frac{2{,}380 \times 2{,}200}{33{,}000} = 158 \text{ horse-power}.$$

The strength of the teeth must now be checked by the Lewis formula to guard against fracture at this load. We find by this method a safe working load of 317 horse-power.

This illustrates that the teeth are capable of carrying 317 horse-power, but as shown by the value W^e they would wear out in about one-half the specified time if such a load were applied. If the example had read "ten hours

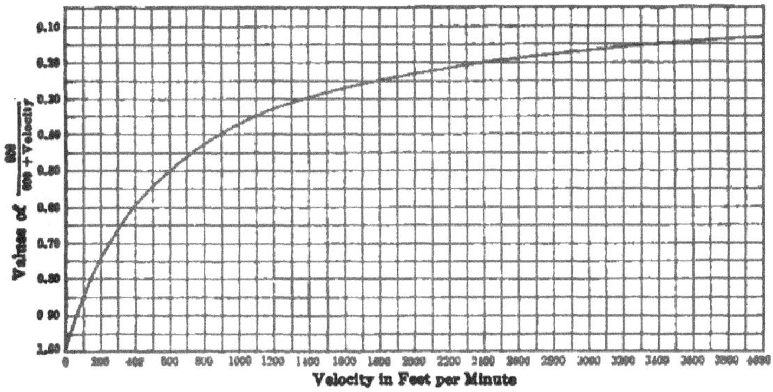

CHART 4. RELATIONS BETWEEN THE QUANTITY $\frac{600}{600 + velocity}$ AND THE VELOCITY.

per day" instead of "continuous," all other conditions remaining the same, we would have:

$$A = 7{,}500 \times 0.40 \times 0.80 \times 1.50 = 3{,}600 \text{ pounds,}$$

$$\frac{3{,}600 \times 2{,}200}{33{,}000} = 240 \text{ horse-power.}$$

For this load, however, the teeth would be liable to fracture.

RELATIVE IMPORTANCE OF STRENGTH AND HARDNESS

It would appear that the actual *strength* of the tooth is to be a secondary consideration, figured only as a preventive against fracture. The real points to be considered are: First, the proper proportion of the gear diameters; second, the hardness of the material and the best combination of hardness for wear. George B. Grant was very near the truth in saying, "It does not proportionately increase the strength of a tooth to double its pitch." With herringbone gears it would seem that the strength need hardly be considered, as it is practically impossible to break out a single tooth of sufficient angle, an entire section must be removed. They must be *worn* out.

Aside from the hardness, the value of the material to avoid crystallization must be considered, as gears in which the teeth are apparently extremely tough will often become brittle and drop off after a comparatively short service from this cause. For this reason high-carbon steel should be avoided and hardness obtained by case-hardening, or by the addition of manganese, nickel, chromium, vanadium, or some other hardening ingredient.

It is hardly necessary to add that proper lubrication adds greatly to the efficiency of a gear drive, except, of course, where they are exposed to brick or cement dust, where it is often advisable to run them dry, as the oil will hold particles of grit and cause the teeth to cut. A jet of air appplied at the point of contact is also found beneficial in such cases, as it will remove particles of grit from the teeth before they enter contact.

The constructions of housings upon which the gears are mounted is of the utmost importance, as the absence of vibration is essential to high efficiency. Where the housings are insecure it is often found that a rawhide pinion will sometimes give better service than one made of iron, as the rawhide will give and absorb vibration that would destroy the harder material.

Another point that naturally suggests itself is the proper value of a suitable lubricant. It is evident that the efficiency of 95 per cent. running dry and 98 per cent. when immersed in oil does represent the total saving. Lubrication means in many cases the difference between a successful drive and a failure which is not apparent from any superficial tests made for power efficiency only.

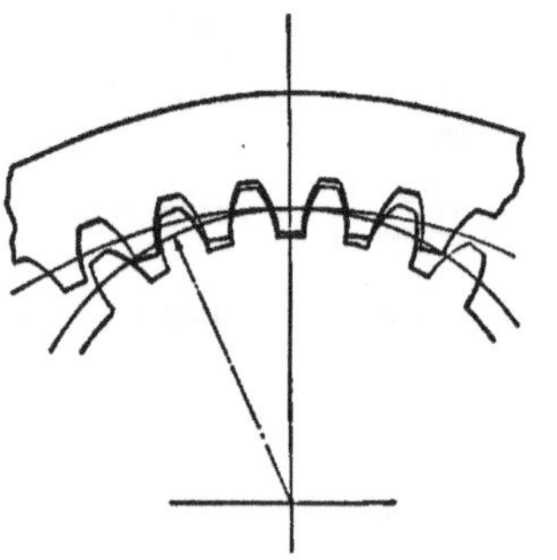

FIG. 60. INTERNAL GEAR MESHING WITH SPUR.

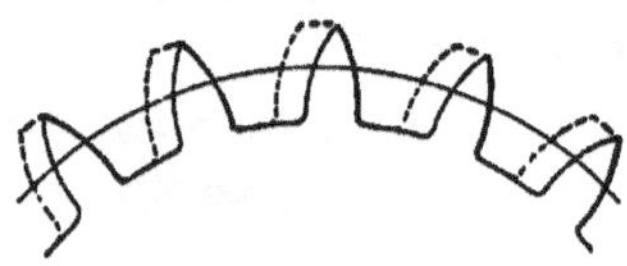

FIG. 61. LIMITING WEAR OF GEAR TEETH.

The question is often asked, "What is the life of a gear?" It is evident that continual service will wear out *any* material no matter how hard. We will say for example that a load of 900 pounds per unit of area will wear out a pair of cast-iron gears in one year's continual service. What pressure must be applied, therefore, to wear out these gears in six months or to allow them to run for five years? It would appear that this could be taken care of nicely by factor A, and as in Table 15, provided, of course, that other conditions are correct and that we have the proper analysis of our materials. We should determine not only what grades of material will wear best together, but also how long they will wear.

A gear may be said to be worn out when the teeth have been reduced to not less than one-half their original thickness—this will subject them to the limit of ultimate stress of the material if allowed their full load according to the Lewis formula. If this is exceeded the teeth would be liable to fracture. See Fig. 61.

It is evident that during the latter part of the life of a gear the teeth will wear more rapidly, as the backlash will allow the teeth to hammer with variations in the load or in reversing.

It is well known that all materials are subject to what is known as fatigue, that is, a piece of steel that will stand an intermittent strain of 2,000 pounds successfully is liable to fail if this load is permanently applied. This should enter into out problem, as it is often required to design a pair of gears to transmit a certain load continuously, with the guarantee that they will render successful service for, say, two years before renewing. The correct solution of this problem would require a thorough knowledge on all the points brought up in this paper, also a proper determination of their efficiency, unless of course, the gears were made amply large, or there had been some precedent upon which to base the calculations.

Another point to be considered is what difference (if any) should be made in the comparative hardness in favor of the pinion, so that it may have wear proportional to the ratio of the pair.

IMPORTANCE OF PROPER DESIGN

Aside from all these conditions, to obtain anything like accurate results the gears must be mounted in such a manner as to obviate practically all vibration. The teeth must be accurately formed and spaced to insure that the impulse received by the driven gear is uniform and without variation, and the thickness of the teeth must be such as to avoid practically all backlash, except just enough to secure free operation, as teeth are often broken by crowding on close centers. The gears must also be properly designed to withstand any strains to which the teeth are subjected. The proper distribution of the material will add greatly to the wear and strength. It is well to have the gear as rigid as possible. It should be remembered that accuracy in cutting the teeth will be of little avail if they are not correctly mounted upon their shafts. If the shaft is a little under size, the key will cause it to run out of true. This will also apply when the shaft is a neat fit and the taper key is driven too tightly.

In the absence of practically all experimental data I have not attempted to put forward anything more than a general outline of the situation and to bring up essential points for consideration. It is trusted that they will be received as such. In view of the growing importance of gears for the transmission of power the points referred to are certainly worthy of attention.

RELATIVE STRENGTH AND HARDNESS OF HOT ROLLED GEARING

The advent of rolled gearing, with its distinctive strength and wear resisting characteristics, has quite materially increased the limits placed upon these important properties for cut gearing and gives promise of greatly increasing the life and modifying the design of all types of commercial gearing.

Tests of similar machine cut and hot rolled gears (6-inch P. D. 6–8 pitch) conducted at the James Herron Laboratories, Cleveland, Ohio, for normal and case-hardened conditions of gears gave the following average results per inch face of gear—the steel of which the gears were made containing 0.20 per cent. carbon and 0.57 per cent. manganese.

	MACHINE CUT GEARS	HOT ROLLED GEARS
Normal Condition		
Yield point	6,470 lbs.	7,918 lbs.
Ultimate strength	12,250 lbs.	13,645 lbs.
Hardness	22	26
Case-hardened Condition		
Yield point	13,529 lbs.	14,750 lbs.
Ultimate strength	17,250 lbs.	19,130 lbs.
Hardness	87	85
Penetration of case	0.027-in.	0.035-in.

In the normal condition—that is, not case-hardened—the hot rolled gears showed a superiority of 22.4 per cent. in yield value, 11.9 per cent. in ultimate strength and 18.2 per cent. in hardness; while in the case-hardened condition the hot rolled gear averaged higher by 9.0 per cent. in yield value, 10.9 per cent. in ultimate strength, 29.6 per cent. in penetration of case and very nearly the same in degree of hardness. The hot rolled gears were also very much the tougher, indicating exceptionally high wear resisting qualities.

Photomicrographs of annealed cut gear tooth sections indicated a coarse structure characteristic of drop forgings with the pearlite gathered together in relatively large islands indiscriminately scattered throughout the metal; while similar photographs of hot rolled gear tooth sections showed a very much more uniform structure with the islands of pearlite small, well broken up and uniformly distributed throughout the metal. In case-hardened specimens, those of cut gears showed more or less free cementite on the outside of the case, but in the specimens of hot rolled gears there was much less free cementite and a more uniform and deeper case penetration, giving the hot rolled gear teeth a very tough and high wear resisting surface.

These laboratory tests indicate that the hot rolling process adds from 20 to 25 per cent. to the strength of the gears, increases their hardness 18 or 20 per cent. and materially betters their wear resisting qualities. These records have been duplicated or bettered repeatedly under normal and exacting working conditions.

Photographs of magnified cross-sections of machine cut and hot rolled gear teeth show plainly that the 10 to 20 ton pressure to which the heated gear blanks are subjected in the process of rolling the teeth gives to the forged teeth not only a much smoother profile surface, but brings about a marked rearrangement of the metal structure, producing a dense almost fiber-like arrangment in a trussed formation about the periphery of the gear which serves to tie the forged teeth to the body of the gear and to equalize any warpage effects in subsequent heat treatment, so eliminating the need of heavy clamping devices for holding the gear true when quenching.

SPEED OF SPUR GEARS

When the question is asked, "What is the greatest circumferential speed at which a spur gear may operate?" we are told, "When 1,200 feet per minute is exceeded either rawhide or herringbone gears must be used, and even when properly mounted 3,000 feet is the limit of speed for any gear."

To illustrate the fallacy of this statement consider a pair of spur pinions, each of 12 teeth, 6 pitch, 2-inch diameter, running at a speed of 1,200 feet per minute. A moment's consideration will show that the noise generated by such a drive would be excessive, as this would mean 2,280 revolutions per minute. These gears would make their presence known at a speed of 400 feet per minute, which represents 761 revolutions per minute.

At first thought it appears that the number of teeth in contact would represent the comparative speed value, but, as it is sometimes possible to obtain as many teeth in contact in a small pair of gears as in a large pair, owing to the difference in pitch, the proportion of gear diameter must also be taken into account. As the proportionate value of the gear diameter is represented by the number of teeth in contact, the pitch remaining constant; this value may be gaged according to the circular pitch. Fig. 62 will illustrate this.

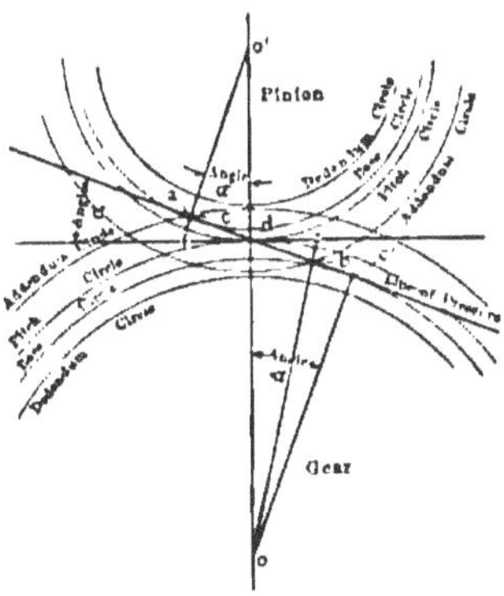

FIG. 62. DIAGRAM ILLUSTRATING METHOD OF DETERMINING NUMBER OF TEETH IN CONTACT.

The relative speed value is found in the product of the number of teeth in contact and the circular pitch, or np'.

n = number of teeth in contact,

p' = circular pitch.

The next step is the determination of a factor from actual practice, ρ, which multiplied by the product of the pitch, ρ', and the number of teeth in contact, n, will give the safe speed. The formula now becomes:

$$\textit{Safe speed} = p'n\rho. \tag{26}$$

The factor ρ may be made to cover almost any type of gear or condition of service. If the safe speed is known for certain machine construction, the value of the factor ρ may be determined as below for similar drives.

$$\textit{Value of } \rho = \frac{\textit{Safe speed}}{p'n}. \tag{27}$$

STYLE OF GEAR	COMMERCIAL CUT GEARS	GENERATED TEETH
Spur gear, pattern molded	0 to 300	....
Spur gear, machine molded	110 to 450	....
Spur gear, commercial cut	600	....
Spur gears, cut with exact cutters, accurately spaced	700	800
Spur gears, cut stepped teeth	820	....
Spur gears, fiber	900	1000
Spur gears, rawhide	1000	....
Herringbone gears, angle of spiral, 10 degrees	700	1100
Herringbone gears, angle of spiral, 20 degrees	800	1400
Herringbone gears, angle of spiral, 30 degrees	1100	1900
Herringbone gears, angle of spiral, 45 degrees	2400	4000

TABLE 14—VALUES OF FACTOR ρ

NUMBER OF TEETH IN CONTACT

As the necessary formula to determine the number of teeth in contact is considered too cumbersome for practical use, a graphical solution is given, as follows:

Referring to Fig. 62, the length of contact is measured between the intersection points of the line of pressure and the addendum circles of the gear and pinion, or between a and b. In case this intersection falls outside the intersection of a line drawn at right angles with the pressure line to the center of the gear as at the point c, the contact is measured from the point c, as this indicates that the distance $a\ c$ must be deducted for interference. The gear tooth is rounded from this point out, or the flank of the pinion tooth is undercut to accomplish the same purpose.

If, on the other hand, this point falls outside the intersection of the pressure line and the addendum circle as at c' this extra length must be deducted, as there is no contact until the point of the mating tooth has passed the addendum circle. In order that the number of teeth in contact can be stepped off, the lines $o'c$ and ob are extended to the pitch or to points e and f. The length of contact is then measured from e to d on the gear, and from d to f on the pinion along this pitch circumference. This distance divided by the circular pitch equals the number of teeth in contact.

LIMITING SPEEDS

Estimate the maximum speed *to avoid danger of fracture* for the best type of gear and condition of service as 500 feet per minute per 1000 pounds safe working stress of the material of which the gear is constructed. Thus the maximum speeds would be as follows:

4,000 feet per minute for cast iron of 8,000 pounds per square inch.
8,000 feet per minute for cast steel of 16,000 pounds per square inch.
10,000 feet per minute for machinery steel of 25,000 pounds per square inch.

To attain anything like these speeds, however, the gears must be exceptionally accurate and well balanced, also the housings must be sufficiently heavy to obviate practically all vibration. The restriction placed by the above limits, however, will avoid the possibility of allowing the higher speeds. According to a series of experiments made by Prof. Charles H. Benjamin, American Machinist, December 28, 1901, page 1421, the bursting speed of a solid, cast-iron gear blank is found to be 24,000 feet per minute. The centrifugal tension at this speed is 15,600 pounds per square inch. The same wheel, split between the arms, burst at an average speed of 11,500 feet per minute.

SPEEDS AND POWERS

The superiority of hot rolled gears in the matter of strength, toughness and uniformity of metal structure permits the attainment of somewhat higher speeds than are safe for cut gears, particularly as they are of relatively small diameter and are, consequently, proportionally of more rugged construction.* The modification in metal structure brought about by the heavy pressure exerted on the semi-plastic metal not only serves to tie the forged teeth to the gear rim but also to form a confining ring of high tension metal about the gear body, increasing the resistance of the gear to the action of centrifugal force. Exactly what is the permissible allowable increase in limiting speed has not as yet been determined, but in the case of unshrouded gears it can be safely taken as 20 per cent. In the case of shrouded rolled gears this increase can be exceeded without danger, the additional increase depending largely upon the height of shroud and probably being limited to another 30 per cent. —a total increase of approximately 50 per cent.

CONSTRUCTION OF THE GEAR

Approximate multipliers should be used for various designs in reference to limiting speeds.

Hot rolled gears (shrouded)	up to 1.50
Hot rolled gears (unshrouded)	1.20
Properly proportioned solid gear	1.00
Gear split through the arms	0.75
Gear split between the arms	0 50
Link flywheel construction	0.60

With the exception of the values given for hot rolled gearing, these values correspond closely to those given by the Fidelity and Casualty Company, which sets the limit of speed for a solid flywheel (cast iron) at 6,000 feet per minute. No value is given for the wheel split through the arms.

The principal cause of failure (as pointed out by Professor Benjamin in the

* Reginald Trautschold.

article above referred to) in gears split between the arms is the necessity of placing bolting lugs on the inside of the rim. These lugs naturally tend to increase the stress at their point of location and fracture the rim in this locality at correspondingly low speeds.

LOCATION OF THE GEAR

The influence of the location on the speed cannot be well determined. In general, however, for gears mounted upon insecure foundations or secured to the wall or ceiling of light buildings more or less allowance must be made. One case in mind is a 10-horse-power motor direct-connected to a line shaft making 150 revolutions per minute. The motor is securely bolted to the joists: The gears have 106 and 22 teeth, respectively, 4 diametral pitch, 5-inch face, cast iron and rawhide; the speed is 1,050 feet per minute. These gears were exceptionally noisy and were replaced by herringbone gears of the same normal pitch, with the angle of spirality 20 degrees. The gear in this instance was cast iron and the pinion machinery steel. These gears were no better than the first pair and were replaced by gears with a spiral angle of 45 degrees. These proved to be but little better, and as a last resort a pair was installed with an angle of 76 degrees. These gears were fairly quiet when operating under full load, but very distressing when running light. This condition is sometimes found in spur gears with excessive backlash. It was noticed that the teeth gave evidence of rapid wear due to the reduced face contact. Rubber cushions placed under the motor and shaft hangers and filling the space between the hub and rim of gear with wood made no perceptible difference in the noise of this drive, which was finally abolished.

In this connection I have knowledge of a pair of herringbone gears, cast iron and machine steel, 80 and 24 teeth, 6 normal pitch, 20- and 6-inch pitch diameters, angle of spiral 45 degrees that are practically noiseless at a speed of 3,150 feet per minute. At this speed, however, the load is comparatively light, These gears are driven by a 10-horse-power motor and are entirely satisfactory.

RELATION OF PRESSURE TO SPEED

The question naturally arises, "In what way does the tooth pressure influence the speed of gears?" From a power standpoint this is ordinarily taken care of by Mr. Barth's expression $\frac{600}{600 + V}$. But this does not fix the speed at which noise may be avoided, which is the subject of this paper.

It has been assumed that the tooth pressure allowable at the speeds given by the formula (26) are within the limits placed by the foregoing formula (24) for *wear*.

As previously pointed out "the *strength* of teeth in herringbone gears need hardly be considered, as it is impossible to break out a single tooth provided the angle be great enough to engage another tooth before the first lets go."

However, the actual contact, which depends upon the angle, should be used instead of the actual face when determining the load for wear, as increasing the angle decreases the noise, but increases the wear.

ILLUSTRATIVE EXAMPLE

Required the safe speed of a pair of solid steel spur gears of the following dimensions:

Gear, 80 teeth, 2-inch pitch, 6-inch face, 50.93-inch pitch diameter, pinion 48 teeth, 2-inch pitch, 6-inch face, 30.558-inch pitch diameter.

Cutters for these gears to be made for the exact number of teeth.

The computations for speed involve the number of teeth in contact = 2.5, value of ρ according to Table 16 = 800, safe speed = $p'n\rho = 2 \times 2\frac{1}{2} \times 800 = 4{,}000$ feet per minute. According to factors given for limiting speeds, these gears would be amply safe to resist fracture, but would be at the extreme limit for cast iron of a safe stress of 8,000 pounds per square inch. The computations for wear involve, assuming that the pinion is made from steel of an elastic limit of 50,000 pounds per square inch, the limiting tooth pressure at this speed would be equal to the value W^e as follows:

$$W^e = CfA \frac{600}{600 + V} = 2.4 \times 6 \times 3{,}000 \times 0.13 = 5{,}600 \text{ pounds},$$

in which

$$C = 2.4,$$
$$f = 6,$$
$$A = 3{,}000,$$

and

$$\frac{600}{600 + V} = 0.13.$$

The computations for strength involve the use of the Lewis formula, based on carrying the entire load on one tooth,

$$W = Sp'fy \frac{600}{600 + V} = 25{,}000 \times 2 \times 6 \times 0.111 \times 0.13 = 4{,}329 \text{ pounds},$$

in which

S = 25,000 pounds per square inch (one-half the elastic limit),
p' = 2 inches,
f = 6 inches,

and

y = 0.111 for 48 teeth.

As the load should not exceed the strength of the tooth, this pressure is the limit to be used. The corresponding horse-power transmitted would equal

$$\frac{WV}{33{,}000} = \frac{4{,}329 \times 4{,}000}{33{,}000} = 524 \text{ horse-power}$$

These gears if properly mounted should be satisfactory at a speed of 4,000 feet per minute, the value W^e indicating that they could be used for 10 hours per day service for two and one-quarter years at a uniform pressure of 4,600 pounds. If longer life or service per day is required the load on the teeth must be reduced accordingly.

HIGH SPEED GEARING*

For the transmission of power it frequently becomes necessary to use toothed gearing, subjected to high peripheral speed conjointly with high pressure per unit of tooth contact, and the object of these remarks is to record what has been successfully done in recent years, as much higher speeds are now successfully attained than formerly. Considered in a static sense, the gear tooth satisfies the condition of stress if it is proportioned to endure forces acting transversely on it, and the pressure per unit of contract is not of such intensity as permanently to deform the curved bearing surface of the teeth. When in motion, the curved surfaces slide upon each other as they enter and leave contact, and when this sliding action is accompanied with high pressure, the limit of endurance is soon reached, and in the case of the inferior materials this occurs at comparatively low speeds and pressures. In addition to this, more or less impact usually occurs, especially when the resistance is of a fluctuating character or the loads are suddenly applied. The effects of this hammering action are discernible by a flattening of the curved faces of the teeth, after which the proper engagement of the teeth ceases and the gear is speedily destroyed.

To prevent this, it is desirable to cut the teeth so accurately that no side clearance or "backlash" exists, and this is now usually done on first-class gearing of even the largest dimensions. Owing to the low elastic limit of cast iron and the bronzes we cannot expect these metals to endure so high a pressure as steel, and steel appears to be the most trustworthy material to endure the highest pressures and speeds. This assertion, however, does not apply to all grades of steel. Soft steel surfaces abrade or cut very readily despite all methods of lubrication, and surfaces of this material should never be allowed to engage in sliding contact. Gearing of soft steel is usually destroyed by abrasion at quite moderate speeds. Rolling-mill pinions of steel, containing 0.3 per cent. carbon, have been destroyed in a few months, whereas the same pattern in steel of 0.6 per cent. carbon has done similar work for several years without distress. Of course it is necessary to shape the teeth to a proper curve to insure proper engagement and uniform angular velocity.

Some years ago there was required suitable gearing to connect the engines to a rolling mill in this vicinity. The diameters of the wheels were 37.6 and 56.4 inches respectively. They were intended to revolve at speeds of 150 and 100 revolutions per minute and expected to transmit about 2,500 horse-

* A paper read before the Engineers' Club of Philadelphia, by James Christie.

power. The character of the service was such that renewal was a serious matter and long endurance very desirable. A high grade of steel was selected especially in the pinion, in which the greatest wear would occur, and which, owing to the location, was the most difficult to replace. The pinion was forged from fluid compressed steel of the following composition:

	Per cent.
Carbon	0.86
Manganese	0.51
Silicon	0.27
Phosphorus and sulphur, both below	0.03

The spur wheel was an annealed steel casting:

	Per cent.
Carbon	0.47
Manganese	0.66
Phosphorus and sulphur, both	0.05

The tooth dimensions were: Pitch, 4.92 inches; face, 24 inches. See Fig. 63.

These were accurately cut with involute curves generated by a rolling tangent of 16 degrees obliquity. No side clearance was allowed. After starting the mill, it was found that a higher speed was practicable than was originally contemplated. Higher pressures on the teeth were also applied, so that ultimately about 3,300 horse-power was transmitted through the gearing, corresponding to a pressure of nearly 2,100 pounds per inch of face. The speed was variable, but occasionally attained a velocity of 260 revolutions per minute for the pinion, corresponding to a peripheral velocity of 2,500 feet per minute. This gearing has been in constant operation for several years and behaves satisfactorily.

The highest recorded speed for gearing that I can recall is that described by Mr. Geyelin in the Club "Proceedings" of June, 1894. The mortise bevels had a peripheral velocity of 3,900 feet per minute, but the pressure per inch of face was only about 680 pounds, the diameter and speed being made high to reduce the pressure on the teeth. I understand that the lifetime of these bevels is not long. If made of a grade of steel, as previously described, their diameter and speed could be considerably reduced and prolonged endurance would be realized.

About the same time No. 63 was installed a similar application was made to another mill, the gear having a different speed ratio, and the angular velocity being lower. See Fig. 64.

	Pinion per cent.	Wheel, per cent.
Carbon	0.90	0.60
Manganese	0.64	0.64

A much larger set had been previously employed, transmitting about 2,400 horse-power at 750 feet per minute peripheral speed, involving a pressure

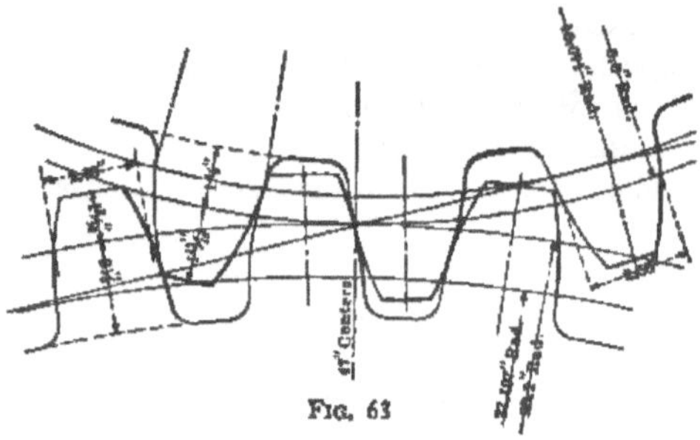

Fig. 63

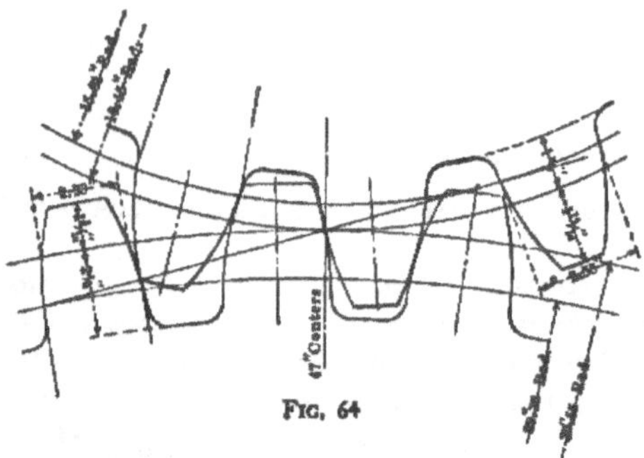

Fig. 64

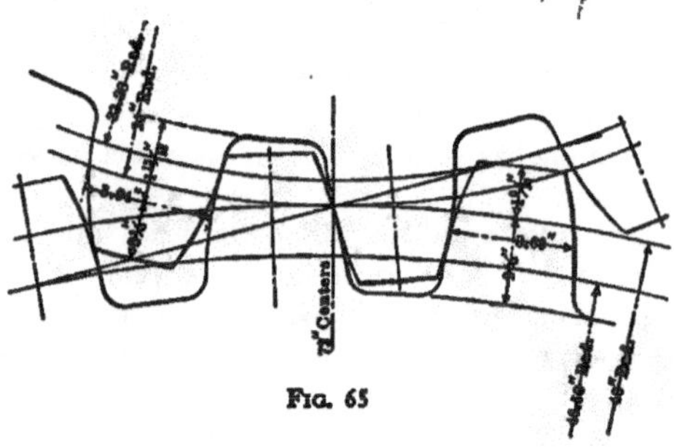

Fig. 65

EXAMPLES OF HIGH SPEED TOOTH GEARING.

per inch of face of 3,500 pounds. This latter pair were 4 feet and 8 feet respectively, 7½-inch pitch, 30-inch face, cut with involute teeth of 14 degrees obliquity. See Fig. 65.

	Pinion	Gear
Carbon	0.52	0.42
Manganese	0.55	0.73
Silicon	0.107	0.279
Phosphorus	0.022	0.078
Sulphur	0.02	0.05

These gears have all rendered excellent service, and to-day are apparently as good as at the beginning.

As considerable expense is involved in cutting large gears of hard steel, it is sometimes practicable to rough-cut the gear after it is made as soft as possible by slow cooling, a higher degree of hardening being imparted before final finishing by air hardening or rapid cooling from the refining heat. This is not infrequently done in the case of screws and gears of moderate dimensions. In this event it is desirable to have the ratio of manganese low—say, not over 0.5 or 0.6 per cent.—as a high manganese content seems to impart a permanent hardness that is not reduced by slow cooling.

It appears to be practicable to maintain sliding surfaces of steel if one of the surfaces is hard, even if the other is comparatively soft, but for steel gearing for ordinary purposes I would suggest the use of steel not less than 0.4 carbon. If the speeds and pressures are unusually high, a much harder grade of steel becomes necessary. When a small pinion engages with a large wheel, the former alone can be made of high grade steel approaching to a carbon content of 1 per cent. When extreme speeds and pressures become necessary, the best results will be found by using in both wheels steel having a carbon content approaching 1 per cent., or an equal hardness, obtained by lower carbon and high manganese or other desirable hardening addition. With gearing accurately cut from steel of this character and securely mounted, it is believed that reasonable endurance will be obtained when the product of speed and pressure, divided by pitch, each within certain limits, does not exceed 1,000,000: For example, a speed of 3,000 feet per minute and 1,600 pounds per inch of face, or *vice versa* for gear of 5-inch pitch, assuming, so far as we know, a maximum speed of 5,000 feet per minute for gear of any pitch, and permissible pressure to be proportional to the pitch.

This statement that speeds and pressures are reciprocal, or as one is increased the other must be reduced, in a fixed ratio, may not strictly be a rational one, but in a broad and general sense it is correct within the usual limits of practice.

It will be understood that such a generalization as herein stated would apply to pinions having a liberal and not the minimum number of teeth.

In the discussion of Mr. Christie's paper, Mr. E. Graves gave particulars of three duplicate sets of cast-steel bevel wheels. The pinions are the drivers and are 57.39-inch pitch diameter and have 36 teeth, 5-inch pitch, 20-inch face. The wheels are 74.8 inches diameter and have 47 teeth. The teeth are carefully cut to involute lay-out and are 3.43 inches high. The normal speed of the pinion is 360 revolutions per minute, giving 200 revolutions per minute to the wheel and nearly 4,000 feet circumferential speed on the pitch line. The horse-power transmitted is 1,300. Assuming the entire load to be distributed along the outer end of one tooth, the fiber strain would be 2,100 pounds per square inch at the root of the tooth.

The pinion is mounted on the upper end of a 10-inch shaft, 148 feet long, with a turbine wheel at the lower end. Both shafts extend through the gears and are supported in a massive bridge casting with adjustable bearings. The gears are enclosed in a casing and are lubricated with oil fed under pressure through several jets applied just in front of the teeth as they mesh together.

The gears have been in service for five years, but have not been entirely satisfactory. Their wearing power in the sense of resisting abrasion is satisfactory, but the teeth break. This breakage is confined to the pinion, the nature of the break being the same in all cases, beginning at the large end, cracking around the root and following along the tooth. The quality of the steel in castings is the ordinary commercial article. The widest variation in analysis observed is, in one instance:

Silicon	0.25
Sulphur	0.036
Phosphorus	0.071
Manganese	0.74
Carbon	0.31

Another:

Silicon	0.27
Sulphur	0.03
Phosphorus	0.032
Manganese	0.80
Carbon	0.23

As is to be expected, the softer metal has resisted breaking the longer. In two sets of these gears the resisting work is of a varying nature with sudden and wide fluctuations; in the third instance the working is more constant. This variation of conditions does not seem to have influenced failure, as the teeth have broken in all the sets.

One of the practical difficulties in operating bevel gears of the nature described is the difficulty of holding them so that they will be in proper contact; longitudinal motion in either shaft throws them out of pitch. The most serious problem, however, is in securing and maintaining shafts so that the extended axis lines of same pass through a common point. The effect of power transmission from pinion to gear is to put these axis lines out of position, moving them in opposite directions and resulting in end contact of teeth and concentrated load instead of evenly distributing the load along the whole length of tooth. In this particular the question of maintaining bevel gears is decidedly more of a problem than that of spur gears. In this latter case small end motions of carrying shafts produce no effect, while the wearing of bearings is only the shifting of pitch line, and, as it occurs slowly, it will, within reasonable limits, adjust itself.

As a matter of further interest, I will mention that in this same room with these gears are three other sets of bevel gears having cut-steel pinions and mortise wheel with cast-iron rims. The diameters and ratios of these—speeds, mountings, and service—are practically the same as those described but the transmission of power is 1,100 instead of 1,300 horse-power. The pinions have 33 teeth, 5½-inch pitch, with 20-inch face, the teeth being planed down to 2½-inch thickness on pitch-line. The wheels have 43 teeth. These gears have been in service some seven years. None of the pinions has ever given way; the wooden teeth in the wheels, however, last only from six weeks to two months, an extra rim being kept on hand for refilling and replacing.

Mr. Lewis, in continuing the discussion, said that in regard to the pressures carried by gear teeth, Mr. Christie seems to lay down a rule making the product of speed and pressure constant. This would reduce the load in proportion to the speed, and it seems to me an open question whether that should be adhered to or not. I do not think it has been demonstrated how the pressure of the teeth should vary with the speed. Some experiments, I think, should be made which would indicate that more clearly than has heretofore been done. It is interesting to note his remarks regarding the influence of the hardness of the metal upon the pressures carried, and instead of reckoning the pressure by the inch as so much per inch of face, it seems to me the pitch should also be included, because the face of a gear tooth is very much like a roller, and the pressure carried by a roller varies with its diameter as well as with the face. Some authorities seem to think that it should vary with the square root of the diameter, others directly with the diameter, and I am inclined to the latter opinion. If gear teeth are proportioned for strength, they are also proportioned for wearing pressure and surface to carry the load.

Mr. W. Trinks said: I wish to call attention to an article on high speed gearing in the November and December numbers of the "Zeitschrift des Vereins deutscher Ingenieure," 1899, by the chief engineer of the General Electric Company, at Berlin, Germany. The experiments show that there is no rule for the relation between pressure and speed, it depends upon accuracy; the load on the teeth may be the higher the more accurately the gears are made. A remarkable method of manufacturing gears was the outcome of the experiment. The curves are laid out on paper three or four times the size of the real tooth, reduced to proper size by photography, transferred on sheet steel, and etched in. Thus the highest degree of accuracy is obtained.

It was found that neither cycloidal nor involute curves gave the best results. Another curve was developed with a view to reducing the sliding motion between the teeth. The article contains very interesting diagrams on this point. By dividing the length of two working teeth into an equal number of parts, the amount of sliding action can be determined and the fact shown that it is reduced to a minimum by these methods.

Another thing shown by the paper is never to place a flywheel close to a gear. If possible, have a good length of shaft between. Slight inaccuracies in the pitch of the wheel require acceleration or retardation of the mass, and in order to do this force is needed. This force causes a hammering on the teeth which may break them—in other words, plenty of elastic material should be between the inertia masses and the gears. I feel pretty sure that all engineers will be much interested in the article; it is a valuable treatise on highspeed gearing.

Mr. Christie added: The bevel gears described by Mr. Graves are very interesting and useful as a record. It is much more difficult to obtain satisfactory results with bevels than with plain spurs, as any deviation from correct alignment is fatal to correct tooth action in the former. In this instance, while speed is very high, the pressure on the teeth is comparatively low—about 750 pounds mean pressure per inch of face. Thus the products of speed and pressure in relation to the pitch are considerably below the quantity assumed as a safe maximum.

Regarding the quality of the material, the manganese is too high. While steel of this composition would be moderately hard and wear fairly well, it would be somewhat brittle. It is not surprising to learn that some teeth gave way by fracture. If the relative proportions of carbon and manganese in the steel were reversed, it would be a much better material for the purpose.

A SPUR GEAR ANGLEMETER*

In the design of spur gears it is desired to give to the teeth such a form that as the gears mesh with each other, the relative motion of the two will be the same as that of two cylinders whose diameters have the same ratio one to the other as have the diameters of the two pitch circles. By the aid of kinematics gear teeth can be so designed as to give exactly this relative motion between the gears. However, in the manufacture of gears, factors enter which make the form of the teeth of the gears as they come from the shop somewhat different from that developed by kinematics. This variation is quite marked in the case of rough-cast gears. Here several errors enter. Because of the difficulty and time required in developing a tooth outline according to kinematics, arcs of circles which approximate the correct outline are used in laying out the gears in the drafting room. From these slightly inaccurate drawings the patternmaker makes the patterns, which are apt to vary slightly in form and the spacing of the teeth from the drawing furnished by the draftsman. These inaccurate patterns then go to the foundry, and from them the molds are made. In making the mold, in order to draw the pattern it is rapped loose, so that the mold is slightly larger than the pattern and still more inaccurate in outline. The casting is poured, and in cooling is warped out of shape because of the cooling stresses leaving a gear with a final error in the form of its teeth made up of several smaller errors, as just enumerated.

* W. M. Wilson, American Machinist, April 13, 1905.

NOTE.—The anglemeter curves for modern well cut generated gears show much smoother lines, and curves of gears rolled by molding generation—Anderson Process—approach closely the ideal of a straight line.

A realization of the presence of these errors suggested that a knowledge of the final inaccuracy in the forms of the teeth of rough-cast gears would be of interest and perhaps not without value. With this idea in mind, an anglemeter for determining the variation in the angular velocity ratio of gears was designed and built.

The anglemeter consists mainly of a responsive frame carrying a drum on which is wrapped a card. If these gears operate without vibration the pointer would draw a straight line around the card. Figs. 66 and 67 will be practically self-explanatory. The pointer was found to multiply the actual variations 9.4 times.

TEST OF A PAIR OF CAST GEARS

The instrument was connected to a pair of spur gears, No. 12020, each having a circular pitch of 1½ inches and 20 teeth. Each gear was mounted on a 2 15/16-inch shaft. As the two gears and the two shafts were the same size, no reducing motion was needed. The guides of the instrument were not long enough to allow a complete revolution of the gears, so the teeth were numbered from 1 to 20 and one card drawn for teeth 1 to 11 and another for teeth from 11 to 1. The points on the curves corresponding to the time when these teeth came into contact were marked and then the cards were cut to these marks and pasted end to end as shown in Fig. 66.

The details of the design of this pair of gears were not known, but judging from the shape of the teeth they were modifications of involutes and the outlines of the teeth were probably laid out according to some empirical method in general use in the manufacture of rough-cast gears. Before the test the gears had been run long enough for the bearing surface to become smooth.

Cards were taken when the gears were adjusted at different distances apart. If the gears had been true involutes the velocity ratio should be constant and the same for the different distances between the centers. Whether the velocity ratio were constant or not is readily seen from the curves in Fig. 66. Due to the uncertainty of the shrinkage of castings, the diameter of the gears was not exactly equal to the computed diameter, but a trifle greater. Judging from the clearance at the ends of the teeth the proper distance between the centers of the gears was 9⅝ inches. Card *A*, Fig. 66, gives a curve corresponding to this distance between centers. Two curves were drawn before the card was removed and the ability of the instrument to duplicate a curve is considered as evidence of its accuracy. It is seen from the curve that the velocity ratio instead of being constant is quite erratic in its variations. While in general the variations in the curve are not related to each other in any way, yet for a portion of the card at least the waves in the curve correspond roughly with the teeth on the gears. This is especially true of the portion of the curve corresponding to the teeth from 11 to 20. The curves on Card *B* were drawn when the distance between centers was 9 13/16 inches. In this case there is a more marked relation between the waves in the curves and the teeth on the gear than in the former case. For Card *C* the distance between centers was 10 1/16 inches, and the regularity of the waves is still more marked. In this case the card was taken for half a revolution only.

For all the cards the arrows above the different curves indicate the general direction of the tangent to the curves which give the angular acceleration of the driven as compared with the driver.

The rough jagged nature of the curves portrays in a vivid manner the wide variance from that uniform positive motion of pitch surfaces necessary for securing high gear efficiency which exists in cast gearing. It also indicates that although the relatively rough profile surfaces of cast teeth are contributory causes of inefficiency they are nowhere as serious as malformation and variations in gear tooth proportions.

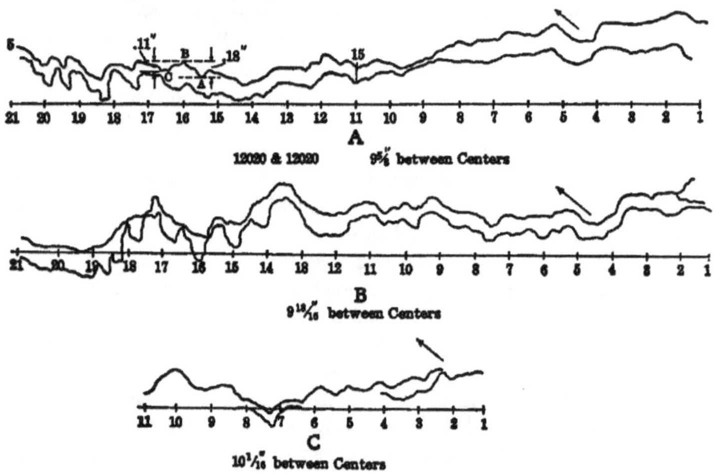

FIG. 66. CURVES FROM CAST GEARS.

From the curves shown the following conclusions have been drawn relative to the forms of the teeth:

(1) The fact that the relation between waves in the curves and the teeth of the gears is increased indicates that all the teeth are subject to a common error due to the empirical method in which the outlines of the teeth were developed, also that this error causes a greater variation in the velocity ratio when the distance between the centers of the gears is increased. (This is in accordance with the empirical method presented in Kent's hand book and attributed to Molsworth. By using the method referred to, the resulting tooth outline would be wider below and narrower above the pitch circle than true involute teeth. When the distance between centers is normal these errors annul each other almost completely, but as the distance is increased the result of the errors is more apparent.)

(2) Irregularities in the curves indicate errors peculiar to the individual teeth, which evidently are involved in the progress of patternmaking and molding.

(3) A shifting of the general vertical position of the curves on the cards, as shown at teeth 4 and 14, indicates errors in the spacing of the teeth.

An effort has been made to analyze the curve in Card *A*. The portion of this curve *A B C* has been chosen as being a wave whose relation to an individual tooth is the most evident. The abscissa *AC* represents the angular space occupied by one tooth on the gear, and *AB* and *BC* each represent one-half of that angular space. The ordinate between *A* and *B* measures 0.18 inch, and that between *B* and *C* measures 0.11 inch. Dividing by 9.4, the constant for the instrument, it is found that the driven gear gains on the driver by an angle whose arc measured on the circumference of the shaft is 0.18 divided by 9.4 equals 0.019 inch, while the gear turns through an angle equal to $\frac{360}{20} \times \frac{1}{2} = 9$ degrees.

An angle whose tangent is

$$\frac{0.019}{1.46885 \text{ (radius of shaft)}} \text{ is 45 minutes.}$$

Computing in the same manner the angle corresponding to the ordinate *BC* is found to be 30 minutes. That is, while the tooth No. 16 is in action the driven gains relative to the driver by an angle of 45 minutes while the latter is turning through an angle of 9 degrees, and then loses by an amount of 30 minutes in the same space. An error of 0.019 inch measured on the circumference of the shaft corresponds to 0.06 inch measured on the pitch circle.

TEST OF SPECIAL GEARS

The second pair of gears to which the instrument was attached consisted of a No. 12016 pinion having 1½-inches circular pitch and 16 teeth and a No. 12050 gear having 50 teeth. The gears were designed to mesh properly when the distance between centers varies by an amount equal to ½ inch. The shrinkage of the casting was not as much as was expected, so that the normal distance between centers is 0.20 inch more than the sum of the computed radii. If the gears were true involutes the velocity ratio should be constant when the distance between centers of the gears varies from 15.7 to 16.2 inches.

The outlines of the teeth of the gears are involutes having an angle of obliquity of 23½ degrees when the distance between centers of the gears is normal. The mold for No. 12016 was made from a pattern, while for No. 12050 it was made from a tooth-block as used in a Walker gear-molding machine. The pattern and tooth-block were both made from drawings laid out in the drafting room. The tooth outlines for these drawings were developed according to kinematics, so that for the exception of any small error which the patternmaker might make, the teeth on the pattern and tooth-block are correct involutes.

How near the castings approached to involute gears is shown by the curves in Fig. 67. As the two gears were not of the same size, it was necessary in

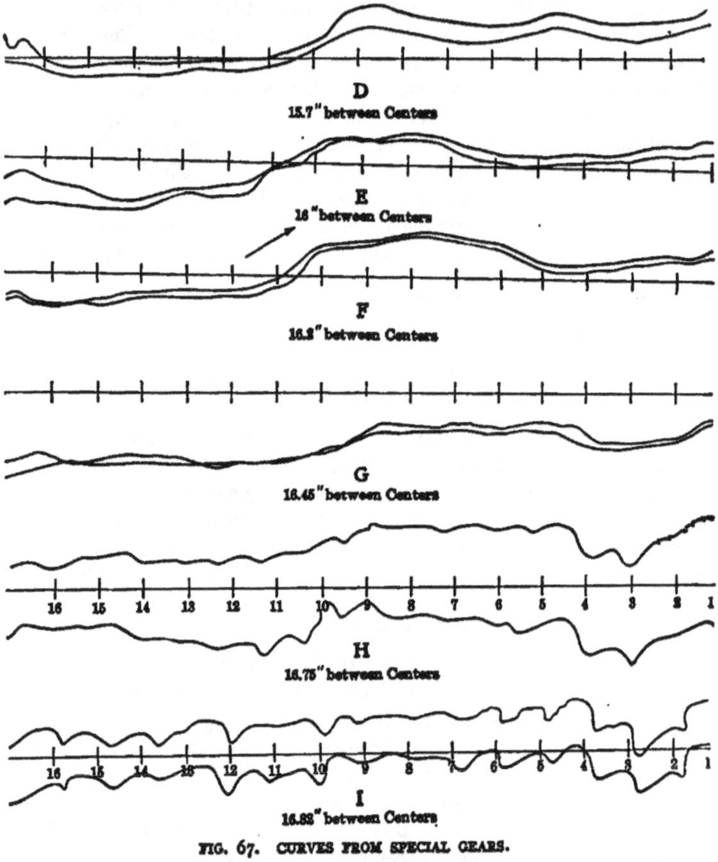

FIG. 67. CURVES FROM SPECIAL GEARS.

taking the cards to use some means of making the average velocity of the two carriages of the instrument the same. To do this the hub of the gear was turned to a smooth surface and the end of the shaft of the pinion was turned down until its diameter bore the same ratio to the diameter of the hub of the gear that the diameter of the pitch circle of the pinion bore to the diameter of the pitch circle of the gear. This gave a rigid and accurate reducing motion. The method of taking the cards was the same as for the other pair of gears except that a curve for a complete revolution of the pinion was obtained upon a single card. The distance between centers of the gears for the different

GEARS LUBRICATED						GEARS NOT LUBRICATED			
Tangential force at pitch circle, 194 lbs.		Tangential force at pitch circle, 330 lbs.		Tangential force at pitch circle, 585 lbs.		Tangential force at pitch circle, 444 lbs.		Tangential force at pitch circle, 148 lbs.	
R. P. M.	Efficiency, %	R. P. M.	Efficiency, %	R. P. M.	Efficiency, %	R. P. M.	Efficiency, %	R. P. M.	Efficiency, %
80	92.00	76	97.33	81	94.97	59	88.58	105	91.40
84	94.05	89	96.66	92	96.10	56	91.40	106	91.50
119	93.55	123	97.75	113	95.28	62	89.17	129	91.80
112	93.31	115	95.99	114	95.45	88	91.18	129	93.50
162	93.60	150	96.27	132	96.28	102	91.03	165	91.00
149	92.57	155	96.57	140	95.29	126	90.21	167	90.80
187	91.56	182	95.81	149	94.81	126	89.29	233	93.30
172	91.08	181	95.05	177	94.57	159	90.60	241	93.50
177	92.04	209	96.16	173	93.87	159	89.94		Av. 92.10
215	91.36	192	95.50		Av. 95.20	202	88.72		
215	91.19	157	95.18				Av. 90.00		
	Av. 91.00	92	97.03						
			Av. 96.30						
Efficiency of Gears alone, 92.60%		Efficiency of Gears alone, 98.00%		Efficiency of Gears alone, 96.80%		Efficiency of Gears alone, 91.50%		Efficiency of Gears alone, 94.80%	

TABLE 15—GEAR NO. 12020 DRIVING 12020—STANDARD INVOLUTES

cards is as indicated in Fig. 67. Cards *D*, *E*, and *F* cover the range of variation in the distance between centers for which the gears were designed. From these curves it is seen that there are no waves on the curves corresponding to the individual teeth, the curves for the most part are smooth, and are practically the same for the different distances between centers of the gears. For card *G* the distance between centers was made 16.45 inches, or 0.75 inch more than the least distance between centers at which the gears were supposed to run. The remarks made in regard to curves *D*, *E*, and *F* apply equally well to this curve also. For cards *H* and *I* the distance between centers was so great that the point of contact of the teeth did not follow the line of obliquity throughout the angle of contact but lay outside of this line for the latter portion of the period of action of the teeth.

This is clearly brought out in the forms of the curves in which waves corresponding to individual teeth are quite evident.

It will be noticed that for all of the curves for the second pair of gears there is an extended wave near the middle of the card and also near each end, so that if several cards were placed end to end so as to form a continuous curve there would be two complete waves on this curve per revolution of the pinion. This indicates that the pinion instead of being exactly round is oval in form. This deformation can be traced to the molder in the foundry who, in order to draw the pattern, raps it loose from the sand and unless great care is taken increases the diameter more in one direction than in the other. The wave in the curve is traced to the pinion instead of to the gear because the spacing of the teeth for the gear was done by means of a molding machine and any errors which might occur would be peculiar to the individual teeth instead of being common to a group of teeth and gradually increasing and decreasing from zero up to a maximum and then back again to zero, as indicated on the cards.

From the curves in Fig. 67 the following conclusions are drawn relative to this pair of gears:

1. The outlines of the individual teeth are true involutes.

2. The spacing of the individual teeth on both the gear and the pinion is accurate.

3. The pinion is slightly oval in form.

4. For the gears used the angular velocity ratio is constant through a range in the variations in the distance between centers equal to 0.75 inch.

GENERAL CONCLUSIONS

If one would be allowed to draw conclusions from the tests of two pairs of gears only, the following statements might be made in regard to the errors in the teeth of rough-cast gears:

1. Of all the errors the greatest is due to laying out the teeth in the drafting room.

2. Given an accurate drawing the patternmaker can make a pattern very accurate in form.

3. In using the patterns the molder is apt to make the mold slightly oval

in form, but in using a molding machine the error induced in the foundry is very small.

4. The surfaces of the teeth of cast gears are so smooth as not to affect the angular velocity ratio.

The difference in the appearance of the curves for the two pairs of gears is very much in accordance with the manner in which they meshed, the second pair running with very much less noise than the first pair. At a speed of 250 revolutions per minute the first pair almost threatened to shake the testing machine to pieces, while at the same speed the second pair ran with but little vibration.

It might be said that the patterns for the second pair of gears were completed before there had been any thought of subjecting the gears to a test, which fact eliminates the possibility of unusual care having been taken in making the patterns.

COMMENTS ON GENERAL CONCLUSIONS

Although the foregoing conclusions are based upon the behavior of cast gears made without the exercise of any exceptional care in their founding, an analysis of these conclusions will throw considerable light on the behavior of cut and generated gearing, indicating as conclusively as the most exacting mechanical tests the conditions in gear fabrication which should be given most attention. In the first place, it is quite evident that the uniform positive motion of pitch surfaces essential for a constant angular velocity ratio is primarily a matter of uniformity in general tooth section and evenness in tooth spacing. In the second place, it is evident that if a constant angular velocity ratio of gears could be attained, the exact form of tooth would be of quite secondary importance.

The modern process of gear cutting, generation, takes care of the uniformity in general tooth section, as each tooth cut on a generated gear is, so far as the mechanical precision of the generating machine permits, a replica of every other tooth and its form is a conjugate of the generating cutter tooth. The efficiency of generated cut gears is, consequently, dependent upon the constancy of the angular velocity ratio between the generation cutter tool and that of the work arbor carrying the gear blank to be generated, and the rigidity with which the gear blank is held on the work arbor. The vital difference between machining generation and molding generation is a question of rigidity of gear blank. Machining generation being of necessity an intermittent process, in order to permit the advance of the gear blank, the gear blank is during generation under a series of intermittent stresses which have a tendency to cause an angular vibration in the gear blank. In molding generation, on the other hand, the stress on the gear blank is a more uniform one and the tendency to cause angular vibration in the gear blank is absent, so the constancy of the angular velocity ratio of hot rolled gears is dependent almost entirely upon the constancy of the angular velocity ratio between the heavy timing gears which actuate the machine.

GEAR EFFICIENCY

No thorough investigation has yet been made of the efficiency of gears, and very few data of any sort has been published, therefore little can be said. One thing is tolerably well known: The efficiency of a gear drive varies with its ratio—that is, a reduction of 2 to 1 will be more efficient than one of 10 to 1, other features governing the efficiency being the accuracy with which the teeth are formed, and spaced, the arc and obliquity of action, and the condition of the engaging surfaces. It is generally conceded that the length of face does not affect the efficiency.

22½ degree involutes. 16″ between centers. Tangential force at pitch line, 361 lbs.		24¼ degree involutes. 16.2″ between centers. Tangential force at pitch line, 355 lbs.		19½ degree involutes. 15.7″ between centers. Tangential force at pitch line, 353 lbs.	
R. P. M.	Efficiency, %	R. P. M.	Efficiency, %	R. P. M.	Efficiency, %
88	92.40	116	91.00	110	90.00
123	90.00	68	96.90	69	94.40
140	89.90	90	91.50	127	86.80
156	91.30	140	89.90	147	88.90
172	91.30	163	90.10	162	88.80
188	92.30	177	90.00	196	88.90
216	92.30	197	99.90	218	89.20
229	92.40	208	90.00		Av. 89.60
	Av. 91.50	214	90.30		
			Av. 91.20		
Efficiency of Gears alone, 92.30%		Efficiency of Gears alone, 92.10%		Efficiency of Gears alone, 90.50%	

TABLE 16—GEAR NO. 12016 DRIVING 12050—LUBRICATED

This loss comes mainly through the sliding action of the teeth when entering and leaving contact; it would seem, therefore, that the arc of action should be as short as possible, carrying the load of that portion of the tooth where it can be best borne, also, as there is less friction in the arc of recess than in the arc of approach, it will increase the efficiency to lengthen the addendum of the driving gear. The contact, however, should be for an equal distance above and below the pitch line to avoid extreme sliding friction. If the entire action was rolling contact, there would be little loss.

Tables 15 and 16 were originally published in AMERICAN MACHINIST, by W. M. Wilson. They give the result of experiments with the same case tooth spur gears mentioned on pages 98 and 100. The gears in Table 16 were made with extra long teeth, the various angles of obliquity as given were obtained by adjusting the gear centers.

The result of these tests is summed up as follows:

1. The efficiency of rough gray iron spur gears is independent of the speed of the gears within the range covered by this report.

2. There is no indication from the limited number of tests that the amount of power transmitted affects the efficiency to any great extent.

3. The use of a heavy grease on the teeth increases the efficiency slightly (average from tests 1.7 per cent.).

EFFICIENCY OF LARGE GEARS

Many of us know things that are not so, and with some of us a part of this useless knowledge may be in regard to the efficiency of large gears.

We believe that we are well within the bounds of truth in saying that the majority believe that large gears are very inefficient.

This misinformation or lack of information is easily explained. The data that are available in regard to the efficiency of gears of any size are few at best and apply to pinions and small gears. Large gears have not been extensively tested, for there are but few technical schools and factories equipped with appliances for gear testing and capable of absorbing several thousand or even several hundred horse-power. Again we do not always distinguish between a large gear with cast teeth and a similar gear with cut teeth.

In the early days of factory engineering, gear drives were common for transmitting power from shaft to shaft throughout the plant. As time went on these drives were replaced—in some cases by belting, in other cases by a change from mechanical transmission to electrical distribution of power. It has been very easy to assume that the reason for discarding the gears was because of their inefficiency. Enthusiastic advocates of electric drives have time and again referred to the substitution of motors for mill gearing with elation and have either stated or implied that the change brought about a great saving of power.

In further support of the common belief that gearing in large sizes is inefficient we quote the following from the presidential address of Mr. Denny to the Institution of Marine Engineers, in England, on October 5, 1908: "It has frequently been suggested that if some inspired engineer could evolve a system of gearing that would be lasting and reliable, not too noisy, and would not absorb in friction more than, say, 10 per cent. of the power, turbine engines would be capable of application to any speed of vessel and to any size of propeller." Here Mr. Denny gives expression to an oft-repeated suggestion, that if large gearing could be made having an efficiency of 90 per cent., a big step forward would be made.

With direct bearing on this belief we have the record of performance of the largest gears in point of horse-power ever made and tested—the Melville-Macalpine reduction gear, through which 6,000 horse-power has been transmitted.

To show the measure of the performance we quote a paragraph from an article by George Westinghouse in the January number of *The Electric Journal:* "Considering the important bearing of the question of efficiency on the ultimate success or failure of the gear, it is peculiarly gratifying to have found by repeated trial and careful measurement that the transmission loss hoped for by Mr. Denny has been divided by seven. To be exact, the efficiency surpasses the more than satisfactory figure of 98.5 per cent."

The hope was for an efficiency of 90 per cent., the fulfilment was an efficiency of 98.5 per cent.

Large bevel-gear drives have perhaps been especially condemned; it is, therefore, of interest to read the following from a paper presented by Prof. C. M. Allen to the American Society of Mechanical Engineers on the testing of water wheels: "The total horse-power delivered to the generator was approximately 700. The driving gear was of the ordinary wood-mortise type, outside diameter 6 feet 5 inches approximately, with 68 teeth 14 inches wide, meshing with a cast-iron pinion which had 48 teeth with planed-tooth outline. At full load the loss of the gear was 3.5 per cent. and 3.4 per cent. for two separate units, or the efficiency of the horizontal-shaft vertical-wheel gear drive was about 96.5 per cent. The gears were well lubricated with a thick grease.

"About nine months later it was necessary to test one of these same units in exactly the same manner. The loss in gears this time was a trifle less, the test giving 3.1 per cent.

"All of the information obtained concerning the loss due to bevel-gear drives leads the writer to conclude that if gears are properly designed, set up, and operated, and are not overloaded intermittently or continuously or left to care for themselves, they should show an efficiency of from 95 to 97 per cent."

Are not these bevel-gear drives efficient compared with the majority of mechanical devices?

As a matter of fact, do not the above figures agree with common sense? The quantity of heat generated by the absorption of one horse-power for an hour is 2,545 British thermal units. If we try to give an expression to the statement that large gears are inefficient by assuming a loss of 10 per cent., in the case cited by Professor Allen, 70 horse-power would be dissipated in heat. A multiplication will show what this means in British thermal units per hour, and the presumption is strong that the gears would heat, cut, and wear under such a condition.

The same reasoning will probably apply to many large-gear drives concerning which there has been speculation in regard to efficiency. Had they been extremely inefficient, they would not have operated satisfactorily for a long period of time.

These figures show that large-gear drives are not necessarily inefficient, but, on the contrary, may be decidedly the reverse.

COMMENTS

As the efficiency of gearing depends upon the uniformity of the angular velocity of the pitch surfaces, it follows that the finer the pitch of the gears and the more numerous the teeth, the more uniform will be the angular velocity of the pitch surfaces and the better the operating efficiency of the gears. Consequently, for any given pitch the efficiency will increase with the number of teeth. This fact explodes the popular belief that large gears are inefficient. They are, on the contrary, apt to be relatively more efficient than smaller gears, providing they are accurately mounted and well aligned, as customarily their pitch is proportionally finer in comparison with their diameters than in the case of small and relatively coarser gears. The inaccuracies which cause variations in angular velocity of pitch surfaces become less potent as the number of teeth increase.

SECTION IV

Gear Proportions and Details of Design

These formulas should not be used indiscriminately, as one design will not meet all conditions. No attempt has been made to proportion arms or rim to the power to be transmitted, as all proportions are derived from the pitch and face of the gear, with the double object of obtaining equal strength and sound castings.

As the smallest section of a casting is, per square inch, its strongest part, this fact should enter more into the question of proportion than has been the custom in the matter of gears. To illustrate the value of this, a test piece cut from the point of a cast gear tooth will often be as much as 30 per cent. stronger than a similar piece cut from its root. Hence the rim of a gear is made thinner than has been the practice, and the central rib deeper, to secure

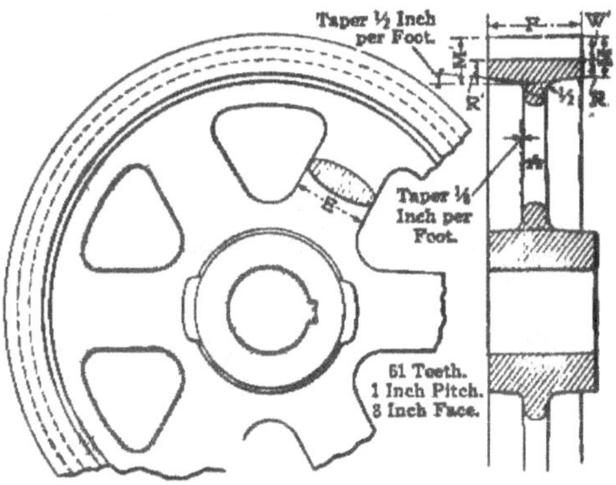

FIG. 68. PLAN AND SECTION OF A SPUR GEAR, SHOWING NOTATION.

the necessary section and thus obtain a stronger casting with the same weight of material. This same rule applies to the hub, the outside diameter being reduced and a deeper rim added, and reinforcements placed over keyways, or, better still, two reinforcements directly opposite, especially if the gear is to be balanced; in fact, this is imperative even at relatively low velocities. For the same reason the teeth should be cored whenever it is possible to do so, as the rim of a casting for a cut gear has always the heaviest section, and, therefore, most subject to blow holes, especially at the junction of the arms and rim. It follows, then, if this be true, that the more uniform the section throughout, the sounder and stronger the casting.

Cored teeth, however, should be machine spaced, as uneven spacing will render it difficult, if not impossible, to cut the teeth in the usual manner. To secure accurate spacing when cutting, it is absolutely necessary that the cutter have an equal amount of stock to remove from each side of the tooth space. Also exposure to hard scale and core sand quickly destroy the cutter. When the teeth are to be planed this point is not of so much importance, as a little unevenness of stock can be readily taken care of, but the cut should be always under the scale if time is of any importance.

FORMULAS

These formulas are based on Brown & Sharpe standard 14½-degree involute tooth.

Thickness of Rim, $M = \frac{3.927}{p}$, or $1.25\ p'$

Mean Thickness of Rim, $M' = \frac{5.026}{p}$, or $1.60\ p'$

Mean Thickness of Rim under Tooth, $R' = \frac{2.868}{p}$, or $0.913\ p'$

Whole Depth of Tooth, $W' = \frac{2.157}{p}$, or $0.6866\ p'$

Minimum Thickness under Tooth, $R' = \frac{1.769}{p}$, or $0.563\ p'$

Area of Rim Total, $MF' = M'F$

Area of Rim under Tooth $= R'F$

Average Area of Arm, $Æ = tF\ 1.3$, or $MF\ 0.52$

Average Thickness of Arm, $A = \sqrt{\frac{Æ\ 1.27}{3}}$

Average Width of Arm, $E = 3A$

Outside Diameter of Hub = Bore $+ \frac{2}{3}\sqrt[3]{NF}$

Number of Arms = 4, 6, 8, 10, etc., according to design and diameter of gear.

p = Diametral pitch.
p' = Circular pitch.
t = Thickness of tooth at pitch line.

DISCUSSION OF FORMULAS

Thickness of Rim, M—The thickness of rim should be equal to 3.927 divided by the diametral pitch, or 1.25 multiplied by the circular pitch. When the gear is small and accurately made it is often good practice to make the dimension 1.12 of the circular pitch, and so secure the same section and additional strength by adding 50 per cent. to the depth of the central rib.

Mean Thickness of Rim, M'—By mean thickness is meant the thickness of one side of a parallelogram necessary to contain the actual area of the rim. This will take care of the central rib and fillets, so that by multiplying the width of the face of the gear by dimension M' the entire area of rim may be obtained. This will be found necessary also for estimating the weight.

Mean Thickness of Rim under Tooth, R'—This dimension $\left(\frac{2.868}{p}\right)$ multiplied by the width of the face will give the area of the entire section of the rim and rib under the teeth.

Minimum Thickness of Rim under Teeth, R—This determines the thickness of the rim under the tooth measured at the edge of the rim.

Whole Depth of Tooth, W'—This gives the whole depth of the tooth as per Brown & Sharpe standard $= 0.6866\, p'$, or $\frac{2.157}{p}$.

Total Area of Rim—The total area of the rim is found by multiplying the mean thickness M' by face of gear F.

Area of Rim under Tooth—The area of the rim under the tooth is determined by multiplying the mean thickness R' by the face F.

Average Area of Arm, $Æ$—The average area of the arm is that area midway between the inside of the rim and the outside of the hub, and is found by adding 30 per cent. to the area of the tooth at the pitch line, or the thickness of tooth at pitch line $t \times F \times 1.3$. The same result may be reached by taking 0.52 of $M' \times F$, although the foregoing is simpler. Taper of arm to be $\frac{1}{8}$ inch per foot above and below this point.

Average Thickness of Arm—The average thickness of arm (A) may be determined by dividing the quotient of the area of arm $Æ \times 1.27$ by 3, and extracting the square root. If the arm was made in the form of a parallelogram it would not be necessary to multiply the area by 1.27, but as it is to be elliptical, this is essential to insure sufficient section, as 27 per cent. of the area of the parallelogram is lost when inscribing an ellipse there in.

NOTE.—If width of arm is desired, 2 or $2\frac{1}{2}$ times the thickness instead of 3 times, as given, 2 or $2\frac{1}{2}$ is to be substituted.

Average Width of Arm, E—To determine the width of an arm multiply its thickness by 3.

Outside Diameter of Hub—As a rule the outside diameter of a gear hub is made double that of its bore, but when keyways are reinforced, or the gear is to carry a load less than proportional to the diameter of its shaft, the hub diameter may be less than this rule prescribes. Thus, if a gear of 30-inch pitch diameter was mounted on a shaft 15 inches in diameter, the hub diameter should only be increased sufficiently to maintain its section and strength proportional to the gear, not to the shaft. The formula given will proportion the hub to easily carry the entire load applied to the gear but should be used with discretion. Generally, however, when the bore is small or proportioned to the diameter of the gear, the outside diameter of the hub may be taken as 1.75 times the diameter of the bore with reinforcement for keyways.

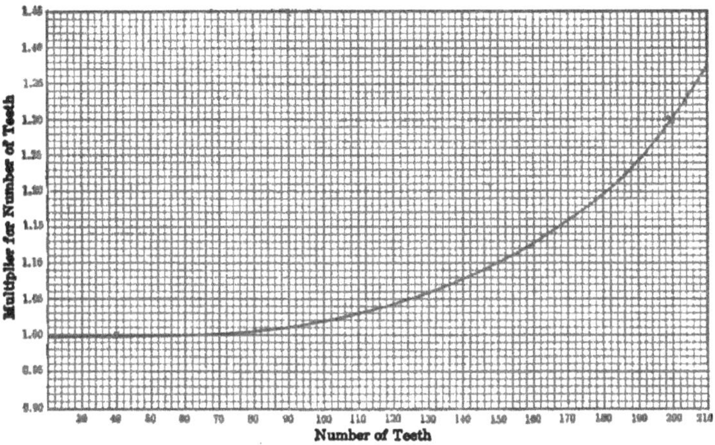

CHART 5. MULTIPLIER FOR INCREASED NUMBER OF TEETH OF SPUR GEARS.

From the foregoing it is obviously almost as important that the hub should not be disproportionately heavy as it should be heavy enough. But if for any reason a materially heavier hub is required, it should be split, by

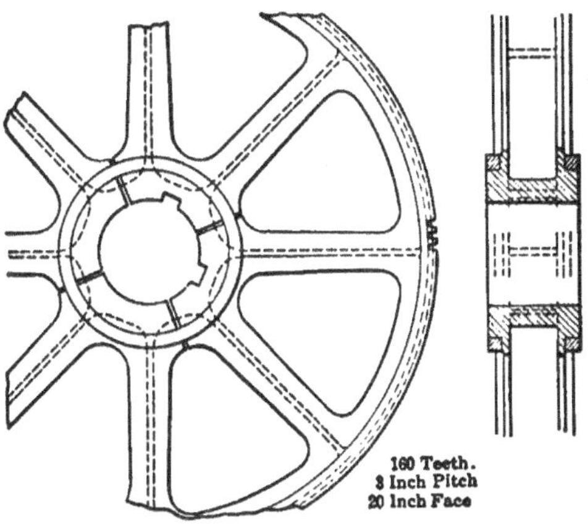

FIG. 69. PLAN AND SECTION OF SPUR GEAR WITH SPLIT HUB.

means of thin cores, into as many equal, radial sections as there are arms in the gear, thus obviating blow-holes, and strains caused by shrinking. Fill the cored spaces with babbitt or lead before machining, and shrink steel bands on the hub for a grip on the shaft. See Fig. 69.

Number of Arms—There is no definite rule for this, as this point depends almost entirely upon the judgment of the designer. In general, however, gears up to 60 inches are either webbed or with four or six arms to suit conditions; over 60 inches eight arms are generally used and over 80 inches in diameter 10 arms. In no case should the greatest distance between the arms exceed the length of the arm measured from the center of the gear to its intersection with the rim.

Width of the Face—The face of spur gear is generally estimated at two or three times its circular pitch, as follows:

1	diametral pitch............	9 inches face
1¼	diametral pitch............	7½ inches face
1½	diametral pitch............	6 inches face
1¾	diametral pitch............	5½ inches face
2	diametral pitch............	5 inches face
2½	diametral pitch............	4 inches face
3	diametral pitch............	3 inches face
4	diametral pitch............	2 inches face
5	diametral pitch............	1¾ inches face
6	diametral pitch............	1½ inches face
8	diametral pitch............	1¼ inches face
10	diametral pitch............	1 inch face
12	diametral pitch............	¾ inch face
14	diametral pitch............	⅝ inch face
16	diametral pitch............	½ inch face
18	diametral pitch............	⅜ inch face
20	diametral pitch............	⅜ inch face

It is becoming better understood, however, that a wider face is more efficient ("increasing the face does not increase the friction of the teeth in proportion"),* and as the wear of the teeth is governed by the diameter of the gear, or rather by the combination of diameters and the width of the face, the face, therefore, should be amply wide, and the pitch just sufficient to resist fracture.

Street-railway gears are made 3-pitch, 5-inch face, with good results. A gear face of five times its circular pitch is now generally considered to be good proportion.

WEBBED SPUR GEARS

No definite rule can be laid down for the design of webbed gears. See Fig. 70. Generally the thickness of the web is made equal to R'', which is the thickness of the rim at its thickest part.

* George B. Grant.

Core holes *H* tend to make a sounder casting, and furnish means to secure the gear while machining. When these holes are made large, or are shaped to follow outlines of the arms and rim, ribs are added on each side. Care

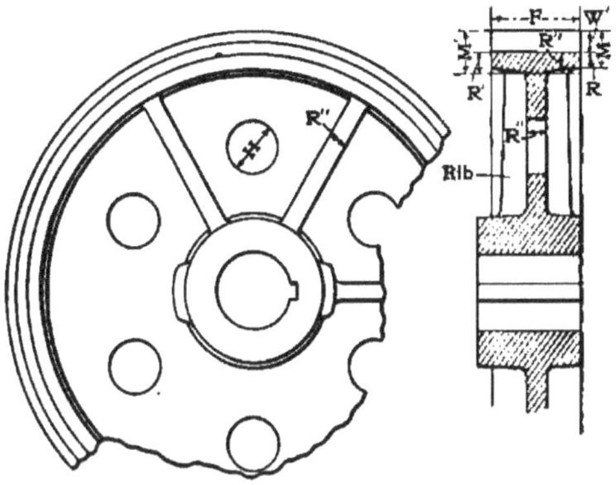

FIG. 70. PLAN AND SECTION OF WEBBED SPUR GEAR.

should be exercised, however, not to make the arm too light, for when the hub is heavy the light section connecting the rim and hub will cool too rapidly, setting up serious shrinkage strains, and causing flaws in the casting that cannot be remedied by annealing. Sharp corners, small fillets, and narrow ribs should be avoided for the same reason.

SPLIT SPUR GEARS

It is not good practice to split a gear between the arms, but when this is necessary, the following points should be kept in mind (see Fig. 71):

Bolts should be placed as close to the rim as possible.

The dimension *b* must in no case be less than the dimension of *a*; otherwise the bolts will be subject to other than the direct tensile stress tending to spread the gear.

Section *C-C* should be stiff enough to resist any strain tending to bend the lugs. By placing the bolt close in the corner of the rim and the lug, the length of the lug may be reduced, as this lug need only be long enough to counteract leverage on bolt.

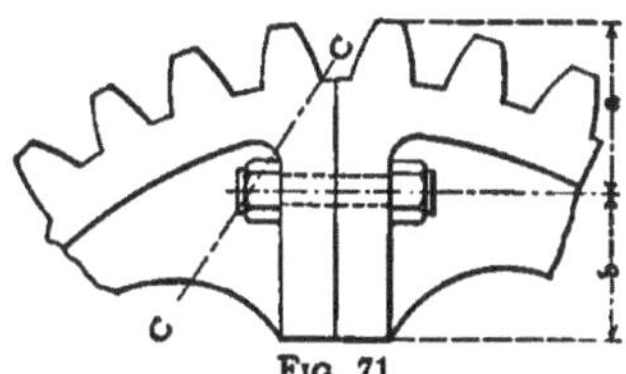

FIG. 71

FIG. 72

DIAGRAM ILLUSTRATING SPLITTING OF GEARS AT THE RIM.

The bolts should be sufficiently heavy to carry the load applied at the pitch line of gear tooth, not neglecting the initial stress set up by the tightening of the nut, which is generally neglected, sometimes disastrously. When a bolt is placed close in the corner it is necessary to use a stud bolt, drawing up the nuts as the gear halves are brought together.

When the dimension *b* is shorter than *a*, Fig. 71, as is generally the case,

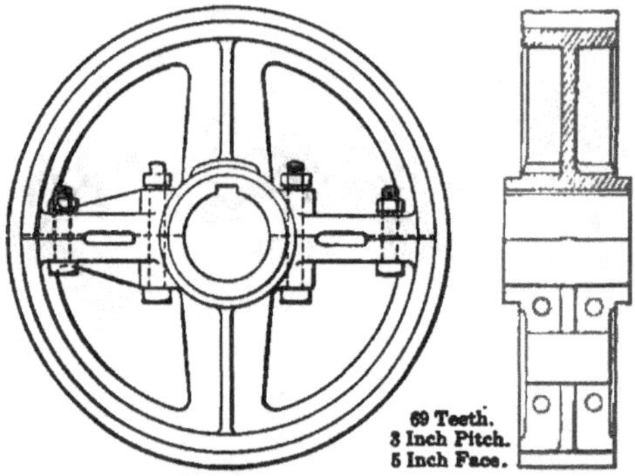

FIG. 73. PLAN AND SECTION OF AVERAGE DESIGN OF SPLIT RAILWAY GEAR.

the load on teeth of gear will cause a fracture of the rim, as illustrated in Fig. 72.

The best method of splitting gears is through the arm, as illustrated in

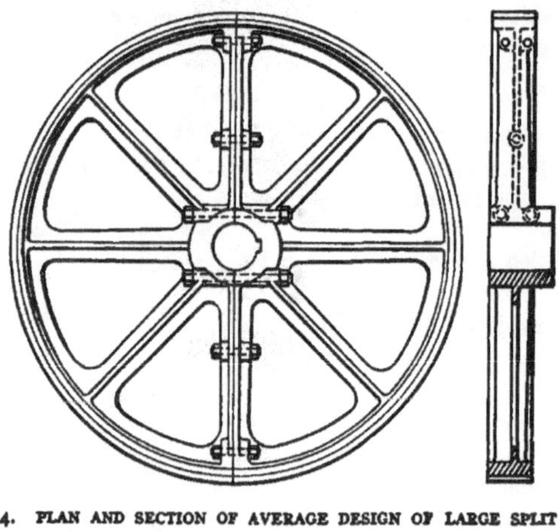

FIG. 74. PLAN AND SECTION OF AVERAGE DESIGN OF LARGE SPLIT GEAR.

Fig. 73, which is a cut of the type used on street railways. When the gear is large, Fig. 74 illustrates a good average design.

When splitting a gear of an odd number of teeth, the split should be made ¼ of the circular pitch off the center line. This will bring the split through center of two tooth spaces. It is good practice to spline the adjoining surfaces, as illustrated in Fig. 73, instead of using fitted bolts or depending on dowel pins. A spline 3/4 inch wide and 1/8 inch high will answer for any but the largest gears.

One point that must be considered in designing split gears for high speed is the fact that weight at any point or part of the rim, and not integral to the rim proper (as lugs for bolting, see Fig. 71), locates the bending moment, and if safe speed is exceeded to the moment of fracture, it will occur at or near such weight.

This fact is pointed and explained by Charles H. Benjamin, in an article in the AMERICAN MACHINIST of December 26, 1901, entitled "The Bursting of Small Cast-iron Flywheels."

I-SHAPED ARMS

For gears such as are shown in Fig. 69, the proportion of rim, arms, and hub may be determined according to formulas given above, the area of arms, of course, being contained in an I instead of an elliptical section. For gears of this size, however, there is more variation in the design of the gear, and the values given cannot be followed so closely.

The I-section arm is much more desirable in heavy gears because it distributes the metal contained in a section of the arm over a greater surface of the rim at their intersection than does the elliptical arm, and, therefore, lessens liability of blow holes at this point. Aside from this, it gives better support to rim in case the face is wide, and makes a stronger section than the elliptical arm of the same weight.

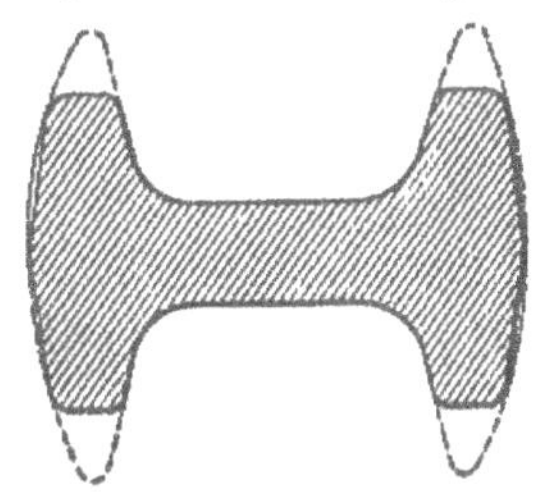

FIG. 75. SUGGESTED CONNECTING-ROD SECTION.

For the above reasons I am disposed to advocate a connecting-rod section, such as is illustrated by Fig. 75, for smaller gears.

CONNECTING-ROD-ARM SPUR GEAR

This design could be applied to gears of all sizes: As the face increased, the section could be changed—as per dotted lines. This, however, would allow a central rib instead of two ribs on each side of the rim as when the ⊢⊣ is turned the other way (see Fig. 69), but would allow the use of bolts instead of links when a split gear had no hub projections. This design is illustrated by Fig. 76.

For gears of an extremely wide face it will be found that the formula for arm will give a section that cannot be contained in the space between hub and rim. This practically means making a web gear. However, when the face is wide it would seem better to use a light web, say, according to dimension on Fig. 70 and extending ribs toward the side, making a cross-sectioned arm, or using the section shown by Fig. 69. This section is generally the most desirable, but this depends upon conditions.

The formulas given here will form a basis for the design of worm and bevel gears, although I believe that the + or cross-shaped section is superior to the oval arm for worm gears on account of the side strain encountered.

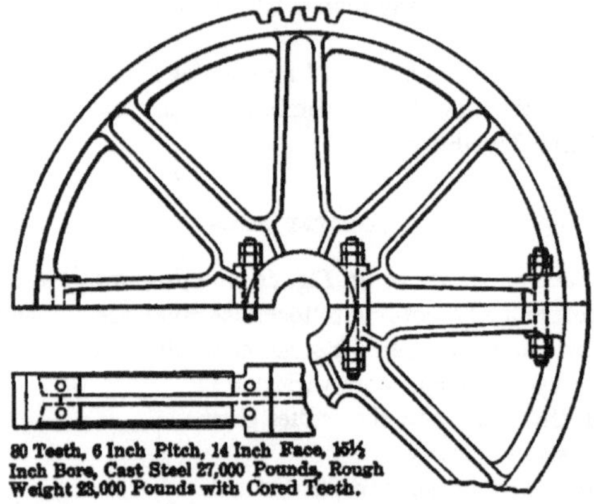

80 Teeth, 6 Inch Pitch, 14 Inch Face, 15½
Inch Bore, Cast Steel 27,000 Pounds, Rough
Weight 23,000 Pounds with Cored Teeth.

FIG. 76. PLAN AND SECTION OF LARGE SPUR GEAR WITH CONNECTING-ROD ARMS.

FOR CALCULATING WEIGHT

The accompanying Chart 6 gives a rapid, approximate method of calculating the weight of a cast-iron spur gear blank for a cut gear, designed according to the above formulas.

This table was derived from a formula by Reuleaux, which gives the

N	0	2	4	6	8
20	5.04	5.60	6.18	6.77	7.38
30	7.99	8.61	9.24	9.89	10.52
40	11.09	11.90	12.59	13.30	14.02
50	14.74	15.48	16.23	17.00	17.77
60	18.55	19.35	20.15	20.97	21.80
70	22.65	23.50	24.36	25.24	26.12
80	27.02	27.93	28.85	29.79	30.73
90	31.69	32.66	33.63	34.62	35.63
100	36.63	37.67	38.70	39.75	40.81
120	47.40	48.54	49.69	50.85	52.03
140	59.30	60.56	61.82	63.10	64.27
160	72.35	73.73	75.10	76.39	77.90
180	86.54	88.03	89.52	91.02	92.54
200	101.88	103.48	104.98	106.70	108.34
320	118.36	120.08	122.15	123.52	125.27

Example: A gear 50 teeth 2 inches pitch, 4 inches face, we have: bc^2 14.74 = $4 \times 2^2 \times$ 14.74 = 235.84, say, 236 pounds.

WEIGHT OF CAST-IRON GEARING. *Reuleaux.*

weight of a cast-tooth gear from the combined product of a constant for the number of teeth, face and square of the circular pitch, as follows:

"The approximate weight of gear wheels, W, may be obtained from the following:

"$W = 0.0357\ bc^2\ (6.25N + 0.04N^2)$, where b = face, c = circular pitch, N = number of teeth, and W = weight of gear.

"The following table will facilitate the application of the formula; it gives the value $\frac{W}{bc^2}$ for the number of teeth which may be given, and the weight may be readily found by multiplying the value in the table by bc^2:"

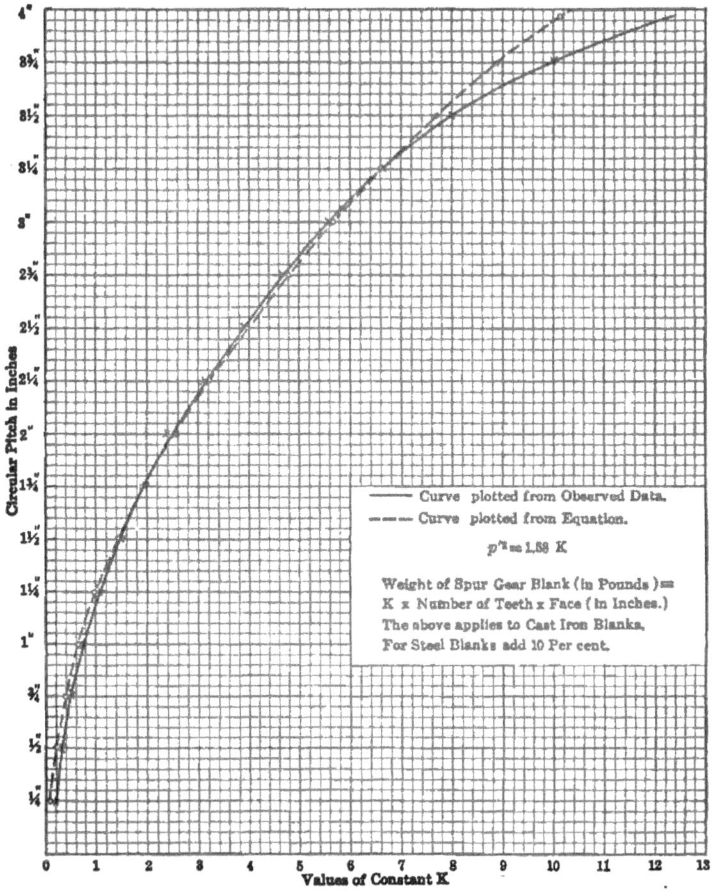

CHART 6. RELATIONSHIP BETWEEN CIRCULAR PITCH AND FACTOR K USED IN ESTIMATING WEIGHTS OF SPUR GEAR BLANKS.

In endeavoring to apply this table to practice it was found that the square of the pitch (C^2) was not a correct factor, and a separate table was made up for it.

After repeated trials and corrections in the value of the constant for the number of teeth, it was noticed (when close results were finally obtained) that the values ran parallel with the number of teeth, and were, therefore, dropped, changing the value of the constants to the square of the pitch, so called, to obtain like results.

To find the weight, therefore, it is only necessary to find the combined product of the number of teeth, face, and a constant given for the pitch:

Weight = *number of teeth* $\times F \times K$.

The accompanying Chart 6 gives the value of K.

This formula has its limitation; it cannot be used for low numbers of teeth, varying with the pitch, also for large gears there is a variation in design that cannot well be covered by one constant. However, the proper constant may be readily determined by trial for different constructions. In general it must be used with discretion, but is invaluable as a check. The finished weight is found by deducting 3 per cent.

The following equation seems to give a close approximation to the curve of Chart 6 throughout the ordinary working range up to 3½ inches circular pitch: $p'^2 = 1.58K$, in which p' is the circular pitch and K the constant previously referred to. Transposing we have $K = \frac{p'^2}{1.58}$, or substituting in equation for weight of spur gears

$$Weight = \frac{p'^2}{1.58} \times \textit{number of teeth} \times \textit{face}.$$

Beyond 3½ inches circular pitch there is considerable variation between the curves from the actual data and from the equation. However, as stated, the great variations in the design of large gears cannot be cared for by a single constant. In using the formula its limitations should be carefully understood. As a matter of interest, it might be stated that for 6 inches circular pitch the curves are again practically in agreement.

ARMS FOR SPUR GEARS*

In deducing a formula for gear arms it is assumed that the thickness of the rim is sufficient to distribute the load between the arms; this assumption is quite justified, as such a depth is necessary to prevent bending of the rim between adjacent arms. By equating the expressions for the tooth strength and that of a beam supported at one end and loaded at the other, the general expression arrived at is

* Henry Hess.

$$Z = \frac{p'^3R(N-7)}{50A} \text{ for circular pitch, and}$$

$$Z = \frac{\pi^3R(N-7)}{50p^3A} \text{ for diametral pitch,}$$

when

Z = modulus resistance of arm cross-section.
p' = circular pitch.
p = diametral pitch.
R = ratio of face width to circular pitch = $\frac{F}{p'}$.
F = face width.
N = number of gear teeth.
A = number of arms.

If it is preferred to use the face width itself, instead of its ratio to the circular pitch, then

$$Z = \frac{p'^2F(N-7)}{50A} \text{ for circular pitch.}$$

$$Z = \frac{F(N-7)}{50p^2A} \text{ for diametral pitch.}$$

By these formulas the dimensions of a gear arm of any section whatever can be determined.

As by far the great majority of cast gear arms are of elliptical cross-section, these expressions are reduced by inserting the terms of the modulus of resistance of an ellipse in which the major axis is double the minor, and the formulas become, when E = the thickness of the arm at its base, and $2E$ = the width of the arm at its base;

$$E = \sqrt{\frac{(N-7)p'^3R}{20A}} = \sqrt{\frac{(N-7)p'^2F}{20A}} \Bigg\} \text{ for circular pitch;}$$

$$E = \sqrt[3]{\frac{(N-7)\pi^3R}{20Ap^3}} = \sqrt[3]{\frac{(N-7)\pi^2F}{20Ap^2}} \Bigg\} \text{ for diameteral pitch.}$$

To reduce the labor involved by the mathematical solution, Charts 7 and 8 have been constructed, one for diametral pitches ranging from 10 to 3, and the other for circular pitches ranging from 1 to 3 inches; as 3 diametral pitch is very nearly equal to 1-inch circular pitch, the second chart extends the range of the first without a break, so that, between the two, any case likely to arise will be taken care of. Diametral pitch is given in Chart 7, as the more general practice uses that for small and medium-sized gears, while for large work; circular pitch is generally employed, and is therefore used as the basis of Chart 8. Two charts are required, as the inclination of the pitch diagonals toward the end values would become too slight to admit of accurate reading on a single one.

By tracing the number of teeth from the bottom scale to the pitch diagonal

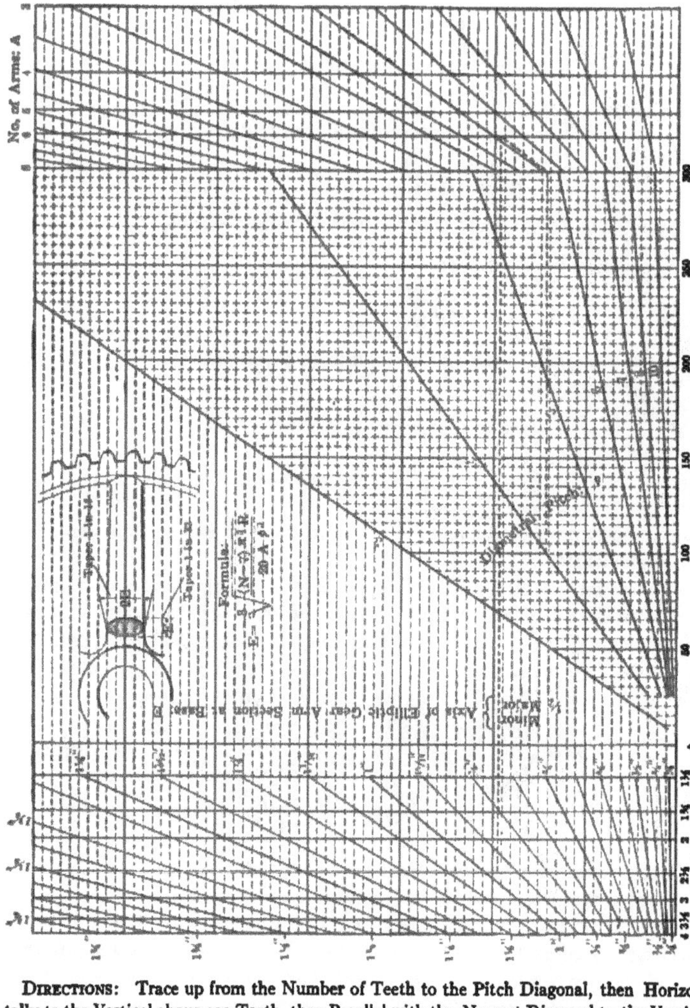

Directions: Trace up from the Number of Teeth to the Pitch Diagonal, then Horizontally to the Vertical above 300 Teeth, then Parallel with the Nearest Diagonal to the Vertical Headed with the Arm Number *E;* then Trace to the left Horizontally to the Vertical *R* representing the Ratio of Face Width to Circular Pitch *R.* Take the Nearest Diagonal as the Arm Base Thickness *E.*

CHART 7. PROPORTIONS OF GEAR ARMS FROM DIAMETRAL PITCH.

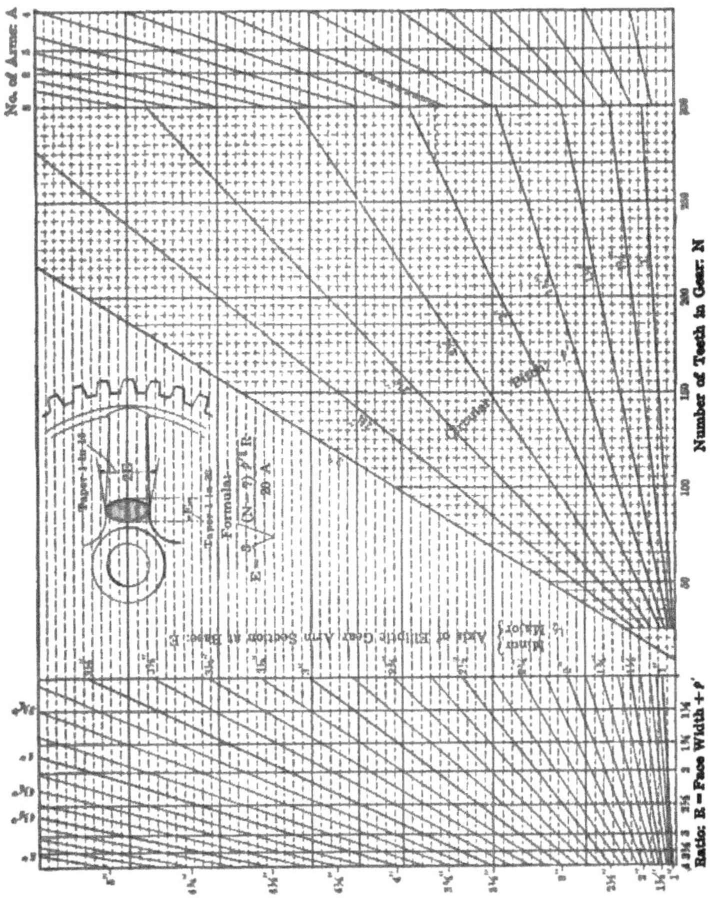

DIRECTIONS: Trace up from the Number of Teeth to the Pitch Diagonal, then Horizontally to the Vertical above 300 Teeth, and Trace Parallel to the Nearest Diagonal to the Vertical Headed with Number of Arms: then Trace to the left Horizontally to the Vertical *R* representing the Ratio of Face Width to Circular Pitch. Take Nearest Diagonal as Base Thickness of Arm *E*.

CHART 8. PROPORTIONS OF GEAR ARMS FROM CIRCULAR PITCH.

in the main portion of the circular pitch chart and referring the intersection to the vertical scale under 8, the value is found of

$$\frac{(N - 7^3 p')}{20A} \text{ for } A = 8 \text{ arms.}$$

For any other number of arms this is modified by employing the auxiliary portion of the chart at the right, referring the value just found along or between the nearest slant lines to intersection with that vertical representing the number of arms actually used. By now tracing this hight horizontally to the left to that vertical R representing the particular ratio of face width to circular pitch employed, and taking a reading from the nearest slant line crossing this vertical, the value first found is multiplied by R and the cube root extracted.

Past practice gives a face width between two and three times the circular pitch, but as the tendency is now toward a wider face, ratios from 1½ to 4 are given.

Concise directions are printed with the charts. Dotted trace lines of the following examples are also drawn in:

Example 1. Given a gear of 100 teeth, 4 diametral pitch, 6 arms and ratio of face width to circular pitch = 2½.

Trace on Chart 7 100 teeth up to diagonal for 4 pitch, horizontally to number of arms 8, slantwise up to number of arms 6, horizontally to the left to ratio 2½, which is intersected between 15/16 inch and 1 inch; therefore thickness of arm is to be taken as 1 inch and width as 2 inches. By calculation the dimensions are 0.96 inch and 1.98 inches.

Example 2. Given a gear of 270 teeth, 2-inch circular pitch, 6 arms and ratio of face width to circular pitch = 2.

Trace on Chart 8 as before and find 3¼ inches full as thickness of gear arm at base, and 6½ inches full as width. The calculated dimensions 3.27 inches and 6.54 inches agree almost absolutely with the much more quickly obtained values by the chart.

In large arms the designer will frequently prefer a cored section. A satisfactory one will be that of Fig. 77, in which major and minor axes of both core and arm are relatively as 2 to 1. By equating the moduli of resistance for solid and hollow elliptical sections of these proportions, it is found that $E^3 = \frac{D^4 - d^4}{D}$, in which E is the thickness of the solid arm as obtained by chart or formula; d and D are dimensions of the cored arm. See Figs. 77 and 78.

In order to lessen the work of making the core box by substituting flat surfaces for curved ones, an approximation like Fig. 78 will add but slightly to the weight, as is shown by the ellipse dotted in for comparison.

The ellipse outlines are formed of circular arcs struck from four centers, which will approximate very closely to the true ellipse. The construction of the core sides is readily apparent from the sketch.

The arm taper is stated as 1 in 32 and 16, respectively, for the arm thickness and width; this gives a pleasing appearance for a moderately long arm,

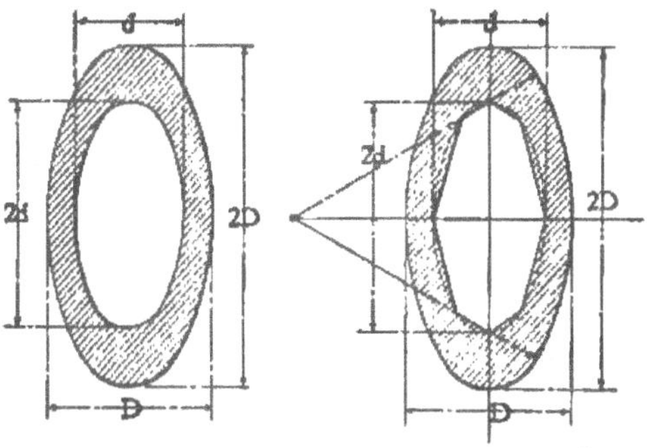

FIG. 77. PROPORTIONS OF HOLLOW ARMS. FIG. 78.

but it is not a hard-and-fast rule, as a greater or lesser taper may be employed to suit the designer's fancy without affecting the strength of the arm, unless the taper is made so excessive as to bring the dimensions at the rim down to one-half of these at the base.

As the tooth and arm are of the same material, the method is satisfactory for all cast gears, but this must not be interpreted to mean that this or any other formula will prevent shrinkage strain due to relatively large hubs or very heavy rims; where these occur, great care must be exercised in the foundry, and it will also not be amiss to add a generous amount of metal to the arms.

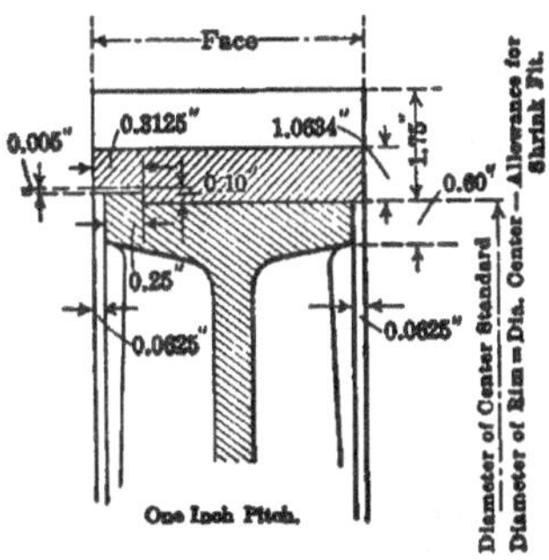

FIG. 79. PROPORTIONS OF RIM GEARS.

RIM GEAR PROPORTIONS

Where steel castings prove inefficient for the work intended, forged steel rims, designed somewhat as illustrated in Fig. 79, are used. The center is made of cast steel of a heavy pattern, the forged rims being shrunk thereon. As this rim material may be obtained of a higher grade of steel than is possible in the casting, and is free from hidden flaws, also as the rim is renewable, it is a much better and, in the end, a cheaper proposition. The rims are made of a forged billet, which is first pierced and then rolled into shape by the same process employed for locomotive tires.

There are many variations of this design, but it is thought best to make the face of the center narrower than the rim on account of possible unevenness in fitting, and turn a shoulder on center to bring rim up true, instead of depending upon parallels or surface plates. Also it will be desirable in many cases to replace rim without removing gear from the shaft. This is accomplished by rapidly heating rim by means of a circular gas or oil burner made to suit diameter of gear and protected by asbestos cover to localize the heat.

DESIGN OF BEVEL GEARS

The rules for the design of spur gears may be applied to helical, herringbone, worm, and spiral gears, using the circular pitch as a basis. Also for bevel gears, taking the proportions from the large end of the tooth. The average design of bevel gears is shown in Fig. 80. The hub should be carried well back of the face and connected to the rim by ribs, 4, 6, 8, or 10 in number, depending upon the diameter of the gear. The hub should not be carried too far in the front, or small end, of the gear, as a long hub will make it impossible in many cases to cut the teeth, for if this hub is carried too far it will interfere with the operation of the machine. See Fig. 81. A small hub, however, should always be put on the front end of the gear, otherwise it will be necessary to counterbore to secure a finished bearing.

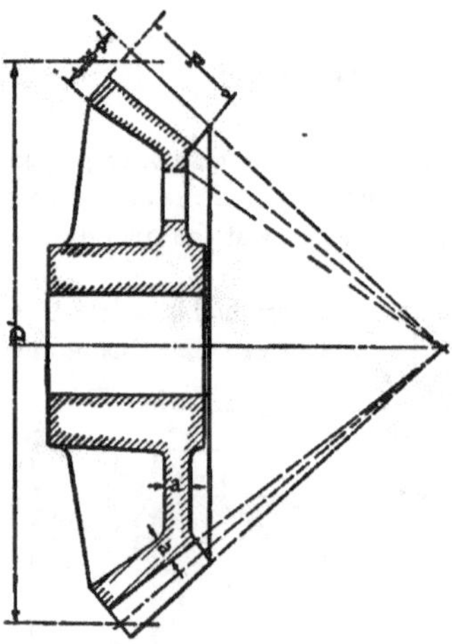

FIG. 80. BEVEL GEAR PROPORTIONS

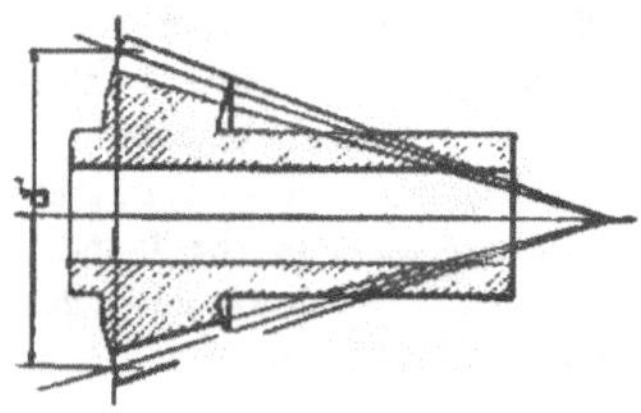

FIG. 81. BEVEL GEAR WITH LONG FRONT HUB EXTENSION.

RAWHIDE GEARS

Rawhide gears are commonly made with brass flanges on either side of the face to hold the rawhide in position and to engage the mating gear, unless rawhide contact alone is desired. It is practice to speak of the face of the rawhide gear as including these flanges. See Fig. 82. There is no great gain in preventing the flanges from coming into contact, and to make a rawhide gear without cutting through the flanges, as per Fig. 83, is unnecessary and expensive.

There is no reason why steel flanges cannot be used in place of brass, especially for the larger sizes; boiler plate will be found excellent for this. The thickness of the flange is something that varies greatly, although $5/16$ of the circular pitch is a fair average.

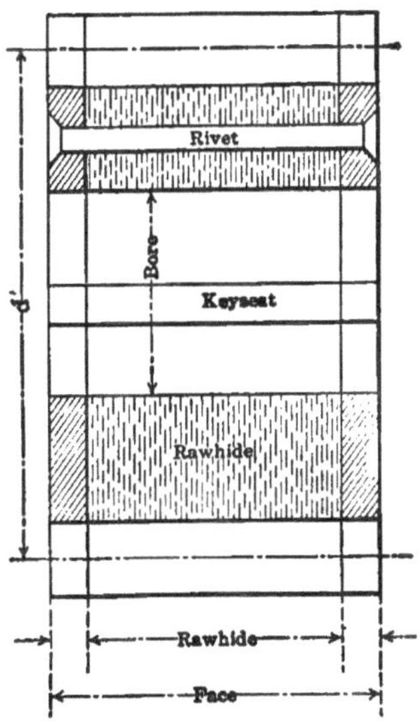

FIG. 82. RAWHIDE GEAR.

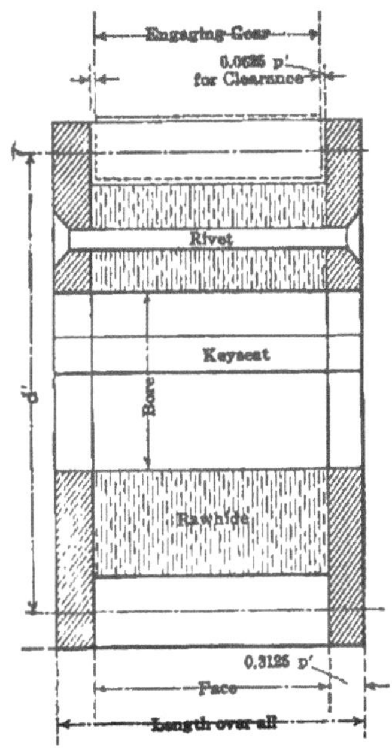

FIG. 83. SHROUDED RAWHIDE GEAR.

For the larger rawhide gears it is recommended that bolts instead of rivets be used, as it is impossible to otherwise draw up a wide-face rawhide gear. The bolt head may be countersunk so that one side of the gear will be flush. It is also sometimes possible to put the nut in a counterbore; this depends, of course, on the design of gear.

Rawhide gears to run loose on the shaft should be bushed. When it is necessary to move a rawhide gear on a spline one of the flanges should be made as part of the bush, as per Fig. 84. The usual design of the larger size rawhide gear is shown in Fig. 85.

Rawhide bevel gears are designed similar to Fig. 86. Both ends of the teeth must be flanged to facilitate the cutting of the teeth.

Fiber is often used in place of rawhide, but is usually more brittle and has a tendency to wear the engaging gear, although made of a harder material.

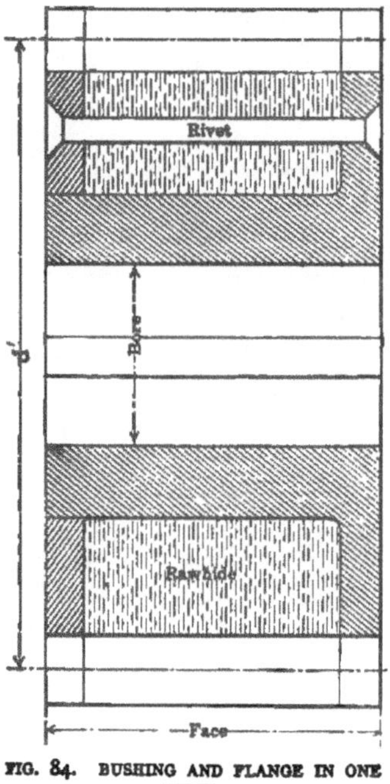

FIG. 84. BUSHING AND FLANGE IN ONE PIECE.

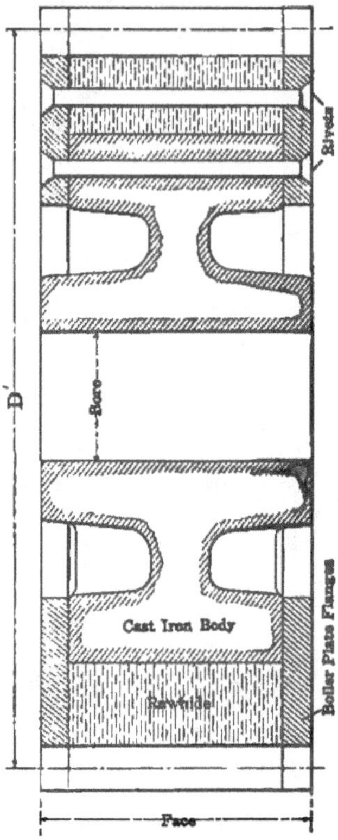

FIG. 85. DESIGN OF RAWHIDE GEAR.

Fiber gears are ordinarily furnished without flanges. Gears are also made of laminations of rawhide and bronze, or fiber and bronze, but not to any great extent.

A fiber stress of 5,000 pounds per square inch is amply safe for calculating the strength of rawhide gears.

MORTISE GEARS

The mortise gear is composed of a cast-iron rim, containing cored slots, into which wooden teeth are driven. See Fig. 87. These gears cannot

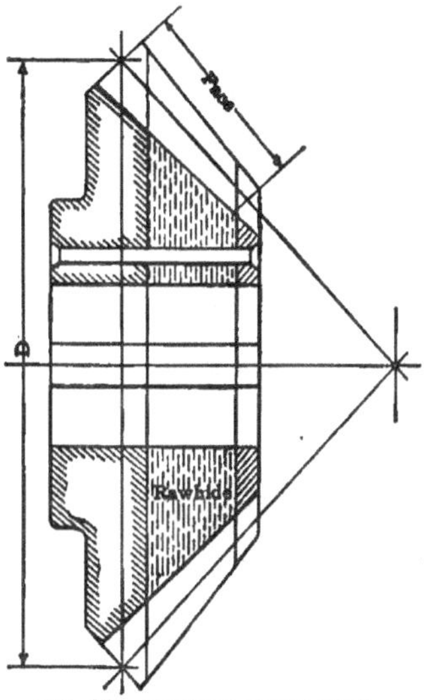

FIG. 86. RAWHIDE BEVEL GEAR.

compare either in efficiency or cost to a properly cut cast-iron gear, but are still used in many places where excessive noise is prohibitive. The wooden

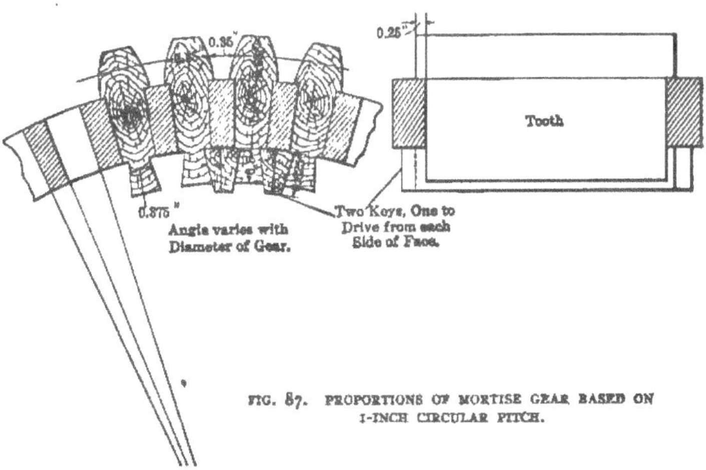

FIG. 87. PROPORTIONS OF MORTISE GEAR BASED ON 1-INCH CIRCULAR PITCH.

teeth are made either of apple or maple, treated in linseed oil and cut to the proper form after being inserted in the cored rim; otherwise the spacing of the teeth would be governed by the spacing of the cores, which can never be very accurate. Replaced teeth must be fitted and shaped by hand. The strength of the mortise gear is governed by the thickness of the iron teeth in the engaging gear, which are made 0.35 of the pitch instead of 0.5 pitch, as for cut gears. The outside diameter of the cored rim is turned to the dedendum diameter that the teeth will be set an equal distance from the center.

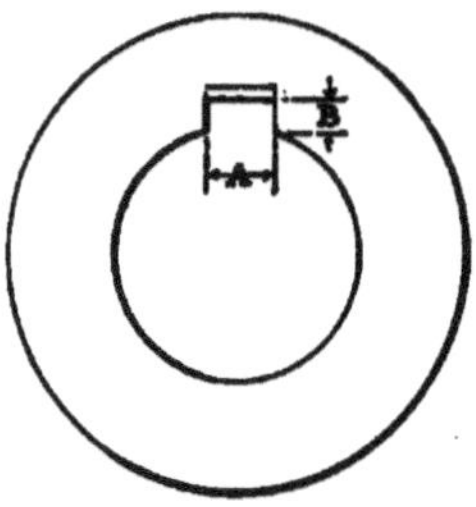

KEYSEATS

The commonly accepted keyseat standard is that of Jones and Laughlin, as per Table 17. In this the width of the key approximates one-quarter the shaft diameter. A square key is used, one-half being in the gear and one-half in the shaft. Dimensions for taper keys are included in this table. A taper keyseat, however, is nothing more than a means of tightening up a poor fit; a gear properly fitted to the shaft will not work loose if a straight key is used with clearance on the top. The best fit can be spoiled by a little carelessness in driving a taper key, as it is sure to make the gear run out if driven

Bore		Keyseat Taper 1/8 Inch per Foot		Straight Keyseat	
		Width "A"	Smallest Hight "B"	Width "A"	Hight "B"
In.	In.	In.	In.	In.	In.
8	to 7 7/8	2	21/32	2	1
7 3/4	" 7 3/8	1 7/8	5/8	1 7/8	15/16
7 1/4	" 6 7/8	1 3/4	19/32	1 3/4	7/8
6 3/4	" 6 3/8	1 5/8	17/32	1 5/8	13/16
6 1/4	" 5 7/8	1 1/2	1/2	1 1/2	3/4
5 3/4	" 5 3/8	1 3/8	29/64	1 3/8	11/16
5 1/4	" 4 7/8	1 1/4	27/64	1 1/4	5/8
4 3/4	" 4 7/16	1 1/8	3/8	1 1/8	9/16
4 3/8	" 4 3/16	1 1/16	11/32	1 1/16	17/32
4 1/8	" 3 15/16	1	21/64	1	1/2
3 7/8	" 3 11/16	15/16	5/16	15/16	15/32
3 5/8	" 3 7/16	7/8	19/64	7/8	7/16
3 3/8	" 3 3/16	13/16	17/64	13/16	13/32
3 1/8	" 2 15/16	3/4	1/4	3/4	3/8
2 7/8	" 2 11/16	11/16	15/64	11/16	11/32
2 5/8	" 2 7/16	5/8	13/64	5/8	5/16
2 3/8	" 2 3/16	9/16	3/16	9/16	9/32
2 1/8	" 1 15/16	1/2	11/64	1/2	1/4
1 7/8	" 1 11/16	7/16	9/64	7/16	7/32
1 5/8	" 1 7/16	3/8	1/8	3/8	3/16
1 3/8	" 1 3/16	5/16	7/64	5/16	5/32
1 1/8	" 1	1/4	3/32	1/4	1/8

Table 17—Standard Keyseats

the least bit too tight. The taper key is applicable to only the heaviest work where the mass of metal will prevent any such distortion.

For gears that are to be hardened it is important that there be a fillet in the corners, the top of the key being beveled off to suit, otherwise a crack is

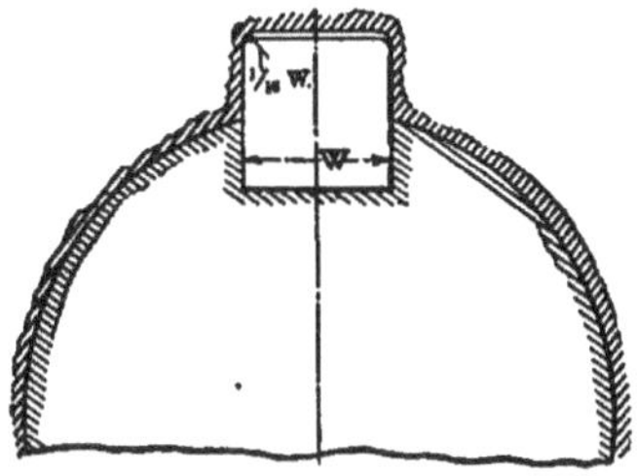

FIG. 88. FILLETED KEYWAY.

very liable to start from the sharp corner of the keyway. For that matter, this would be an excellent plan to adopt for all keyways. See Fig. 88.

For machine tools and automobiles the Woodruff key is generally used.

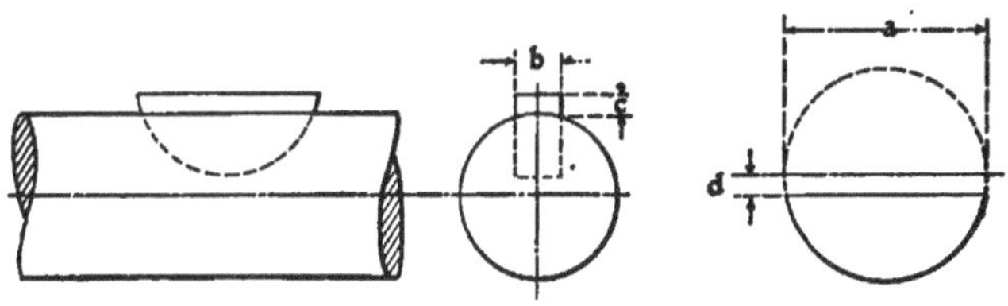

NO. OF KEY	DIAMETER OF KEY	THICKNESS OF KEY	DEPTH OF KEY-WAY	CENTER OF STOCK, FROM WHICH KEY IS MADE, TO TOP OF KEY	NO. OF KEY	DIAMETER OF KEY	THICKNESS OF KEY	DEPTH OF KEY-WAY	CENTER OF STOCK, FROM WHICH KEY IS MADE, TO TOP OF KEY
	a	*b*	*c*	*d*		*a*	*b*	*c*	*d*
1	1/2	1/16	1/32	3/64	*B*	1	5/16	5/32	1/16
2	1/2	3/32	3/64	3/64	16	1 1/8	3/16	3/32	5/64
3	1/2	1/8	1/16	3/64	17	1 1/8	7/32	7/64	5/64
4	5/8	3/32	3/64	1/16	18	1 1/8	1/4	1/8	5/64
5	5/8	1/8	1/16	1/16	*C*	1 1/8	5/16	5/32	5/64
6	5/8	5/32	5/64	1/16	19	1 1/4	3/16	3/32	5/64
7	3/4	1/8	1/16	1/16	20	1 1/4	7/32	7/64	5/64
8	3/4	5/32	5/64	1/16	21	1 1/4	1/4	1/8	5/64
9	3/4	3/16	3/32	1/16	*D*	1 1/4	5/16	5/32	5/64
10	7/8	5/32	5/64	1/16	*E*	1 1/4	3/8	3/16	5/64
11	7/8	3/16	3/32	1/16	22	1 3/8	1/4	1/8	3/32
12	7/8	7/32	7/64	1/16	23	1 3/8	5/16	5/32	3/32
A	7/8	1/4	1/8	1/16	*F*	1 3/8	3/8	3/16	3/32
13	1	3/16	3/32	1/16	24	1 1/2	1/4	1/8	7/64
14	1	7/32	7/64	1/16	25	1 1/2	5/16	5/32	7/64
15	1	1/4	1/8	1/16	*G*	1 1/2	3/8	3/16	7/64

TABLE 18—WOODRUFF STANDARD KEYS

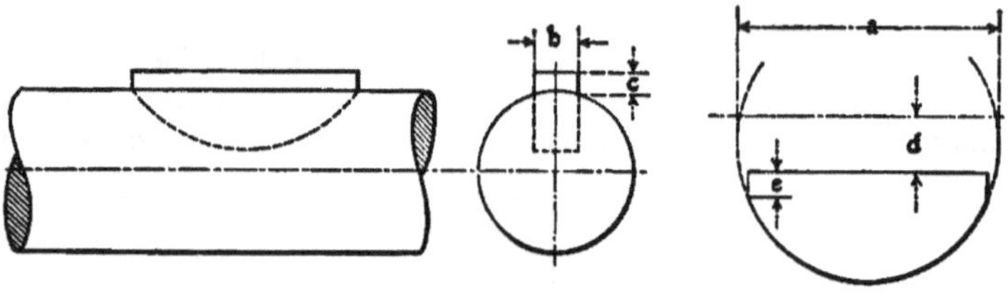

NUMBER OF KEY	DIAMETER OF KEY	THICKNESS OF KEY	DEPTH OF KEYWAY	CENTER OF STOCK, FROM WHICH KEY IS MADE, TO TOP OF KEY	WIDTH OF FLAT	NUMBER OF KEY	DIAMETER OF KEY	THICKNESS OF KEY	DEPTH OF KEYWAY	CENTER OF STOCK, FROM WHICH KEY IS MADE, TO TOP OF KEY	WIDTH OF FLAT
	a	*b*	*c*	*d*	*e*		*a*	*b*	*c*	*d*	*e*
26	2⅛	3/16	3/32	17/32	3/32	31	3½	7/16	7/32	13/16	3/16
27	2⅛	¼	⅛	17/32	3/32	32	3½	½	¼	13/16	3/16
28	2⅛	5/16	5/32	17/32	3/32	33	3½	9/16	9/32	13/16	3/16
29	2⅛	⅜	3/16	17/32	3/32	34	3½	⅝	5/16	13/16	3/16
30	3½	⅜	3/16	13/16	3/16						

TABLE 19—WOODRUFF SPECIAL KEYS

DIAMETER OF SHAFT	NUMBER OF KEYS	DIAMETER OF SHAFT	NUMBER OF KEYS	DIAMETER OF SHAFT	NUMBER OF KEYS
5/16–⅜	1	⅞–15/16	6, 8, 10	1⅜–1 7/16	14, 17, 20
7/16–½	2, 4	1	9, 11, 13	1½–1⅝	15, 18, 21, 24
9/16–⅝	3, 5	1 1/16–1⅛	9, 11, 13, 16	1 11/16–1¾	18, 21, 24
11/16–¾	3, 5, 7	1 3/16	11, 13, 16	1 13/16–2	23, 25
13/16	6, 8	1¼–1 5/16	12, 14, 17, 20	2 1/16–2½	25

WOODRUFF STANDARD KEYS TO USE WITH VARIOUS DIAMETER SHAFTS.

For heavy work the ordinary key will not answer: it is often necessary to put in two keys diametrically opposite, and for extremely heavy work, what is known as the "Kennedy" key, Fig. 89, is used. This is the only key that will answer the requirements of rolling-mill work. At the armor-plate mill of the Carnegie plant this type of key is used in the 22-inch shaft of rolls that reverse on an average of 20 times per minute. No other key would stand up to this work. These keys are made approximatly one-quarter of the shaft diameter, and located in the gear so that the corners of the keys intersect the bore. It is not necessary for the bottoms of the keys to be on a vertical line. The keys are made to a taper of ⅛ inch per foot on the

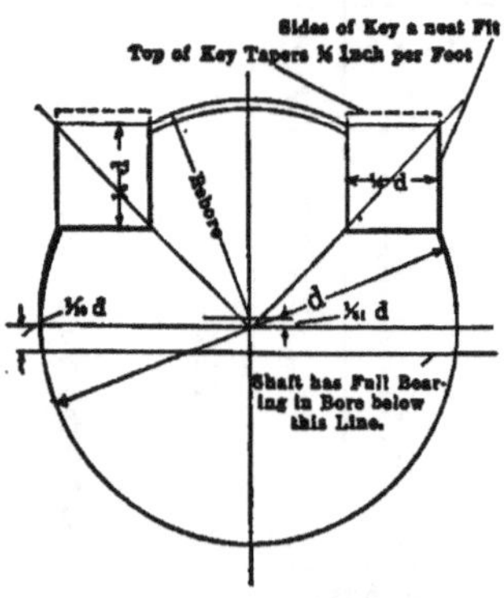

FIG. 89. THE KENNEDY KEY.

top for a driving fit, the sides being just a neat fit. The shaft is first bored for a press fit, then rebored about $\frac{1}{64}$ of the shaft diameter off center; the keyways are cut in the eccentric side. That portion of the bore opposite the keys remains as originally bored to within one-tenth of the shaft diameter below the center line.

The "Kennedy" key is especially desirable where it is necessary to move the gear for any considerable distance on the shaft before securing.

For street-railway work the gear is often pressed on the axle, not using a key of any description.

Where a sliding gear is used for heavy work, three keys, having radial sides as illustrated by Fig. 90, are generally employed. An example of this is the

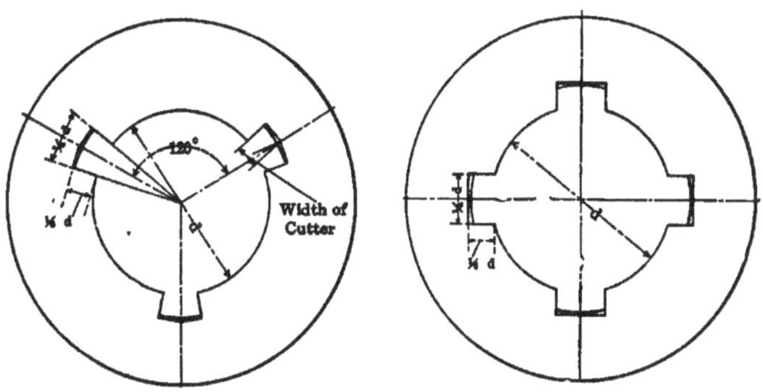

FIG. 90. KEYWAYS FOR HEAVY SLIDING GEAR. FIG. 91. FOUR-KEYED SLIDING GEAR.

gear drive for vertical rolls. This style of key has a distinct advantage over the four-key type with parallel sides as illustrated by Fig. 91.

BORE

The bore of a gear is supposed to be standard, any allowance for a fit being made in the shaft. This follows out the practice of the manufacturers of cold-rolled shafting, that is, to make the shaft enough under size for a sliding fit in a gear or pulley which is bored standard.

An exception to this is when press or shrink fits are desired. In this case the allowance is made in the gear, the shaft being turned standard.

PRESSURES AND ALLOWANCES FOR FORCE FITS

The Lane & Bodley Company of Cincinnati, O., furnishes the following data. For several years this firm has been keeping a record of observations on press fits with a view to making an analysis of them when a sufficient body of data had been accumulated, and thus obtaining a guide for future practice. Hundreds of cases of such press fits have been recorded, forming a body of data which is probably unequaled. See Chart 9.

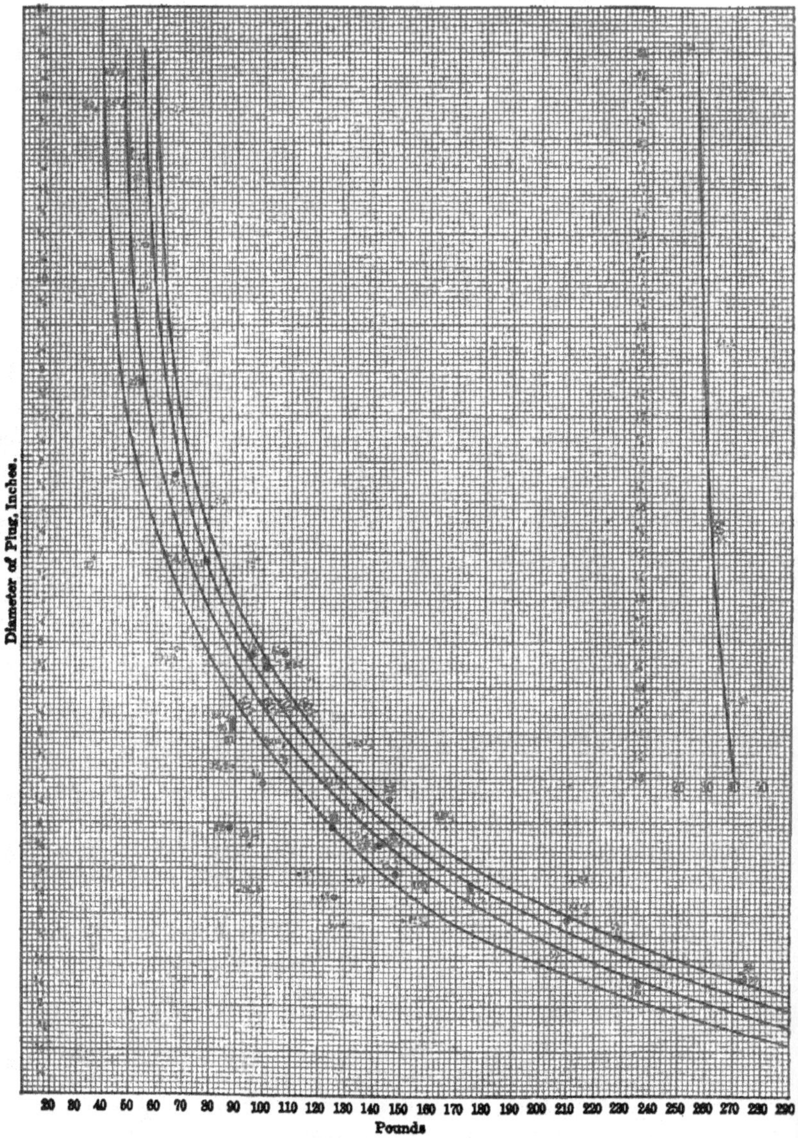

CHART 9. FORCE-FIT RELATIONS SHOWING LANE & BODLEY PRACTICE.

MEASUREMENTS AND PRESSURE READINGS

In these records the measurements have been made with great thoroughness. Both plug and hole have been measured on two diameters and at both ends, the average of these micrometer readings being taken as the true diameter. The pressures have been read at the beginning, middle, and end of the length of the fit; the material of both plug and ring, the length of the fit, the radial thickness of the hub, the areas and volumes of the fitted surfaces, and some other minor points have been noted, 24 entries being regularly made for each case. The resulting chart will thus be seen to have a very broad foundation.

In ordinary cases the quantities which are fixed by the conditions are the nominal diameter and length of the fit, the radial thickness of the hub, and the material. With these given it is required to find the press allowance for a given pressure to force the plug home.

HOW THE PRESSURE VARIES

Regarding the influence of these various factors, the pressure varies:

1. Directly as the surface of the fit for a given diameter.

2. Directly as the press-fit allowance, this allowance being such as not to stretch the metal beyond the elastic limit.

3. As some function of the radial thickness of the hub, which, while not determined mathematically, is shown in the chart.

4. With the materials used in a manner not yet determined owing to insufficient data. The chart is for steel or iron shafts and cast-iron cranks. Cast-steel cranks require much heavier pressures for the same press-fit allowances, but how much heavier cannot at present be said.

5. With other conditions, which cannot be formulated, and which lead to erratic results in the observations make it impossible to formulate a rule or construct a chart which shall give other than approximate results.

VARYING CONDITIONS

Among these are the nature of the surfaces as regards smoothness, the varying character of materials going under the same name, the shape of the crank and the speed with which the work is done. The term cast iron includes materials of widely varying hardness and other properties, and it is apparent that the web of a disk crank would have an influence not expressed by the radius of the hub. If a counterbalance were cast in the disk, this and the crank arm would naturally produce an effect on the effective thickness of the hub which would be different from the effect of the arm alone on a plain crank. Again, with a plain crank, the arm, being taper, would reinforce a greater arc of the pin eye than of the shaft eye.

STILL ANOTHER FACTOR

Another factor, which no doubt introduces some of the discrepancies of the diagram, is that while most of the shafts were of steel, some were of wrought iron, and no discrimination between these materials has been made in the analysis. These considerations explain the erratic results obtained, but it is nevertheless plain that the observations follow the general direction of the curves in a very marked manner.

AVERAGES SHOWN BY THE DIAGRAM

The plotted observations, it should be remarked, are in most cases the averages of many. The figures attached give the percentage of radial hub thickness to plug diameter. It will be observed in this connection that the discrepancies grow less as the diameters increase. This is doubtless chiefly due to the fact that the percentage of error is always greater with small experiments than with large. Its effect is to give increased value to the diagram when used with large sizes where it is most needed.

THE PROBLEM SOLVED BY THE CURVES

Of course, the holding power of these fits is the real thing desired, but it is obvious that the probability of adequate experiments being undertaken to determine this in large sizes is slight. The problem as it presents itself in the shop calls for the determination of the press-fit allowance to give a required pressure in forcing the parts home, and this the present diagram solves with a degree of accuracy sufficient for most purposes.

In order to reduce the size of the diagram, the portion applying to diameters above 13 inches has been detached and placed at the right. For these large sizes the observations are too few to justify the drawing of more than one curve. The lubricant used in all cases was linseed oil.

DIRECTIONS FOR USE

Select the curve which gives the ratio of the radial thickness of the hub divided by the diameter of the plug. Below the point of intersection of the plug diameter line with the selected curve, read pounds. Multiply this reading by the area of the fitted surface in square inches, and by the number of thousandths of an inch allowed for the press fit. The result will be the pressure in pounds carried to force the plug home.

AN EXAMPLE

The following example illustrates the use of the diagram: Diameter of plug, 8 inches, length of fit, 6 inches, diameter of hub, 16 inches, press-fit allowance, 0.020 inch. Required, the pressure to force the parts together.

$$\frac{\textit{Radial thickness of hub} = 4 \textit{ inches}}{\textit{Diameter of plug} = 8 \textit{ inches}} = 0.5.$$

Finding a point on the 50 per cent. curve opposite 8 inches, and tracing downward, we find 52 pounds. Area of fitted surface = 8 × 3.1416 × 6 = 150 square inches, 52 × 150 × 20 = 156,000 pounds = 78 tons.

SECTION V

Bevel Gears *

Cut bevel gears may be divided into three classes: First, that in which the teeth are of uniform depth throughout their entire length and of exact profile; second, that in which the depth of teeth correctly decreases toward the apex, but in which the teeth are originally cut with a correct profile at one point only, necessitating subsequent work on the teeth; and third, that in which the true taper type of teeth with correct profile at all points is generated by a process reproducing the rolling action of a pair of gears in mesh. These three classes may then be referred to according to the form of tooth cut: First, a modification of the correct tooth; second, an approximation of the correct tooth; and third, the correct form of tooth.

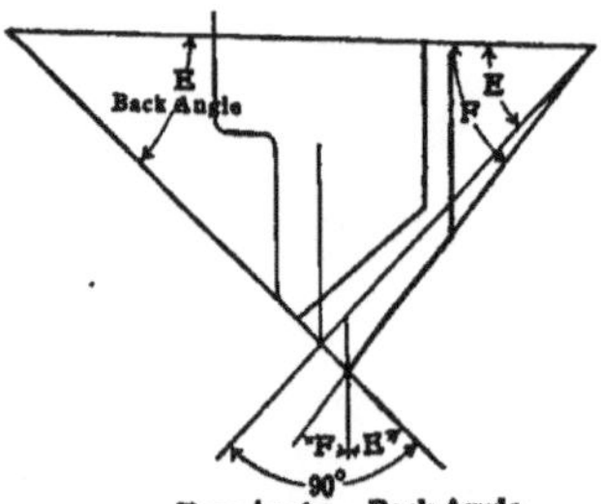

FIG. 92. LOCATION OF FACE AND BACK ANGLES.

The principal relationships of the various dimensions, angles, etc., of standard bevel gears of true form, in accordance with the notations and formulas on the succeeding pages, are as follows:

For gears meshing at right angles, the tangent of the center angle of a bevel gear is found by dividing the number of teeth in the gear by the number of teeth in the pinion; obviously the complement of the center angle of the gear is the center angle of the pinion. If the axes of the gears do not meet at right angles more complicated calculations are necessary. See the succeeding formulas.

The face angle is found by adding the angle increment, the angle between the pitch line and the face of the tooth, to the center angle.

The back angle is the complement of the center angle. See Fig. 92.

The cutting angle is found by subtracting the angle decrement, the angle between the pitch line and the bottom plane of the tooth, from the center angle.

When bevel gears are to be milled with parallel depth teeth, the back angle becomes the complement of the face angle, and the cutting angle is obtained by subtracting the angle increment instead of the angle decrement from the face angle, making the face angle of the gear the same as the cutting

* See Section XVI for Rolled Bevel Gears and Section XIV for Special Bevel Gears.

angle of the pinion and making the clearance the same at both ends of the teeth. The back and center angles are the same.

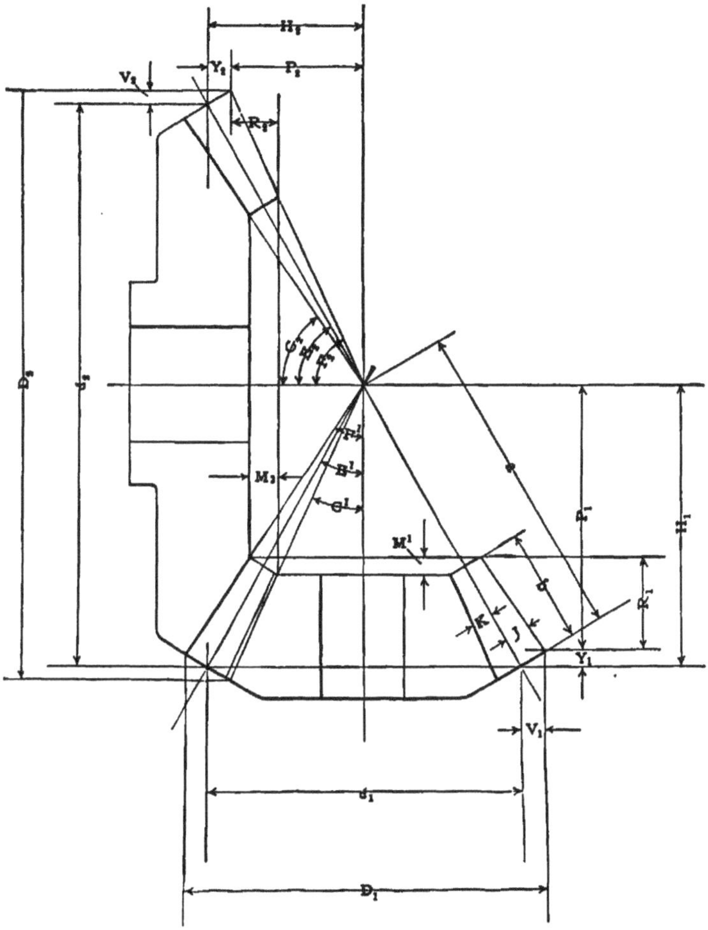

FIG. 93. DIAGRAM AND NOTATION FOR BEVEL GEAR.

The tangent of the angle decrement is found by dividing twice the sine of the center angle by the number of teeth. This holds true, however, only for gears having an addendum of $\frac{1}{p}$ or $0.3183p'$ (Brown & Sharpe standard).

When special teeth are used, the angle increment equals the addendum divided by the apex distance $\left(\frac{s}{a}\right)$, and the angle decrement equals the addendum plus the clearance divided by the apex distance $\left(\frac{s+f}{a}\right)$. When the addendum equals $\frac{1}{p}$, the angle increment is found by dividing the product of 2.314 and the sine of the center angle by the number of teeth.

The diameter increment is found by multiplying the addendum by the cosine of the center angle. Twice this diameter increment added to the pitch diameter of a bevel gear gives its outside diameter.

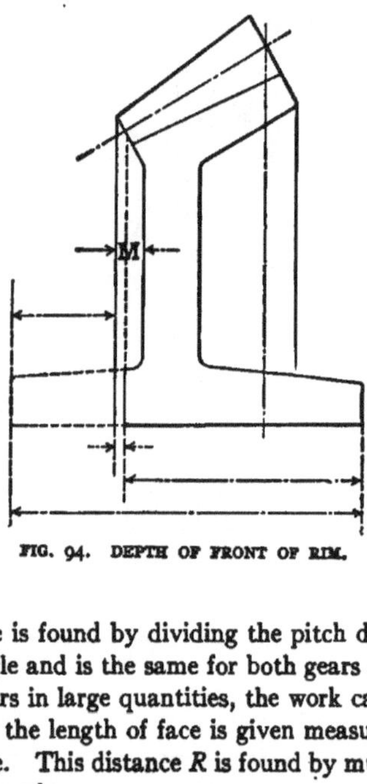

FIG. 94. DEPTH OF FRONT OF RIM.

The apex distance is found by dividing the pitch diameter by twice the sine of the center angle and is the same for both gears of a pair.

When turning gears in large quantities, the work can be carried forward more expeditiously if the length of face is given measured parallel with the center line of the bore. This distance R is found by multiplying the face by the cosine of the face angle.

Time may also be saved in turning bevel gears if the depth of rim at the front end is given. This will allow the workman to finish the face, front end, and bore of the gear during the first operation, while it is held in the chuck by the back hub. Ordinarily the depth is made 0.2 inch per inch circular pitch, but this can be varied to suit requirements, provided the rim depth is sufficient to allow for a full tooth at the small end. See Fig. 94.

The pitch diameter at the small end of the teeth is found by subtracting twice the product of the face and the sine of the center angle from the pitch diameter at the large end of the gear.

The outside diameter at the small end bears a similar relationship to the outside diameter at the large end.

The number of teeth for which the cutter should be selected is equal to the number of teeth in the gear or pinion divided by the cosine of its center angle.

The thickness of the tooth at the small end is found by dividing the product of the thickness of the tooth at the large end and the difference between the apex distance and the face by the apex distance. Or:

$$a : t :: a - b : t'. \quad \text{Therefore } t' = \frac{t(a-b)}{a}.$$

All dimensions for the small end of the tooth may be determined in this proportion.

NOTATION AND FORMULAS FOR MITRE GEARS

Center angle.	$E = 45$ degrees.
Face angle.	$F = 45$ deg. $+ J$.
Cutting angle.	$C = 45$ deg. $- K$.
Angle increment.	$Tan.\ J = \frac{s}{a}$, or $\frac{1.414}{N}$ when $s = \frac{1}{p}$.
Angle decrement.	$Tan.\ K = \frac{s+f}{a}$, or $\frac{1.636}{N}$ when $s = \frac{1}{p}$.
Diameter increment.	$V = s\ 0.70711$.
Backing.	$Y = s\ 0.70711$.
Outside diameter.	$D = d + 2V$.
Apex distance.	$a = \frac{d}{0.1414}$, or $d\ 0.70711$.
Distance from apex to point of tooth— large end. Parallel with axis.	$P = \frac{d}{2} - Y$.
Face, measured parallel with axis.	$R = b\ cos.\ F$.
Depth of rim at small end.	$M = \frac{W + 0.2\,p'\,(a-b)}{a}\ 0.70711$.
Pitch diameter at small end.	$d^{s\prime} = d - 2\,(b\ 0.70711)$.
Outside diameter at small end.	$D^{s} = D - 2\,(b\ sin.\ F)$.
Number of teeth for which cutter should be selected.	$= \frac{N}{0.70711}$.
Thickness of tooth at small end.	$t' = \frac{t\,(a-b)}{a}$.

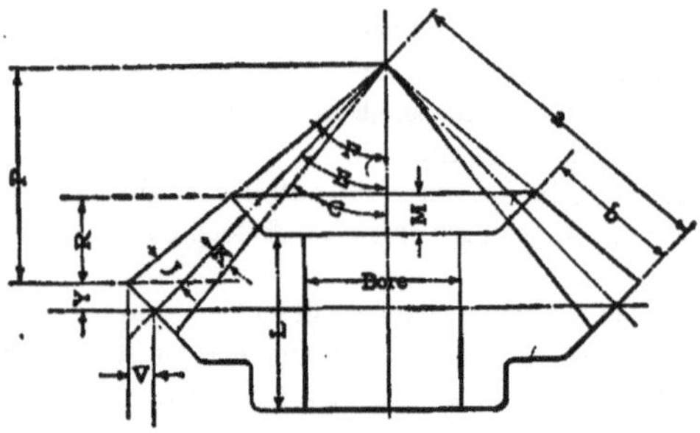

FIG. 95. MITRE GEAR.

NOTATION FOR BEVEL GEARS

AXIS AT 90 DEGREES

E = center angle.
F = face angle.
C = cutting angle.
J = angle increment.
K = angle decrement.
V = diameter increment.
Y = backing, or distance from point of tooth to pitch line.
d = pitch diameter.
D = outside diameter.
p' = circular pitch.
p = diametral pitch.
d^s = pitch diameter at small end.
D^s = outside diameter at small end.
t = thickness of tooth at largest pitch diameter.
t' = thickness of tooth at smallest pitch diameter.
a = apex distance.
H = distance from pitch line to apex.
P = distance from point of tooth to axis of mating gear.
R = face measured parallel with axis.
M = depth of rim at front end.
b = face.
s = addendum.
W' = whole depth.
N = number of teeth.

FORMULAS FOR BEVEL GEARS AT ANY ANGLE

Many formulas have heretofore been published for determining the angles and dimensions of bevel gears not at right angles, all of which are more or less confusing. It is simply a matter of finding the center angles; all other angles and dimensions are obtained as for ordinary bevel gears at right angles, each gear being figured separately. This also applies to the use of tables, the table

GEAR		PINION		REMARKS
REQ.	FORMULA	REQ.	FORMULA	
E_2	$Tan.\ E_2 = \frac{N_2}{N_1}$	E_1	$90° - E_2$	
F_2	$E_2 + J$	F_1	$E_1 + J$	Subtract from 90° for use in lathe.
C_2	$E_2 - K$	C_1	$E_1 - K$	
J	$Tan.\ J = \frac{2\ sine\ E_2}{N_2}$	J	$Tan.\ J = \frac{2\ sine\ E_1}{N_1}$	Same for both gear and pinion.
K	$Tan.\ K = \frac{2.314\ sine\ E_2}{N_2}$	K	$Tan.\ K = \frac{2.314\ sine\ E_1}{N_1}$	Same for both gear and pinion.
V_2	$s\ cos.\ E_2$	V_1	$S\ cos.\ E_1$	V_1 or pinion same as Y_2 for gear.
Y_2	$s\ sine\ E_2$	Y_1	$S\ sine\ E_1$	Y_1 for pinion same as V_2 for gear.
D_2	$d_2 + 2V_2$	D_1	$d_1 + 2V_1$	
a	$\frac{d_2}{2\ sine\ E_2}$	a	$\frac{d_1}{2\ sine\ E_1}$	Same per both gear and pinion.
P_2	$\frac{d_1}{2} - Y_2$	P_1	$\frac{d_2}{2} - Y_1$	
R_2	$b\ cos.\ F_2$	R_1	$b\ cos.\ F_1$	
H_2	$\frac{d_1}{2}$	H_1	$\frac{d_2}{2}$	
M_2	$\frac{W' + 0.2\ p'\ (a - b)}{a}\ sine\ E_2$	M_1	$\frac{W' + 0.2\ p'\ (a - b)}{a}\ sine\ E_1$	
d_2'	$d_2 - 2\ (b\ sine\ E_2)$	d_1'	$d_1 - 2\ (b\ sine\ E_1)$	
D_2'	$D_2 - 2\ (b\ sine\ F_2)$	D_1'	$D_1 - 2\ (b\ sine\ F_1)$	
Cutter$_2$	$\frac{N_2}{cos.\ E_2}$	Cutter$_1$	$\frac{N_1}{cos.\ E_1}$	
t'	$\frac{t\ (a - b)}{a}$			Same for both gear and pinion.

FORMULAS FOR BEVEL GEARS, SHAFTS AT 90 DEGREES

being entered for each gear separately, according to its center angle and independent of its mate.

When the axes are not at right angles there are four other combinations: Axes less than 90 degrees, see Fig. 96; axes greater than 90 degrees, Fig. 97;

crown bevel gears, Fig. 98; and internal bevel gears, Fig. 99. The center angles for these gears are found as follows:

N_2 = number of teeth in gear.
N_1 = number of teeth in pinion.

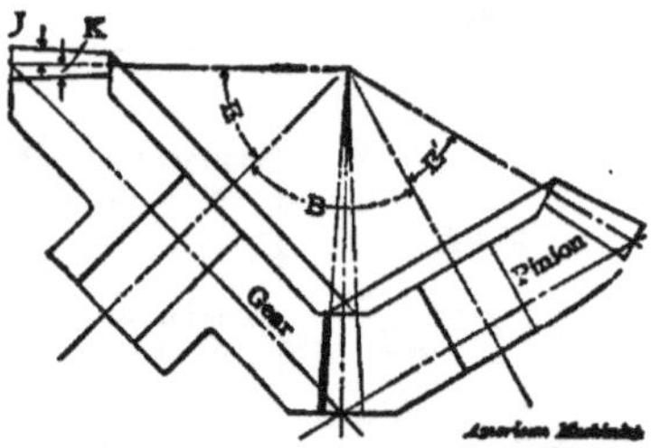

FIG. 96. BEVEL GEARS WITH AXES LESS THAN 90 DEGREES.

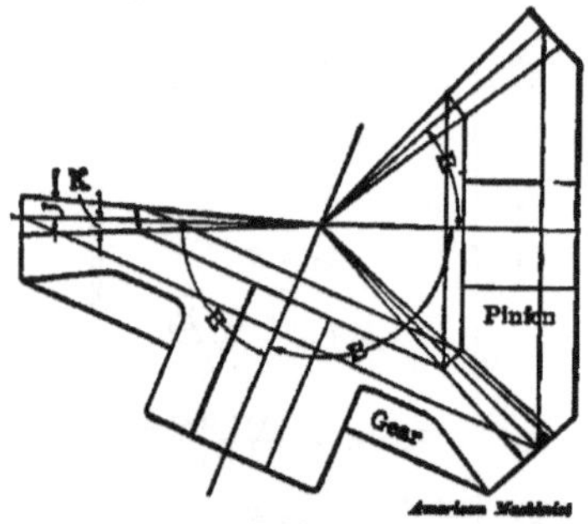

FIG. 97. BEVEL GEARS WITH AXES GREATER THAN 90 DEGREES.

$$Tan.\ E = \frac{sin.\ B}{\frac{N_1}{N_2} + cos.\ B}.$$

$$E' = B - E.$$

$$Tan.\ E = \frac{sin.\ (180 - B)}{\frac{N_1}{N_2} - cos.\ (180 - B)}.$$

$$E' = B - E.$$

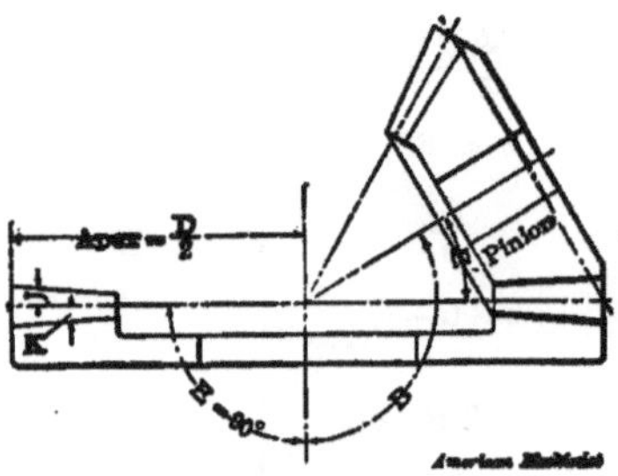

FIG. 98. CROWN BEVEL GEARS.

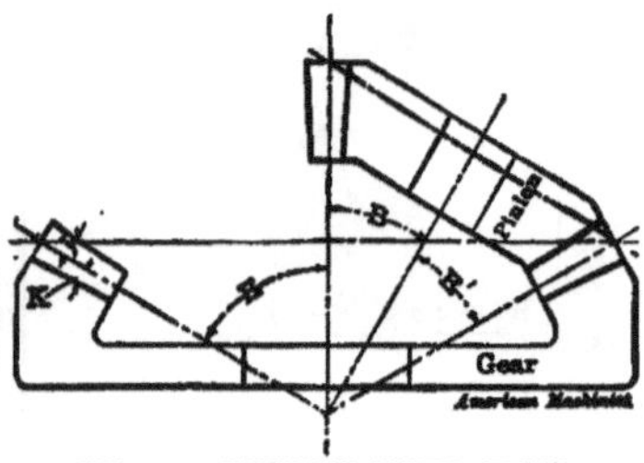

FIG. 99. INTERNAL BEVEL GEARS.

$$E = 90°.$$

$$E' = B - 90.$$

$$Tan.\ E = \frac{sin.\ B}{sin.\ B - \frac{N_1}{N_2}}$$

$$E' = E - B.$$

USE OF BEVEL GEAR TABLE

To use the bevel gear table, Table 21, divide the number of teeth in pinion by the number of teeth in gear; the quotient will equal the tangent of center angles. Find the nearest number in the table of tangents and on the same line on the left will be found the degrees, and at the top of the column hundredth degrees for the center angle of pinion. On the same line of the right will be found the degrees, and at the bottom of the column hundredth degrees for gear.

On the same line on the left will be found the angle increase, which, when divided by the number of teeth in the pinion, will give as a quotient the angle increase for either wheel.

To obtain the face angles add the angle increase to the center angle, and to obtain the cutting angle subtract the angle decrement from the center angle. Now on the same line on the left will be found the diameter increase for the pinion, and on the same line on the right will be found the diameter increase for the gear. These when divided by the required diametral pitch equal the diametral increase for that pitch, which, added to the pitch diameters, give the outside diameters.

AN ILLUSTRATIVE EXAMPLE

In a pair of bevel gears, 24 and 27 teeth, 8 diametral pitch; 24 divided by 72 = 0.3333, which is the tangent of the center angle. The nearest tangent in the table is 0.3346, which gives:

Center angle of pinion, 18.50 degrees.
Center angle of gear, 71.50 degrees.

On the same line at the left will be found the angle increase, 36, which divided by the number of teeth in the pinion will give the angle increase $\frac{36}{24} = 1.5$ degrees. This angle added to the center angle will give the face angle.

The cutting angle is found by subtracting this angle, plus 16 per cent., from the center angle, which in this example would be $1.50 \times 0.16 = 0.24$, therefore the angle increment would be $1.50 + 0.24 = 1.74$ degrees.

Face angle of pinion $= 18.50 + 1.50 = 20.00$ degrees.
Cutting angle of pinion $= 18.50 - 1.74 = 16.76$ degrees.
Face angle of gear $= 71.50 + 1.50 = 73.00$ degrees.
Cutting angle of gear $= 71.50 - 1.74 = 69.76$ degrees.

On the same line to the left will be found the diameter increase for the pinion, 1.90, which divided by the pitch, or $\frac{1.90}{8}$, $= 0.237$ inch. On the right of this same line will be found the diameter increase for the gear, 0.63, which divided by the pitch, or $\frac{0.65}{8}$, $= 0.081$ inch.

The pitch diameter of the pinion is $\frac{24}{8} = 3$ inches. The pitch diameter of the gear is $\frac{72}{8} = 9$ inches.

Outside diameter of gear $= 9 + 0.081 = 9.081$ inches.
Outside diameter of pinion $= 3 + 0.237 = 3.237$ inches.

For gears not at right angles, first determine the center angle, and enter the table for each gear separately, using the formulas given on pages 145 and 146. Except in the case of crown bevel gears, the formulas give tangents of the center angle, so the angles themselves may be most conveniently found from Table 21.

Number of Teeth in Wheel

		42	41	40	39	38	37	36	35	34	33	32	31	30	29	28
12	l........	2° 38'	2° 42'	2° 45'	2° 48'	2° 53'	2° 57'	3° 2'	3° 6'	3° 11'	3° 15'	3° 21'	3° 27'	3° 33'	3° 38'	3° 45'
	k........	3° 2'	3° 6'	3° 10'	3° 15'	3° 19'	3° 24'	3° 29'	3° 35'	3° 40'	3° 46'	3° 52'	3° 59'	4° 6'	4° 13'	4° 21'
13	l........	2° 37'	2° 40'	2° 44'	2° 47'	2° 52'	2° 56'	3° 0'	3° 4'	3° 9'	3° 14'	3° 19'	3° 24'	3° 29'	3° 36'	3° 43'
	k........	3° 1'	3° 5'	3° 9'	3° 13'	3° 18'	3° 23'	3° 28'	3° 33'	3° 38'	3° 44'	3° 50'	3° 56'	4° 3'	4° 10'	4° 17'
14	l........	2° 35'	2° 38'	2° 43'	2° 46'	2° 50'	2° 54'	2° 58'	3° 3'	3° 7'	3° 12'	3° 17'	3° 22'	3° 27'	3° 33'	3° 39'
	k........	2° 59'	3° 4'	3° 7'	3° 12'	3° 16'	3° 21'	3° 26'	3° 31'	3° 36'	3° 42'	3° 47'	3° 54'	4° 0'	4° 7'	4° 13'
15	l........	2° 35'	2° 38'	2° 41'	2° 44'	2° 48'	2° 53'	2° 56'	3° 1'	3° 4'	3° 10'	3° 14'	3° 19'	3° 24'	3° 30'	3° 36'
	k........	2° 58'	3° 2'	3° 6'	3° 10'	3° 14'	3° 19'	3° 24'	3° 29'	3° 34'	3° 39'	3° 45'	3° 51'	3° 57'	4° 4'	4° 10'
16	l........	2° 34'	2° 37'	2° 40'	2° 43'	2° 47'	2° 51'	2° 55'	2° 58'	3° 3'	3° 7'	3° 12'	3° 17'	3° 22'	3° 27'	3° 33'
	k........	2° 57'	3° 1'	3° 4'	3° 8'	3° 13'	3° 17'	3° 22'	3° 26'	3° 31'	3° 37'	3° 42'	3° 48'	3° 54'	4° 0'	4° 6'
17	l........	2° 33'	2° 35'	2° 38'	2° 42'	2° 45'	2° 49'	2° 53'	2° 57'	3° 1'	3° 5'	3° 9'	3° 14'	3° 18'	3° 24'	3° 29'
	k........	2° 56'	2° 59'	3° 3'	3° 7'	3° 11'	3° 15'	3° 20'	3° 24'	3° 27'	3° 34'	3° 39'	3° 45'	3° 50'	3° 56'	4° 2'
18	l........	2° 31'	2° 33'	2° 37'	2° 40'	2° 43'	2° 47'	2° 51'	2° 55'	2° 58'	3° 3'	3° 7'	3° 12'	3° 16'	3° 21'	3° 26'
	k........	2° 54'	2° 58'	3° 1'	3° 5'	3° 9'	3° 13'	3° 17'	3° 22'	3° 26'	3° 31'	3° 36'	3° 42'	3° 47'	3° 53'	3° 57'
19	l........	2° 30'	2° 32'	2° 35'	2° 38'	2° 42'	2° 45'	2° 49'	2° 53'	2° 57'	3° 0'	3° 4'	3° 9'	3° 13'	3° 18'	3° 23'
	k........	2° 52'	2° 56'	2° 59'	3° 3'	3° 7'	3° 12'	3° 16'	3° 19'	3° 24'	3° 29'	3° 33'	3° 38'	3° 44'	3° 49'	3° 55'
20	l........	2° 29'	2° 30'	2° 33'	2° 37'	2° 40'	2° 43'	2° 47'	2° 50'	2° 54'	2° 58'	3° 2'	3° 6'	3° 10'	3° 15'	3° 19'
	k........	2° 50'	2° 54'	2° 58'	3° 1'	3° 5'	3° 9'	3° 14'	3° 17'	3° 21'	3° 26'	3° 31'	3° 35'	3° 40'	3° 45'	3° 51'
21	l........	2° 28'	2° 28'	2° 32'	2° 35'	2° 38'	2° 42'	2° 45'	2° 48'	2° 52'	2° 56'	3° 0'	3° 3'	3° 8'	3° 12'	3° 17'
	k........	2° 49'	2° 52'	2° 56'	2° 59'	3° 3'	3° 7'	3° 11'	3° 15'	3° 19'	3° 23'	3° 27'	3° 32'	3° 36'	3° 41'	3° 47'
22	l........	2° 26'	2° 27'	2° 30'	2° 33'	2° 37'	2° 40'	2° 43'	2° 46'	2° 49'	2° 53'	2° 57'	3° 0'	3° 4'	3° 8'	3° 13'
	k........	2° 48'	2° 51'	2° 54'	2° 58'	3° 1'	3° 5'	3° 8'	3° 12'	3° 16'	3° 20'	3° 24'	3° 29'	3° 32'	3° 38'	3° 43'
23	l........	2° 25'	2° 26'	2° 28'	2° 32'	2° 35'	2° 38'	2° 42'	2° 44'	2° 47'	2° 51'	2° 54'	2° 58'	3° 2'	3° 5'	3° 9'
	k........	2° 46'	2° 49'	2° 52'	2° 56'	2° 59'	3° 2'	3° 6'	3° 10'	3° 14'	3° 18'	3° 22'	3° 26'	3° 30'	3° 34'	3° 39'
24	l........	2° 24'	2° 24'	2° 27'	2° 30'	2° 33'	2° 36'	2° 39'	2° 42'	2° 45'	2° 48'	2° 52'	2° 55'	2° 58'	3° 2'	3° 6'
	k........	2° 44'	2° 47'	2° 51'	2° 54'	2° 58'	3° 1'	3° 4'	3° 7'	3° 11'	3° 15'	3° 19'	3° 23'	3° 27'	3° 31'	3° 35'
25	l........	2° 22'	2° 23'	2° 26'	2° 28'	2° 31'	2° 34'	2° 37'	2° 40'	2° 43'	2° 46'	2° 49'	2° 53'	2° 56'	2° 59'	3° 3'
	k........	2° 43'	2° 46'	2° 49'	2° 52'	2° 56'	2° 58'	3° 1'	3° 4'	3° 8'	3° 12'	3° 16'	3° 20'	3° 24'	3° 28'	3° 32'
26	l........	2° 20'	2° 22'	2° 24'	2° 26'	2° 29'	2° 32'	2° 35'	2° 38'	2° 40'	2° 43'	2° 47'	2° 49'	2° 53'	2° 56'	3° 0'
	k........	2° 41'	2° 44'	2° 46'	2° 49'	2° 54'	2° 56'	2° 59'	3° 2'	3° 6'	3° 9'	3° 13'	3° 16'	3° 20'	3° 24'	3° 28'
27	l........	2° 19'	2° 19'	2° 22'	2° 24'	2° 27'	2° 30'	2° 33'	2° 36'	2° 38'	2° 41'	2° 44'	2° 47'	2° 50'	2° 53'	2° 57'
	k........	2° 39'	2° 42'	2° 44'	2° 47'	2° 51'	2° 53'	2° 57'	3° 0'	3° 3'	3° 6'	3° 10'	3° 13'	3° 17'	3° 20'	3° 24'
28	l........	2° 17'	2° 18'	2° 21'	2° 23'	2° 26'	2° 28'	2° 31'	2° 33'	2° 36'	2° 39'	2° 42'	2° 44'	2° 47'	2° 51'	2° 53'
	k........	2° 37'	2° 40'	2° 43'	2° 45'	2° 49'	2° 51'	2° 54'	2° 58'	3° 0'	3° 3'	3° 7'	3° 10'	3° 14'	3° 17'	3° 21'
29	l........	2° 16'	2° 17'	2° 19'	2° 21'	2° 24'	2° 26'	2° 29'	2° 31'	2° 34'	2° 37'	2° 39'	2° 42'	2° 44'	2° 47'	
	k........	2° 36'	2° 38'	2° 41'	2° 44'	2° 47'	2° 49'	2° 52'	2° 54'	2° 57'	3° 1'	3° 5'	3° 7'	3° 10'	3° 14'	
30	l........	2° 14'	2° 15'	2° 17'	2° 19'	2° 22'	2° 24'	2° 27'	2° 29'	2° 32'	2° 34'	2° 36'	2° 39'	2° 42'		
	k........	2° 34'	2° 36'	2° 39'	2° 42'	2° 44'	2° 46'	2° 49'	2° 52'	2° 55'	2° 58'	3° 1'	3° 4'	3° 7'		
31	l........	2° 13'	2° 13'	2° 16'	2° 18'	2° 20'	2° 22'	2° 24'	2° 27'	2° 29'	2° 32'	2° 34'	2° 37'			
	k........	2° 32'	2° 34'	2° 37'	2° 39'	2° 42'	2° 44'	2° 47'	2° 50'	2° 53'	2° 56'	2° 58'	3° 1'			
32	l........	2° 11'	2° 12'	2° 14'	2° 16'	2° 18'	2° 20'	2° 23'	2° 25'	2° 27'	2° 29'	2° 32'				
	k........	2° 31'	2° 33'	2° 35'	2° 37'	2° 40'	2° 43'	2° 45'	2° 48'	2° 51'	2° 53'	2° 56'				
33	l........	2° 10'	2° 11'	2° 12'	2° 14'	2° 16'	2° 18'	2° 21'	2° 23'	2° 25'	2° 27'					
	k........	2° 29'	2° 31'	2° 33'	2° 36'	2° 38'	2° 40'	2° 42'	2° 45'	2° 48'	2° 51'					
34	l........	2° 9'	2° 8'	2° 11'	2° 12'	2° 14'	2° 17'	2° 18'	2° 21'	2° 23'						
	k........	2° 27'	2° 29'	2° 31'	2° 34'	2° 36'	2° 37'	2° 40'	2° 42'	2° 45'						
35	l........	2° 7'	2° 7'	2° 9'	2° 11'	2° 13'	2° 15'	2° 17'	2° 19'							
	k........	2° 25'	2° 27'	2° 30'	2° 32'	2° 34'	2° 35'	2° 38'	2° 40'							
36	l........	2° 6'	2° 6'	2° 8'	2° 9'	2° 11'	2° 13'	2° 15'								
	k........	2° 24'	2° 26'	2° 28'	2° 30'	2° 32'	2° 34'	2° 36'								
37	l........	2° 5'	2° 4'	2° 6'	2° 8'	2° 9'	2° 12'									
	k........	2° 22'	2° 24'	2° 26'	2° 28'	2° 30'	2° 32'									
38	l........	2° 3'	2° 3'	2° 4'	2° 6'	2° 7'										
	k........	2° 20'	2° 22'	2° 24'	2° 25'	2° 27'										
39	l........	2° 1'	2° 1'	2° 3'	2° 5'											
	k........	2° 18'	2° 20'	2° 22'	2° 23'											
40	l........	1° 59'	1° 59'	2° 1'												
	k........	2° 17'	2° 18'	2° 20'												
41	l........	1° 58'	1° 58'													
	k........	2° 15'	2° 16'													
42	l........	1° 57'														
	k........	2° 13'														

TABLE 20—ADDENDUM AND DEDENDUM ANGLES FOR BEVEL GEARS. ANGLE BETWEEN

Number of Teeth in Wheel

Number of Teeth in Pinion	27	26	25	24	23	22	21	20	19	18	17	16	15	14	13	12
12	3° 53' / 4° 29'	4° 0' / 4° 37'	4° 7' / 4° 46'	4° 16' / 4° 56'	4° 25' / 5° 6'	4° 34' / 5° 16'	4° 43' / 5° 28'	4° 55' / 5° 40'	5° 5' / 5° 53'	5° 16' / 6° 6'	5° 30' / 6° 22'	5° 42' / 6° 36'	5° 56' / 6° 52'	6° 10' / 7° 9'	6° 27' / 7° 27'	6° 43' / 7° 46'
13	3° 49' / 4° 25'	3° 57' / 4° 33'	4° 3' / 4° 42'	4° 12' / 4° 51'	4° 20' / 5° 0'	4° 29' / 5° 10'	4° 38' / 5° 21'	4° 48' / 5° 32'	4° 58' / 5° 43'	5° 8' / 5° 57'	5° 20' / 6° 10'	5° 32' / 6° 24'	5° 45' / 6° 39'	5° 57' / 6° 54'	6° 13' / 7° 10'	
14	3° 46' / 4° 21'	3° 54' / 4° 29'	3° 59' / 4° 37'	4° 7' / 4° 46'	4° 15' / 4° 54'	4° 24' / 5° 4'	4° 32' / 5° 14'	4° 40' / 5° 25'	4° 50' / 5° 36'	5° 0' / 5° 48'	5° 11' / 6° 0'	5° 22' / 6° 12'	5° 37' / 6° 24'	5° 45' / 6° 43'		
15	3° 43' / 4° 16'	3° 49' / 4° 24'	3° 55' / 4° 32'	4° 2' / 4° 41'	4° 9' / 4° 49'	4° 18' / 4° 58'	4° 26' / 5° 7'	4° 34' / 5° 17'	4° 43' / 5° 27'	4° 52' / 5° 38'	5° 2' / 5° 50'	5° 12' / 6° 2'	5° 27' / 6° 14'			
16	3° 39' / 4° 13'	3° 45' / 4° 21'	3° 52' / 4° 27'	3° 58' / 4° 36'	4° 5' / 4° 44'	4° 14' / 4° 52'	4° 20' / 5° 0'	4° 28' / 5° 10'	4° 36' / 5° 19'	4° 45' / 5° 29'	4° 53' / 5° 40'	5° 3' / 5° 50'				
17	3° 35' / 4° 9'	3° 41' / 4° 16'	3° 47' / 4° 23'	3° 53' / 4° 30'	4° 0' / 4° 37'	4° 8' / 4° 45'	4° 13' / 4° 54'	4° 22' / 5° 2'	4° 30' / 5° 11'	4° 37' / 5° 21'	4° 45' / 5° 30'					
18	3° 32' / 4° 4'	3° 37' / 4° 10'	3° 43' / 4° 18'	3° 49' / 4° 25'	3° 55' / 4° 32'	4° 2' / 4° 39'	4° 8' / 4° 47'	4° 15' / 4° 55'	4° 22' / 5° 3'	4° 29' / 5° 12'						
19	3° 28' / 4° 1'	3° 35' / 4° 7'	3° 38' / 4° 13'	3° 44' / 4° 19'	3° 50' / 4° 26'	3° 56' / 4° 33'	4° 2' / 4° 40'	4° 8' / 4° 48'	4° 15' / 4° 55'							
20	3° 24' / 3° 56'	3° 29' / 4° 2'	3° 34' / 4° 8'	3° 40' / 4° 14'	3° 45' / 4° 20'	3° 51' / 4° 27'	3° 56' / 4° 34'	4° 3' / 4° 41'								
21	3° 21' / 3° 53'	3° 25' / 3° 58'	3° 30' / 4° 3'	3° 35' / 4° 8'	3° 40' / 4° 14'	3° 46' / 4° 21'	3° 52' / 4° 27'									
22	3° 17' / 3° 48'	3° 22' / 3° 52'	3° 26' / 3° 58'	3° 31' / 4° 4'	3° 36' / 4° 9'	3° 41' / 4° 15'										
23	3° 14' / 3° 44'	3° 18' / 3° 49'	3° 22' / 3° 54'	3° 27' / 3° 59'	3° 32' / 4° 4'											
24	3° 10' / 3° 39'	3° 14' / 3° 44'	3° 18' / 3° 49'	3° 22' / 3° 54'												
25	3° 7' / 3° 36'	3° 10' / 3° 40'	3° 14' / 3° 45'													
26	3° 3' / 3° 32'	3° 7' / 3° 36'														
27	3° 0' / 3° 28'															
28																
29																
30																
31																
32																
33																
34																
35																
36																
37																
38																
39																
40																
41																
42																

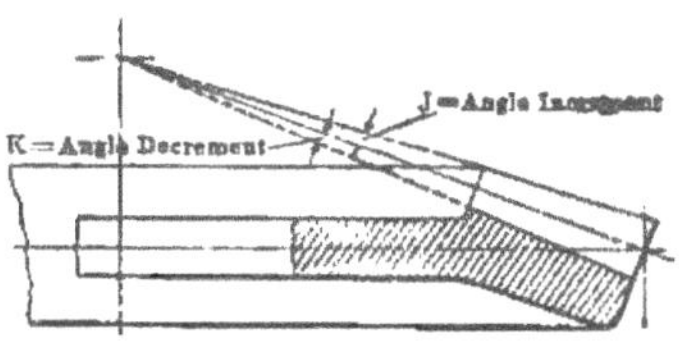

AXES 90°, TOOTH PROPORTIONS BROWN & SHARPE STANDARD. BY F. WITHERS

Number of Teeth in Wheel

		72	71	70	69	63	67	66	65	64	63	62	61	60	59
12	J..............	1° 34'	1° 35'	1° 37'	1° 38'	1° 40'	1° 41'	1° 43'	1° 45'	1° 46'	1° 48'	1° 50'	1° 51'	1° 53'	1° 55'
	K..............	1° 49'	1° 50'	1° 52'	1° 54'	1° 55'	1° 57'	1° 58'	2° 0'	2° 2'	2° 4'	2° 6'	2° 8'	2° 10'	2° 12'
13	J..............	1° 34'	1° 35'	1° 37'	1° 38'	1° 40'	1° 41'	1° 43'	1° 44'	1° 46'	1° 48'	1° 50'	1° 51'	1° 53'	1° 55'
	K..............	1° 49'	1° 50'	1° 52'	1° 53'	1° 55'	1° 56'	1° 58'	2° 0'	2° 2'	2° 4'	2° 5'	2° 7'	2° 9'	2° 11'
14	J..............	1° 34'	1° 35'	1° 37'	1° 38'	1° 39'	1° 41'	1° 43'	1° 44'	1° 45'	1° 47'	1° 50'	1° 50'	1° 52'	1° 54'
	K..............	1° 48'	1° 50'	1° 51'	1° 53'	1° 55'	1° 56'	1° 58'	2° 0'	2° 1'	2° 3'	2° 5'	2° 7'	2° 9'	2° 11'
15	J..............	1° 34'	1° 35'	1° 36'	1° 38'	1° 39'	1° 40'	1° 42'	1° 44'	1° 45'	1° 47'	1° 49'	1° 50'	1° 51'	1° 54'
	K..............	1° 48'	1° 50'	1° 51'	1° 53'	1° 54'	1° 56'	1° 57'	1° 59'	2° 0'	2° 3'	2° 5'	2° 7'	2° 9'	2° 11'
16	J..............	1° 33'	1° 35'	1° 36'	1° 37'	1° 38'	1° 40'	1° 42'	1° 43'	1° 45'	1° 45'	1° 48'	1° 50'	1° 51'	1° 53'
	K..............	1° 47'	1° 49'	1° 51'	1° 52'	1° 54'	1° 55'	1° 57'	1° 58'	2° 0'	2° 2'	2° 4'	2° 6'	2° 8'	2° 10'
17	J..............	1° 33'	1° 34'	1° 35'	1° 37'	1° 38'	1° 40'	1° 41'	1° 43'	1° 45'	1° 45'	1° 48'	1° 50'	1° 50'	1° 53'
	K..............	1° 47'	1° 49'	1° 50'	1° 51'	1° 53'	1° 55'	1° 57'	1° 58'	2° 0'	2° 2'	2° 4'	2° 6'	2° 8'	2° 10'
18	J..............	1° 33'	1° 34'	1° 35'	1° 37'	1° 38'	1° 39'	1° 41'	1° 42'	1° 44'	1° 45'	1° 47'	1° 49'	1° 50'	1° 52'
	K..............	1° 47'	1° 49'	1° 50'	1° 51'	1° 53'	1° 55'	1° 56'	1° 58'	1° 59'	2° 1'	2° 3'	2° 5'	2° 7'	2° 9'
19	J..............	1° 33'	1° 34'	1° 35'	1° 36'	1° 37'	1° 39'	1° 40'	1° 42'	1° 43'	1° 45'	1° 47'	1° 48'	1° 50'	1° 52'
	K..............	1° 47'	1° 48'	1° 50'	1° 51'	1° 53'	1° 54'	1° 56'	1° 57'	1° 59'	2° 1'	2° 2'	2° 4'	2° 6'	2° 8'
20	J..............	1° 32'	1° 33'	1° 34'	1° 36'	1° 37'	1° 38'	1° 40'	1° 41'	1° 42'	1° 45'	1° 46'	1° 48'	1° 49'	1° 51'
	K..............	1° 46'	1° 48'	1° 49'	1° 51'	1° 52'	1° 54'	1° 55'	1° 57'	1° 59'	2° 0'	2° 2'	2° 4'	2° 6'	2° 8'
21	J..............	1° 32'	1° 33'	1° 34'	1° 35'	1° 37'	1° 38'	1° 40'	1° 41'	1° 42'	1° 44'	1° 45'	1° 47'	1° 49'	1° 50'
	K..............	1° 46'	1° 47'	1° 49'	1° 50'	1° 52'	1° 53'	1° 55'	1° 56'	1° 58'	2° 0'	2° 1'	2° 3'	2° 5'	2° 7'
22	J..............	1° 32'	1° 33'	1° 34'	1° 35'	1° 36'	1° 38'	1° 39'	1° 40'	1° 42'	1° 44'	1° 45'	1° 47'	1° 48'	1° 50'
	K..............	1° 46'	1° 47'	1° 48'	1° 49'	1° 51'	1° 52'	1° 54'	1° 55'	1° 57'	1° 59'	2° 1'	2° 2'	2° 4'	2° 6'
23	J..............	1° 31'	1° 33'	1° 33'	1° 35'	1° 36'	1° 37'	1° 39'	1° 40'	1° 41'	1° 43'	1° 45'	1° 46'	1° 48'	1° 50'
	K..............	1° 45'	1° 46'	1° 48'	1° 49'	1° 51'	1° 52'	1° 54'	1° 55'	1° 57'	1° 58'	2° 0'	2° 2'	2° 4'	2° 5'
24	J..............	1° 31'	1° 32'	1° 33'	1° 34'	1° 35'	1° 37'	1° 38'	1° 39'	1° 41'	1° 42'	1° 44'	1° 45'	1° 47'	1° 49'
	K..............	1° 45'	1° 46'	1° 48'	1° 49'	1° 50'	1° 51'	1° 53'	1° 54'	1° 56'	1° 58'	2° 0'	2° 1'	2° 3'	2° 5'
25	J..............	1° 30'	1° 32'	1° 33'	1° 34'	1° 35'	1° 36'	1° 38'	1° 39'	1° 40'	1° 42'	1° 43'	1° 45'	1° 46'	1° 48'
	K..............	1° 44'	1° 45'	1° 47'	1° 48'	1° 50'	1° 51'	1° 52'	1° 54'	1° 56'	1° 57'	1° 59'	2° 0'	2° 2'	2° 4'
26	J..............	1° 30'	1° 31'	1° 32'	1° 33'	1° 34'	1° 35'	1° 37'	1° 38'	1° 40'	1° 41'	1° 43'	1° 44'	1° 45'	1° 47'
	K..............	1° 44'	1° 45'	1° 46'	1° 47'	1° 49'	1° 50'	1° 52'	1° 53'	1° 55'	1° 56'	1° 58'	1° 59'	2° 1'	2° 3'
27	J..............	1° 30'	1° 31'	1° 32'	1° 33'	1° 34'	1° 35'	1° 37'	1° 38'	1° 39'	1° 40'	1° 42'	1° 43'	1° 45'	1° 46'
	K..............	1° 44'	1° 45'	1° 46'	1° 47'	1° 49'	1° 50'	1° 51'	1° 53'	1° 55'	1° 56'	1° 58'	1° 59'	2° 1'	2° 2'
28	J..............	1° 29'	1° 30'	1° 31'	1° 33'	1° 33'	1° 35'	1° 36'	1° 37'	1° 38'	1° 40'	1° 41'	1° 43'	1° 44'	1° 45'
	K..............	1° 43'	1° 44'	1° 45'	1° 46'	1° 48'	1° 49'	1° 51'	1° 52'	1° 54'	1° 55'	1° 57'	1° 58'	2° 0'	2° 1'
29	J..............	1° 29'	1° 30'	1° 31'	1° 32'	1° 33'	1° 34'	1° 35'	1° 36'	1° 38'	1° 39'	1° 40'	1° 42'	1° 44'	1° 45'
	K..............	1° 42'	1° 43'	1° 45'	1° 46'	1° 47'	1° 48'	1° 50'	1° 51'	1° 53'	1° 54'	1° 56'	1° 57'	1° 59'	2° 1'
30	J..............	1° 28'	1° 29'	1° 30'	1° 32'	1° 33'	1° 34'	1° 35'	1° 35'	1° 37'	1° 38'	1° 40'	1° 41'	1° 43'	1° 45'
	K..............	1° 42'	1° 42'	1° 44'	1° 45'	1° 47'	1° 48'	1° 50'	1° 51'	1° 53'	1° 54'	1° 55'	1° 56'	1° 58'	2° 0'
31	J..............	1° 28'	1° 29'	1° 30'	1° 31'	1° 32'	1° 33'	1° 35'	1° 35'	1° 37'	1° 38'	1° 40'	1° 40'	1° 42'	1° 44'
	K..............	1° 41'	1° 42'	1° 44'	1° 45'	1° 46'	1° 47'	1° 49'	1° 50'	1° 52'	1° 53'	1° 55'	1° 56'	1° 58'	1° 59'
32	J..............	1° 27'	1° 28'	1° 30'	1° 30'	1° 32'	1° 33'	1° 34'	1° 34'	1° 36'	1° 38'	1° 39'	1° 40'	1° 41'	1° 43'
	K..............	1° 41'	1° 42'	1° 43'	1° 44'	1° 45'	1° 47'	1° 48'	1° 50'	1° 51'	1° 52'	1° 54'	1° 55'	1° 57'	1° 58'
33	J..............	1° 27'	1° 28'	1° 29'	1° 30'	1° 31'	1° 32'	1° 33'	1° 34'	1° 35'	1° 37'	1° 38'	1° 40'	1° 40'	1° 42'
	K..............	1° 40'	1° 41'	1° 43'	1° 44'	1° 45'	1° 46'	1° 48'	1° 49'	1° 50'	1° 51'	1° 53'	1° 54'	1° 56'	1° 57'
34	J..............	1° 26'	1° 27'	1° 29'	1° 30'	1° 30'	1° 32'	1° 33'	1° 33'	1° 35'	1° 36'	1° 38'	1° 39'	1° 40'	1° 41'
	K..............	1° 40'	1° 41'	1° 42'	1° 43'	1° 45'	1° 46'	1° 47'	1° 48'	1° 50'	1° 51'	1° 52'	1° 53'	1° 55'	1° 56'
35	J..............	1° 26'	1° 27'	1° 28'	1° 29'	1° 30'	1° 31'	1° 32'	1° 33'	1° 35'	1° 35'	1° 37'	1° 38'	1° 39'	1° 40'
	K..............	1° 39'	1° 40'	1° 42'	1° 43'	1° 44'	1° 45'	1° 46'	1° 47'	1° 49'	1° 50'	1° 52'	1° 53'	1° 55'	1° 56'
36	J..............	1° 25'	1° 26'	1° 28'	1° 28'	1° 30'	1° 30'	1° 32'	1° 32'	1° 34'	1° 35'	1° 36'	1° 37'	1° 38'	1° 40'
	K..............	1° 39'	1° 40'	1° 41'	1° 42'	1° 43'	1° 44'	1° 46'	1° 47'	1° 48'	1° 49'	1° 51'	1° 52'	1° 54'	1° 55'
37	J..............	1° 25'	1° 26'	1° 27'	1° 28'	1° 29'	1° 30'	1° 31'	1° 32'	1° 33'	1° 34'	1° 35'	1° 36'	1° 38'	1° 39'
	K..............	1° 38'	1° 39'	1° 40'	1° 41'	1° 43'	1° 44'	1° 45'	1° 46'	1° 47'	1° 48'	1° 50'	1° 51'	1° 53'	1° 54'
38	J..............	1° 24'	1° 25'	1° 26'	1° 27'	1° 28'	1° 30'	1° 30'	1° 31'	1° 32'	1° 33'	1° 35'	1° 35'	1° 37'	1° 38'
	K..............	1° 38'	1° 39'	1° 40'	1° 41'	1° 42'	1° 43'	1° 45'	1° 46'	1° 47'	1° 48'	1° 49'	1° 50'	1° 52'	1° 53'
39	J..............	1° 24'	1° 25'	1° 25'	1° 27'	1° 28'	1° 29'	1° 30'	1° 31'	1° 32'	1° 33'	1° 34'	1° 35'	1° 36'	1° 37'
	K..............	1° 38'	1° 38'	1° 39'	1° 40'	1° 41'	1° 42'	1° 44'	1° 45'	1° 46'	1° 47'	1° 48'	1° 49'	1° 51'	1° 52'
40	J..............	1° 23'	1° 24'	1° 25'	1° 26'	1° 27'	1° 28'	1° 30'	1° 31'	1° 32'	1° 33'	1° 33'	1° 34'	1° 35'	1° 37'
	K..............	1° 37'	1° 37'	1° 38'	1° 39'	1° 40'	1° 41'	1° 43'	1° 44'	1° 45'	1° 46'	1° 48'	1° 49'	1° 50'	1° 51'
41	J..............	1° 23'	1° 24'	1° 25'	1° 26'	1° 26'	1° 27'	1° 29'	1° 30'	1° 31'	1° 32'	1° 33'	1° 34'	1° 35'	1° 36'
	K..............	1° 36'	1° 37'	1° 38'	1° 39'	1° 40'	1° 41'	1° 42'	1° 43'	1° 44'	1° 45'	1° 47'	1° 48'	1° 49'	1° 50'
42	J..............	1° 23'	1° 24'	1° 24'	1° 25'	1° 26'	1° 27'	1° 28'	1° 29'	1° 30'	1° 31'	1° 32'	1° 33'	1° 34'	1° 35'
	K..............	1° 35'	1° 36'	1° 37'	1° 38'	1° 39'	1° 40'	1° 41'	1° 42'	1° 43'	1° 44'	1° 46'	1° 47'	1° 48'	1° 49'

TABLE 20—*Continued.* ADDENDUM AND DEDENDUM ANGLES FOR BEVEL GEARS

Number of Teeth in Wheel

58	57	56	55	54	53	52	51	50	49	48	47	46	45	44	43	Number of Teeth in Pinion
1° 56' 2° 14'	1° 58' 2° 16'	2° 0' 2° 19'	2° 2' 2° 21'	2° 4' 2° 24'	2° 6' 2° 26'	2° 8' 2° 29'	2° 11' 2° 31'	2° 13' 2° 35'	2° 16' 2° 38'	2° 19' 2° 41'	2° 21' 2° 44'	2° 24' 2° 47'	2° 27' 2° 51'	2° 30' 2° 54'	2° 34' 2° 58'	12
1° 56' 2° 14'	1° 58' 2° 16'	2° 0' 2° 18'	2° 2' 2° 21'	2° 4' 2° 23'	2° 5' 2° 26'	2° 8' 2° 28'	2° 10' 2° 31'	2° 14' 2° 34'	2° 15' 2° 37'	2° 18' 2° 40'	2° 21' 2° 43'	2° 24' 2° 46'	2° 27' 2° 50'	2° 30' 2° 53'	2° 33' 2° 57'	13
1° 55' 2° 13'	1° 58' 2° 15'	2° 0' 2° 18'	2° 1' 2° 20'	2° 3' 2° 22'	2° 5' 2° 25'	2° 7' 2° 28'	2° 10' 2° 30'	2° 13' 2° 33'	2° 15' 2° 36'	2° 17' 2° 39'	2° 20' 2° 42'	2° 23' 2° 45'	2° 26' 2° 49'	2° 29' 2° 52'	2° 32' 2° 56'	14
1° 55' 2° 13'	1° 57' 2° 15'	1° 59' 2° 17'	2° 0' 2° 19'	2° 3' 2° 21'	2° 5' 2° 24'	2° 6' 2° 27'	2° 9' 2° 29'	2° 12' 2° 32'	2° 14' 2° 35'	2° 16' 2° 38'	2° 19' 2° 41'	2° 22' 2° 44'	2° 25' 2° 48'	2° 28' 2° 51'	2° 31' 2° 54'	15
1° 55' 2° 12'	1° 56' 2° 14'	1° 58' 2° 17'	2° 0' 2° 19'	2° 2' 2° 21'	2° 4' 2° 24'	2° 5' 2° 26'	2° 8' 2° 28'	2° 10' 2° 31'	2° 13' 2° 34'	2° 15' 2° 37'	2° 18' 2° 40'	2° 22' 2° 43'	2° 24' 2° 46'	2° 27' 2° 50'	2° 29' 2° 53'	16
1° 54' 2° 12'	1° 56' 2° 14'	1° 57' 2° 16'	2° 0' 2° 18'	2° 1' 2° 20'	2° 3' 2° 23'	2° 5' 2° 25'	2° 8' 2° 27'	2° 10' 2° 30'	2° 13' 2° 33'	2° 15' 2° 36'	2° 17' 2° 39'	2° 21' 2° 42'	2° 23' 2° 45'	2° 25' 2° 49'	2° 28' 2° 52'	17
1° 54' 2° 11'	1° 55' 2° 13'	1° 56' 2° 15'	1° 59' 2° 17'	2° 0' 2° 20'	2° 3' 2° 22'	2° 5' 2° 24'	2° 7' 2° 25'	2° 9' 2° 27'	2° 12' 2° 32'	2° 14' 2° 35'	2° 16' 2° 38'	2° 19' 2° 41'	2° 21' 2° 44'	2° 25' 2° 47'	2° 27' 2° 50'	18
1° 53' 2° 10'	1° 55' 2° 12'	1° 56' 2° 14'	1° 59' 2° 16'	2° 0' 2° 19'	2° 2' 2° 21'	2° 4' 2° 23'	2° 6' 2° 25'	2° 8' 2° 28'	2° 11' 2° 31'	2° 13' 2° 34'	2° 15' 2° 37'	2° 18' 2° 40'	2° 20' 2° 43'	2° 23' 2° 46'	2° 26' 2° 49'	19
1° 53' 2° 10'	1° 54' 2° 12'	1° 55' 2° 14'	1° 58' 2° 16'	2° 0' 2° 18'	2° 1' 2° 20'	2° 3' 2° 23'	2° 5' 2° 24'	2° 7' 2° 27'	2° 10' 2° 30'	2° 12' 2° 33'	2° 14' 2° 36'	2° 17' 2° 39'	2° 19' 2° 41'	2° 22' 2° 44'	2° 25' 2° 47'	20
1° 52' 2° 9'	1° 54' 2° 11'	1° 55' 2° 13'	1° 57' 2° 15'	1° 59' 2° 17'	2° 0' 2° 19'	2° 2' 2° 22'	2° 4' 2° 24'	2° 6' 2° 25'	2° 9' 2° 29'	2° 11' 2° 31'	2° 13' 2° 35'	2° 15' 2° 37'	2° 18' 2° 39'	2° 21' 2° 43'	2° 24' 2° 46'	21
1° 51' 2° 8'	1° 53' 2° 10'	1° 54' 2° 12'	1° 56' 2° 14'	1° 58' 2° 16'	2° 0' 2° 18'	2° 2' 2° 21'	2° 3' 2° 23'	2° 5' 2° 25'	2° 8' 2° 28'	2° 10' 2° 30'	2° 12' 2° 33'	2° 15' 2° 35'	2° 17' 2° 38'	2° 19' 2° 42'	2° 22' 2° 45'	22
1° 50' 2° 7'	1° 53' 2° 9'	1° 54' 2° 11'	1° 56' 2° 13'	1° 57' 2° 15'	1° 59' 2° 17'	2° 0' 2° 19'	2° 3' 2° 22'	2° 4' 2° 24'	2° 7' 2° 26'	2° 9' 2° 29'	2° 11' 2° 31'	2° 13' 2° 34'	2° 15' 2° 37'	2° 18' 2° 40'	2° 21' 2° 43'	23
1° 50' 2° 7'	1° 52' 2° 8'	1° 53' 2° 10'	1° 55' 2° 12'	1° 56' 2° 14'	1° 57' 2° 16'	2° 0' 2° 19'	2° 1' 2° 21'	2° 3' 2° 23'	2° 6' 2° 25'	2° 8' 2° 28'	2° 10' 2° 30'	2° 12' 2° 33'	2° 15' 2° 36'	2° 17' 2° 39'	2° 19' 2° 41'	24
1° 49' 2° 6'	1° 51' 2° 8'	1° 52' 2° 10'	1° 54' 2° 12'	1° 55' 2° 14'	1° 57' 2° 16'	1° 59' 2° 19'	2° 1' 2° 20'	2° 2' 2° 22'	2° 5' 2° 24'	2° 7' 2° 27'	2° 9' 2° 29'	2° 11' 2° 32'	2° 13' 2° 34'	2° 15' 2° 37'	2° 18' 2° 40'	25
1° 49' 2° 5'	1° 50' 2° 7'	1° 51' 2° 9'	1° 53' 2° 11'	1° 55' 2° 13'	1° 56' 2° 15'	1° 58' 2° 17'	2° 0' 2° 19'	2° 1' 2° 21'	2° 4' 2° 23'	2° 6' 2° 26'	2° 8' 2° 28'	2° 10' 2° 30'	2° 12' 2° 32'	2° 14' 2° 35'	2° 16' 2° 38'	26
1° 48' 2° 4'	1° 50' 2° 6'	1° 51' 2° 8'	1° 53' 2° 10'	1° 54' 2° 12'	1° 55' 2° 14'	1° 57' 2° 16'	1° 59' 2° 18'	2° 0' 2° 20'	2° 3' 2° 22'	2° 5' 2° 24'	2° 7' 2° 26'	2° 9' 2° 29'	2° 11' 2° 31'	2° 13' 2° 34'	2° 15' 2° 36'	27
1° 47' 2° 5'	1° 49' 2° 5'	1° 50' 2° 7'	1° 52' 2° 9'	1° 53' 2° 11'	1° 55' 2° 13'	1° 56' 2° 15'	1° 58' 2° 17'	2° 0' 2° 19'	2° 2' 2° 21'	2° 3' 2° 23'	2° 5' 2° 25'	2° 8' 2° 28'	2° 10' 2° 30'	2° 12' 2° 32'	2° 14' 2° 34'	28
1° 46' 2° 3'	1° 48' 2° 4'	1° 50' 2° 6'	1° 51' 2° 8'	1° 52' 2° 10'	1° 54' 2° 11'	1° 55' 2° 14'	1° 57' 2° 15'	1° 59' 2° 17'	2° 1' 2° 19'	2° 3' 2° 22'	2° 4' 2° 24'	2° 6' 2° 26'	2° 8' 2° 28'	2° 10' 2° 31'	2° 12' 2° 33'	29
1° 45' 2° 2'	1° 47' 2° 3'	1° 49' 2° 5'	1° 50' 2° 7'	1° 51' 2° 9'	1° 53' 2° 10'	1° 55' 2° 13'	1° 56' 2° 14'	1° 58' 2° 16'	2° 0' 2° 18'	2° 2' 2° 20'	2° 3' 2° 22'	2° 5' 2° 25'	2° 7' 2° 27'	2° 9' 2° 30'	2° 11' 2° 33'	30
1° 44' 2° 1'	1° 46' 2° 2'	1° 48' 2° 4'	1° 50' 2° 6'	1° 51' 2° 8'	1° 52' 2° 9'	1° 54' 2° 11'	1° 55' 2° 13'	1° 57' 2° 15'	1° 59' 2° 17'	2° 0' 2° 19'	2° 2' 2° 21'	2° 4' 2° 23'	2° 5' 2° 25'	2° 8' 2° 28'	2° 9' 2° 30'	31
1° 44' 2° 0'	1° 45' 2° 2'	1° 46' 2° 3'	1° 47' 2° 5'	1° 50' 2° 7'	1° 51' 2° 8'	1° 53' 2° 10'	1° 54' 2° 12'	1° 55' 2° 14'	1° 58' 2° 16'	1° 59' 2° 18'	2° 1' 2° 20'	2° 3' 2° 22'	2° 4' 2° 24'	2° 6' 2° 26'	2° 8' 2° 28'	32
1° 43' 1° 50'	1° 45' 2° 0'	1° 45' 2° 2'	1° 43' 2° 4'	1° 49' 2° 6'	1° 50' 2° 7'	1° 52' 2° 9'	1° 53' 2° 11'	1° 55' 2° 13'	1° 57' 2° 15'	1° 58' 2° 17'	2° 0' 2° 18'	2° 1' 2° 20'	2° 3' 2° 22'	2° 5' 2° 25'	2° 7' 2° 27'	33
1° 42' 1° 58'	1° 44' 1° 59'	1° 45' 2° 1'	1° 47' 2° 3'	1° 48' 2° 5'	1° 50' 2° 6'	1° 51' 2° 8'	1° 53' 2° 9'	1° 54' 2° 11'	1° 55' 2° 13'	1° 57' 2° 15'	1° 59' 2° 17'	2° 0' 2° 19'	2° 2' 2° 21'	2° 4' 2° 23'	2° 5' 2° 25'	34
1° 41' 1° 57'	1° 43' 1° 58'	1° 45' 2° 0'	1° 45' 2° 2'	1° 47' 2° 4'	1° 49' 2° 5'	1° 50' 2° 7'	1° 51' 2° 8'	1° 53' 2° 10'	1° 54' 2° 11'	1° 56' 2° 13'	1° 58' 2° 15'	1° 59' 2° 17'	2° 2' 2° 19'	2° 3' 2° 21'	2° 4' 2° 23'	35
1° 40' 1° 56'	1° 42' 1° 57'	1° 44' 1° 59'	1° 45' 2° 1'	1° 46' 2° 3'	1° 48' 2° 4'	1° 47' 2° 6'	1° 50' 2° 7'	1° 52' 2° 9'	1° 53' 2° 10'	1° 54' 2° 12'	1° 57' 2° 14'	1° 58' 2° 16'	1° 59' 2° 18'	2° 1' 2° 20'	2° 3' 2° 22'	36
1° 40' 1° 56'	1° 41' 1° 57'	1° 43' 1° 58'	1° 44' 2° 0'	1° 45' 2° 2'	1° 47' 2° 3'	1° 45' 2° 5'	1° 47' 2° 6'	1° 50' 2° 8'	1° 51' 2° 9'	1° 53' 2° 11'	1° 55' 2° 13'	1° 57' 2° 15'	1° 58' 2° 16'	2° 0' 2° 18'	2° 1' 2° 20'	37
1° 39' 1° 55'	1° 40' 1° 56'	1° 42' 1° 57'	1° 43' 1° 55'	1° 44' 2° 0'	1° 46' 2° 1'	1° 47' 2° 3'	1° 48' 2° 4'	1° 49' 2° 6'	1° 50' 2° 7'	1° 52' 2° 9'	1° 54' 2° 11'	1° 55' 2° 13'	1° 57' 2° 14'	1° 58' 2° 16'	1° 59' 2° 18'	38
1° 38' 1° 51'	1° 39' 1° 54'	1° 41' 1° 56'	1° 42' 1° 57'	1° 43' 1° 59'	1° 45' 2° 0'	1° 45' 2° 2'	1° 47' 2° 3'	1° 48' 2° 5'	1° 49' 2° 6'	1° 51' 2° 8'	1° 52' 2° 9'	1° 53' 2° 11'	1° 55' 2° 12'	1° 57' 2° 14'	1° 58' 2° 16'	39
1° 38' 1° 53'	1° 39' 1° 53'	1° 40' 1° 55'	1° 41' 1° 56'	1° 42' 1° 58'	1° 44' 1° 50'	1° 45' 2° 1'	1° 45' 2° 2'	1° 47' 2° 4'	1° 48' 2° 5'	1° 50' 2° 7'	1° 51' 2° 8'	1° 52' 2° 10'	1° 54' 2° 11'	1° 55' 2° 13'	1° 57' 2° 15'	40
1° 36' 1° 51'	1° 38' 1° 52'	1° 39' 1° 54'	1° 40' 1° 55'	1° 41' 1° 57'	1° 43' 1° 58'	1° 44' 2° 0'	1° 45' 2° 1'	1° 46' 2° 5'	1° 47' 2° 4'	1° 49' 2° 5'	1° 50' 2° 6'	1° 51' 2° 8'	1° 53' 2° 10'	1° 54' 2° 11'	1° 56' 2° 13'	41
1° 36' 1° 50'	1° 37' 1° 51'	1° 38' 1° 53'	1° 39' 1° 54'	1° 40' 1° 56'	1° 42' 1° 57'	1° 41' 1° 59'	1° 43' 2° 0'	1° 45' 2° 1'	1° 46' 2° 2'	1° 47' 2° 4'	1° 49' 2° 5'	1° 50' 2° 7'	1° 52' 2° 8'	1° 51' 2° 10'	1° 55' 2° 11'	42

ANGLE BETWEEN AXES 90°, TOOTH PROPORTIONS BROWN & SHARPE STANDARD

ANGLE INCREASE. DIVIDE BY TEETH IN PINION	DIAMETER INCREASE. DIVIDE BY PITCH OF PINION	CENTER ANGLE FOR PINION	CENTER-ANGLE-HUNDREDTH-DEGREES LEFT-HAND COLUMN READ HERE							CENTER ANGLE FOR GEAR	DIAMETER INCREASE. DIVIDE BY PITCH OF GEAR
			.00	.17	.33	.50	.67	.83	1.00		
1	2.00	0	.0000	.0029	.0058	.0087	.0116	.0145	.0175	89	.03
2	2.00	1	.0175	.0204	.0233	.0262	.0291	.0320	.0349	88	.07
4	2.00	2	.0349	.0378	.0407	.0437	.0466	.0495	.0524	87	.10
6	2.00	3	.0524	.0553	.0582	.0612	.0641	.0670	.0699	86	.14
8	1.99	4	.0699	.0729	.0758	.0787	.0816	.0846	.0875	85	.17
10	1.99	5	.0875	.0964	.0934	.0963	.0992	.1022	.1051	84	.21
12	1.99	6	.1051	.1080	.1110	.1139	.1169	.1198	.1228	83	.24
14	1.98	7	.1228	.1257	.1278	.1317	.1346	.1376	.1405	82	.28
16	1.98	8	.1405	.1435	.1465	.1495	.1524	.1554	.1584	81	.31
18	1.98	9	.1584	.1614	.1644	.1673	.1703	.1733	.1763	80	.34
20	1.97	10	.1763	.1793	.1823	.1853	.1883	.1914	.1944	79	.38
22	1.96	11	.1944	.1974	.2004	.2035	.2065	.2095	.2126	78	.41
24	1.96	12	.2126	.2156	.2186	.2217	.2247	.2278	.2309	77	.45
26	1.95	13	.2309	.2339	.2370	.2401	.2431	.2462	.2493	76	.48
28	1.94	14	.2493	.2524	.2555	.2586	.2617	.2648	.2679	75	.51
30	1.93	15	.2679	.2711	.2742	.2773	.2805	.2836	.2867	74	.55
32	1.92	16	.2867	.2899	.2931	.2962	.2994	.3026	.3057	73	.58
34	1.91	17	.3057	.3089	.3121	.3153	.3185	.3217	.3249	72	.62
36	1.90	18	.3249	.3281	.3314	.3346	.3378	.3411	.3443	71	.65
37	1.89	19	.3443	.3476	.3508	.3541	.3574	.3607	.3640	70	.68
39	1.88	20	.3640	.3673	.3706	.3739	.3772	.3805	.3839	69	.71
41	1.86	21	.3839	.3872	.3906	.3939	.3973	.4006	.4040	68	.75
43	1.85	22	.4040	.4074	.4108	.4142	.4176	.4210	.4245	67	.78
45	1.84	23	.4245	.4279	.4314	.4348	.4383	.4417	.4452	66	.81
47	1.82	24	.4452	.4487	.4522	.4557	.4592	.4628	.4663	65	.84
49	1.81	25	.4663	.4699	.4734	.4770	.4806	.4841	.4877	64	.88
50	1.79	26	.4877	.4913	.4950	.4986	.5022	.5059	.5095	63	.91
52	1.78	27	.5095	.5132	.5169	.5206	.5243	.5280	.5317	62	.93
54	1.76	28	.5317	.5354	.5392	.5430	.5467	.5505	.5543	61	.97
56	1.74	29	.5543	.5581	.5619	.5658	.5696	.5735	.5774	60	1.00
57	1.73	30	.5774	.5812	.5851	.5890	.5930	.5969	.6009	59	1.03
59	1.71	31	.6009	.6048	.6088	.6128	.6168	.6208	.6249	58	1.05
61	1.69	32	.6249	.6289	.6330	.6371	.6412	.6453	.6494	57	1.08
63	1.67	33	.6494	.6536	.6577	.6619	.6661	.6703	.6745	56	1.11
64	1.65	34	.6745	.6787	.6830	.6873	.6916	.6959	.7002	55	1.14
66	1.63	35	.7002	.7046	.7089	.7133	.7177	.7221	.7265	54	1.17
68	1.61	36	.7265	.7310	.7355	.7400	.7445	.7490	.7536	53	1.20
69	1.59	37	.7536	.7581	.7627	.7673	.7720	.7766	.7813	52	1.23
71	1.57	38	.7813	.7860	.7907	.7954	.8002	.8050	.8098	51	1.25
72	1.55	39	.8098	.8146	.8195	.8243	.8292	.8342	.8391	50	1.28
73	1.53	40	.8391	.8441	.8491	.8541	.8591	.8642	.8693	49	1.31
75	1.51	41	.8693	.8744	.8796	.8847	.8899	.8952	.9004	48	1.33
77	1.48	42	.9004	.9057	.9110	.9163	.9217	.9271	.9325	47	1.36
79	1.46	43	.9325	.9380	.9435	.9490	.9545	.9601	.9657	46	1.39
80	1.43	44	.9657	.9713	.9770	.9827	.9884	.9942	1.0000	45	1.41
81	1.41	45	1.0000	1.0058	1.0117	1.0176	1.0235	1.0295	1.0355	44	1.43
			1.00	.83	.67	.50	.33	.17	.00		
			RIGHT-HAND COLUMN READ HERE								

TABLE 21—DIAMETERS AND ANGLES OF BEVEL GEAR—SHAFT ANGLES 90 DEGREES

Becker Milling Machine Company

MACHINING BEVEL GEARS

Cutting bevel-gear teeth necessitates quite intricate operations if a correctly proportioned tooth is to be finished on other than a generating machine. In fact, so intricate would be the required adjustments that quite radical departures from a true form of tooth are frequently resorted to—modifications that are almost considered standard, so universally are they employed.

The recognized standard tooth for bevel gears, after which all modifications are modeled, is of the octoid form. This form of tooth, which was described in a preceding section, is simply a modification of the involute and can be accurately conjugated only by a generating machine.

MACHINES FOR CUTTING BEVEL GEARS

Milling machines are the ones most generally employed in cutting bevel gears unless extreme accuracy in the form of teeth is essential. These machines readily cut teeth of uniform depth of correct profile, and are also used for the cutting operations of bevel gears with teeth of approximately correct form—the type of bevels which are finished by filing the teeth to approximately correct profile.

Planing machines are employed in similar operations and also for cutting teeth which closely resemble the true octoid form. Somewhat special planers, known as bevel-gear planing machines, are employed for this more exacting work. Three templets are used which guide the simple planing tools: A straight-faced one for gashing, and two formed ones, one for each side of the tooth space. All the tooth spaces are first gashed, one side of each tooth finished, and the gear completed by finishing up the other side of the teeth.

Generating machines are the third type of machines used for cutting the teeth of bevel gears. These machines all use the generating principle, but either the shaping or milling process may be employed. The generating machines operating on the shaping process have a crown gear or its equivalent provided with cutting tools representing its sides in successive positions. Meshing with this is a large master bevel gear carrying on its arbor the gear to be cut. The master and crown gears rotate in mesh, thus presenting the work to the cutting tools in successive positions. The blank and the cutting tools virtually roll together to conjugate the teeth. Generating machines employing the milling process differ in having the sides of the teeth of the imaginary crown gear represented by the plane faces of milling cutters. The crown gear equivalent is stationary and the master bevel gear is rolled over it, thus presenting the work to the cutting tools. The action in the two machines is identical, the conjugation of true octoid teeth.

Of these types of machines for cutting bevel gears, the milling machine is generally employed on account of the speed with which the work can be performed. Nevertheless the operations required are quite intricate and deserve special attention. The meager description of the bevel-gear planer, on the other hand, should suffice to explain its operations and, as generating machines are fully automatic, little explanation of their action is necessary.

MILLING BEVEL GEARS

Though rotary cutters do not lend themselves to the production of really high-class bevel gears, in view of the many conditions governing the shape of the correctly proportioned tooth, yet gears cut on milling machines are nevertheless entirely satisfactory for many purposes, and until the time arrives when all gear teeth are really standardized, will continue to be used quite extensively.

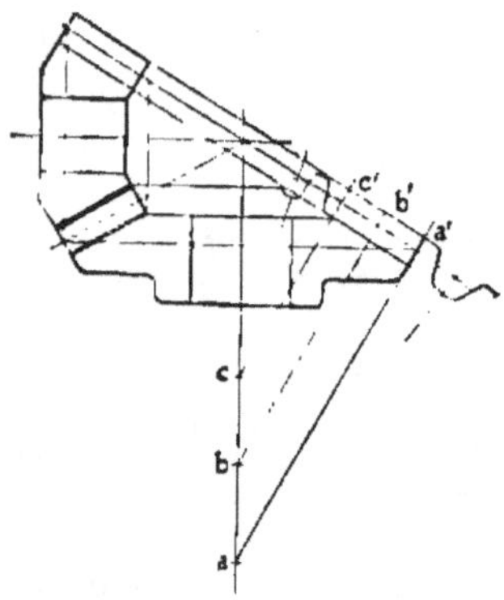

FIG. 100. PARALLEL DEPTH BEVEL GEARS.

The fact that the depth of the true *octoid* tooth decreases, as well as the outside and pitch diameters of a bevel gear, as the cone apex of the gear is approached, makes sole reliance upon a rotary cutter impossible. Notwithstanding a correct profile to the cutter, the tooth must be gone over with a file to remove surplus material, etc. This drawback led to the development of the *Pentz Parallel Depth Gear* in which the depth of the tooth is constant throughout its entire length. See Fig. 100.

By this radical departure from the true *octoid*, the necessity of finishing tne teeth with a file is avoided—the form of tooth being correct at one point, it is correct at every other. This form of tooth has one possible drawback, however, although it is quite generally considered as the equal of the true *octoid* tooth of varying depth.

The cutter for the parallel depth tooth is selected for the *inner* or *smaller* pitch circle, so there is always a certain deficiency in the amount of metal at the outer ends of the teeth, just as there is a surplus of metal at the inner ends of the teeth which has to be filed away when the cutter is selected for the outer pitch circle—the method when milling the standard varying depth tooth.

The *Pentz Parallel Depth* bevel gears are quite distinctive, owing to the stubby appearance of the teeth at the outer edge of the gear, but as this form of tooth is somewhat easier to mill than the one in which the depth of tooth

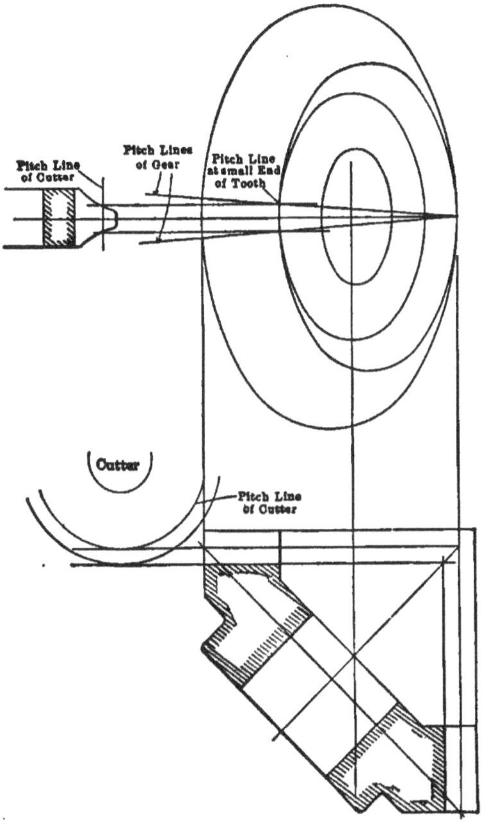

FIG. 101. FIRST CUT IN MILLING BEVEL GEARS.

varies, a detailed explanation of the steps required in machining such a gear will be more readily grasped than if the more complicated method necessitating more adjustments be first considered. The more involved method can then be quite easily comprehended.

The pitch of the cutter is determined from the small end of the gear, and the form of the cutter from the average back cone distance, as at *b-b* (Fig. 100). This will give the average form of tooth, as it is apparent that the true form cannot be maintained the entire length of face. The only change in form is due to the reduced back cone distance from *a* to *c*; there is no change due to reducing the depth of cut, as is found in the ordinary methods of bevel gear milling, the pitch line of the bevel gear being cut, and the pitch line of the cutter coinciding during the entire cutting operation.

The milling of a parallel or, for that matter, tapered tooth bevel gears will be better understood if the pitch is considered at the small end of the tooth. When taking side cuts, this pitch alone should be kept in mind and the matter will appear in a new light.

After the first central cut has been taken as illustrated by Fig. 101 the blank is rolled to the right, bringing the pitch line of the left-hand side of space parallel with the travel of cutter, as shown by Fig. 102. This movement is accomplished by indexing the blank one-quarter as many holes in the index plate as are used altogether to space the teeth. The table is then moved toward the nose of the spindle a distance equal to one-half the tooth thickness at *small end*, or one-quarter the pitch at small end. This will bring the pitch line of gear to the pitch line of the cutter, and the blank in position for first side cut, as shown in Fig. 103.

After this cut has been made, roll the gear blank to the left, one-half as many holes in index plate as are used altogether to space the tooth, as per Fig. 104, and move the table *away* from the nose of spindle the thickness of tooth *at small end;* this will bring the other side of the tooth into the same relative position, and the blank is in position for the second side cut, as in Fig. 105.

It will be noted that the pitch is figured at the small end of the tooth, an ordinary spur gear cutter corresponding to the pitch at this point being used.

The half tone, Fig. 106, shows a pair of parallel tooth gears, sent with W. Allen's original article. Their operation was entirely satisfactory and the teeth gave no evidence of being filed. These samples were 25 teeth, 10 pitch at small end and 1-inch face, No. 2, 10 pitch cutter used.

Referring to Figs. 101 to 105; in moving the table forward and back one-half the thickness of space at the small end there is a small error due to the fact that instead of these moves being made in the direction of the pitch line they are made tangent to it, as illustrated by Fig. 107, which is exaggerated to show this clearly, *c* representing the distance actually moved and *f* the theoretical distance. This error, however, is on the safe side, so that setting the machine as directed will allow a little clearance, depending, of course, upon the diameter and pitch of gear, but will not be noticed except in extreme cases.

In milling the taper-tooth type of bevels, very similar steps are taken, but, of course, the face and cutting angles are not the same. This naturally complicates the holding and adjusting of the work to a considerable extent.

The cutter should be selected with a correct profile for the outer ends of the teeth, but its thickness must be such as to allow it to pass between the teeth at the inner edge of the gear. The necessary adjustments of the blank

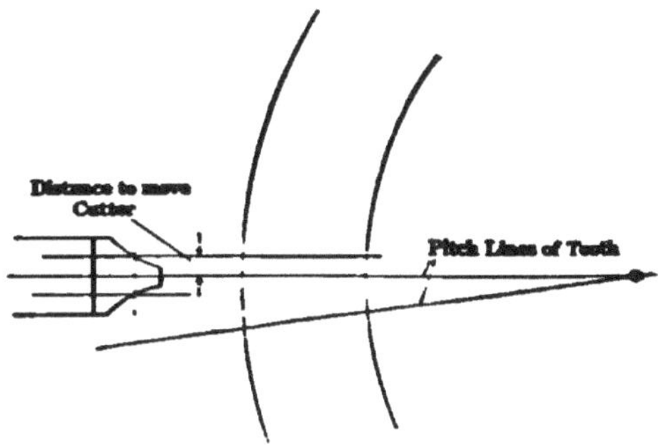

FIG. 102. POSITION OF BLANK WHEN ROTATED ONE-QUARTER OF THE INDEX.

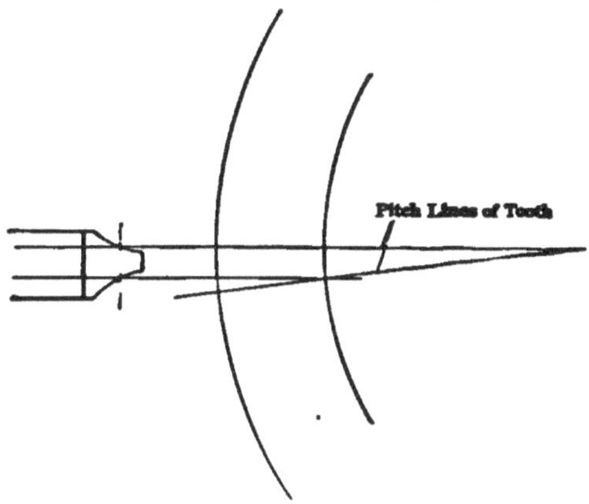

FIG. 103. CUTTER IN POSITION FOR THE FIRST SIDE CUT.

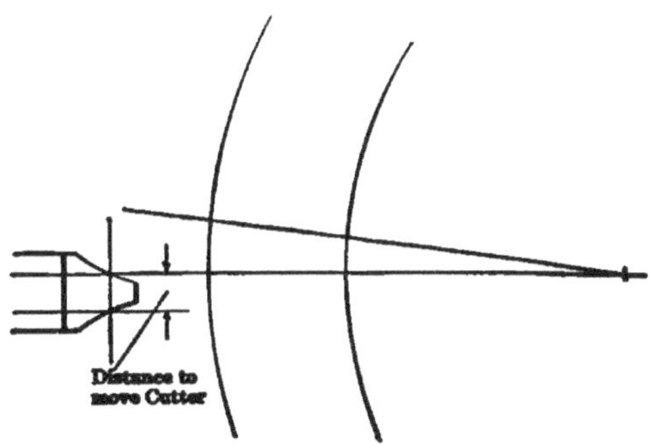

FIG. 104. BLANK IN POSITION FOR SECOND SIDE CUT.

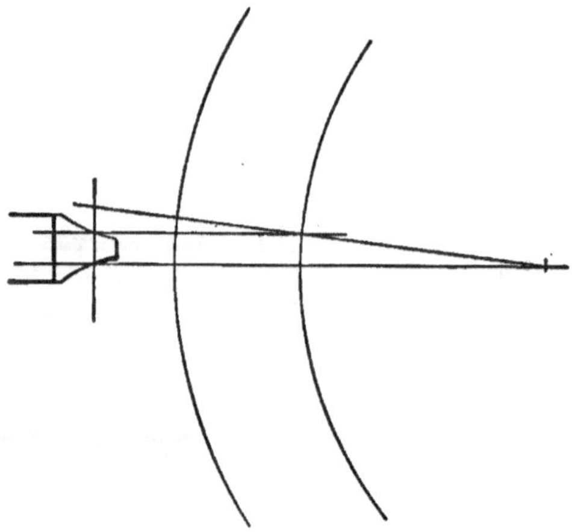

FIG. 105. CUTTER IN POSITION FOR SECOND SIDE CUT.

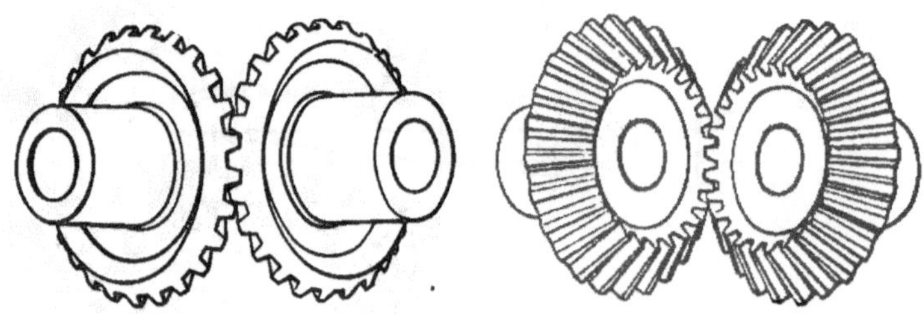

FIG. 106. A PAIR OF PARALLEL DEPTH BEVEL GEARS IN MESH.

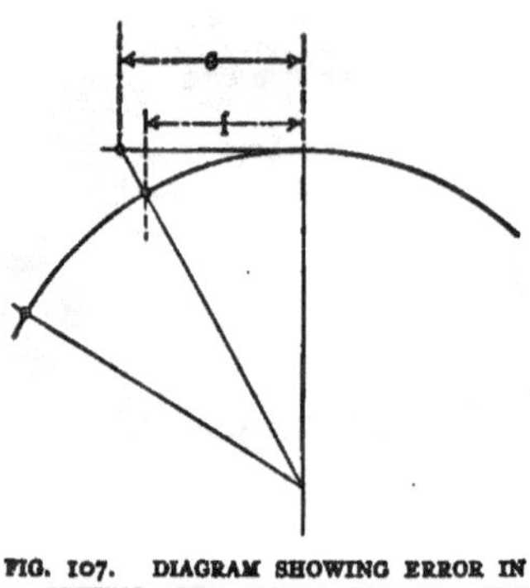

FIG. 107. DIAGRAM SHOWING ERROR IN SETTING OVER CUTTER FOR SIDE CUT ON MILLED BEVEL GEARS.

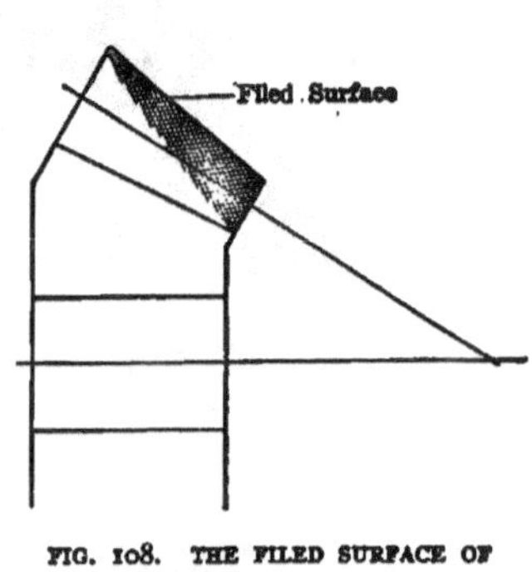

FIG. 108. THE FILED SURFACE OF MILLED BEVEL GEARS.

are made, usually by trial, and the cuts taken in a manner very similar to that described for the parallel-depth method. The teeth are milled to a correct outer end thickness and the surplus metal toward the inner edge removed by a file. See Fig. 108.

The reason that this excess metal is left by the cutter is that, though the set of teeth as cut are correct throughout their entire length at the pitch line, their curvature is correct only at the outer end of the teeth. In the octoid form of tooth, for which the cutters were selected, the radius of curvature grows less and less as the pitch diameter of the gear decreases—*i.e.*, toward its inner edge. Rotary cutters, being unable to alter their curvature, leave a surplus of metal outside the pitch line which gradually increases in thickness as the inner edge of the gear is approached. This surplus can only be removed by filing. In performing such an operation, great care is necessary not to reduce the thickness of the teeth at the pitch line. Particularly is this so if the gear is one of wide face, as the greater the face of the teeth, the more filing required.

THE USE OF GENERATING MACHINES

If the gears to be finished are stocked out before being mounted on the generating machine, these machines are really fully automatic and when once set up require no further attention until the job is completed. Stocking out on generating machines is not to be recommended ordinarily, although the larger machines are equipped for this operation. Generating machines are essentially finishing machines with delicate adjustment of cutting tools that finish a stocked out tooth in one generating operation, so that the rougher and heavier work of stocking out should advisably always be performed on some other and more rugged machine. This will enable the capacity of the generating machine to be realized fully, and operations in quantities carried through rapidly and efficiently.

EFFICIENCY OF BEVEL GEARS

The chief cause of decreased efficiency in bevel gearing is due to inaccuracies in the shape and form of the teeth, whereby a lateral thrust is produced which tends to force the gears out of mesh. This is particularly noticeable in gears finished on milling machines or on planers unless the gears have teeth of the parallel-depth type, owing to the fact that it is seldom that the center angles of the gear and pinion coincide as they roll together. The tooth pressure cannot be normal between all points of engaging teeth and in such case a decided lateral thrust tending to separate the teeth is produced. Furthermore, the pinion, and usually the gear also, must be overhung so that the thrust is greatly increased at the bearing by leverage, materially increasing friction, twisting the shaft, etc.

Even if it were possible to produce a perfect tooth on a generating machine, so that the tooth pressure is transmitted normally, inefficiencies that cannot be avoided arise, due to the angularity of the gear and pinion shafts. The greater the center angle of a bevel gear, the greater, as a rule, the loss in efficiency. The most efficient center angle to employ is that of 45 degrees, for then both the gear and the pinion suffer equally in respect to angularity.

Correctly proportioned teeth and adequate shaft diameters with minimum backing are particularly essential for the satisfactory operation of bevel gears. Unsatisfactory tooth speed and wide differences between the number of teeth on the gear and on the pinion are also more serious drawbacks in bevel gearing than in spur gearing, owing to the fact that a bevel gear has pitch circles and corresponding tooth speeds varying from that of the inner or smaller end of the gear through to that of the larger end.

HARDENING BEVEL GEARS

The abrasive wear on gear teeth is always appreciable if the profile of the teeth varies even slightly from the correct form, and once such deterioration commences, it rapidly becomes serious. This is more noticeable in bevel gearing than it is in spur gearing, owing to the greater difficulty of securing the correct tooth profile in the former. To overcome or minimize this destruction, case-hardening of the gears is the logical procedure.

The great demand for satisfactory gears, bevels as well as other types of toothed wheels, in automobile construction has probably been the main reason for the knowledge gained in recent years of the art of case-hardening gears. The transmission of the mechanically propelled vehicle must possess exceptional wearing qualities to give satisfaction, and, though other intricate and delicate machinery also requires gears capable of resisting the abrasive action of teeth slipping upon one another, it is the automobile that has made this subject one of such general interest.

Processes of hardening gears, of course, differ in nearly every shop, but the general requirements and results secured are similar. The accurately cut gear should first be slowly and uniformly preheated to avoid undue distortion and to bring the metal to as homogeneous a condition as possible. The gear should then be subjected to the chemical treatment by which the skin of the metal is prepared for the hardening quench. Finally the quenching operation is performed, usually in an oil bath, and the gears rapidly but uniformly cooled.

One process of hardening automobile gears which has proved very successful and satisfactory is to preheat the gears in a commodious gas forge or oven until they become dull red (at a temperature of from 1,300 to 1,400 degrees Fahrenheit) and then boil them in a solution of cyanide of potassium for about an hour. The gears are then immediately plunged into a bath of fish oil which is kept in constant circulation to guard against localized overheating of the oil.

Gears hardened by this process show remarkably little warping and are uniformly hard—the case-hardening penetrating an average depth of 0.025

inch—and are practically free from any change in volume or distribution of metal. The average shrinkage of the diameter of shaft holes, in a large number of automobile transmission gears so treated, is reported to have been only about 0.0005 inch (total shrinkage).

GRINDING BEVEL GEARS

Gears that have been most carefully case-hardened usually show some deformation, however, and no matter how slight this is their smooth and proper action is impossible. Thus the benefits derived from hardening are partially counteracted. The distortion that is noticeable in a gear that has been carefully hardened, however, is so slight that it can be localized in the hub and there corrected. That is, the gear may be mounted on a rotary machine so that the teeth run true and any deformity thereby concentrated in the hub, which may be rebored, the bore straightened and the hub correctly faced.

A simple method of trueing up hardened bevel gears for grinding, and holding them in such position, was described in AMERICAN MACHINIST, July 20, 1911. Briefly summarized, the salient points in this description follow: The gears are first clamped to a face plate by a bolt through their bore, set true by using a test indicator, and kept in position by four perfectly round and straight pins placed equidistant on the periphery of the gear and held in place by rubber bands. The gear is then firmly strapped to the face plate, the bolt through the bore removed, and the bore and back of the hub properly ground.

Draw-in chucks are also used for holding gears for grinding. The gears are held by the roots of the teeth, by the pitch line, or by the top of the teeth.

OTHER TYPES OF BEVEL GEARS

The unprecedented demand for bevel gears in automotive devices has led to the development of a number of meritorious types of bevels and to new methods of gear production which have greatly added to the scope of bevel gearing. Among the type developments may be mentioned the Gleason Spiral Bevel Gear and the Williams "Master Form" Gearing which will be described briefly in a subsequent section.

The Anderson process of rolling gears from heated metal blanks—Section XVI—has also contributed a new form of bevel gear in the Herringbone bevel.

SECTION VI

Worm Gears

An interesting model of half a dozen sets of worm gears is shown in Fig. 109. All the gears are of the same diameter with teeth of the same normal pitch, though the respective speed ratios of the varius sets differ. The hori-

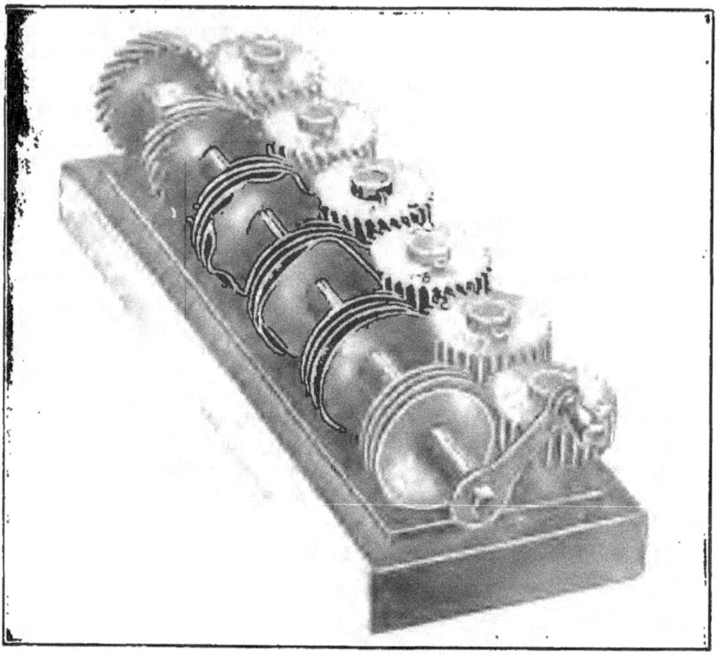

FIG. 109. MODEL OF SPIRAL GEARS OF VARIOUS RATIOS.

zontal shaft in making 32 revolutions causes the vertical gear in the foreground to make but one complete revolution, the second, two complete turns, and each of the succeeding vertical gears twice the number of revolutions made by the gear immediately in front of it, the furthermost gear making 32 revolutions to 32 turns of the horizontal shaft The consecutive speed ratios, commencing with the set in the foreground, are respectively 32 : 1, 16 : 1, 8 : 1, 4 : 1, 2 : 1 and 1 : 1, the revolutions of the horizontal shaft being named first.

The first three drives are evidently worms of single, double and quadruple thread respectively, but they are also spiral gears as the driven gears are cut with spiral teeth, not simply hobbed as is customary in laying out worm gears. The balance of the drivers resemble spiral gears even more, particularly the most remote, showing that the so-called "worm" is in reality a toothed gear of the spiral type.

NOTATION FOR WORM GEARS

N = number of teeth in worm wheel.
n = number of threads in worm.
p' = circular pitch (distance from center to center of teeth).
L = lead (advance of worm in one revolution).
D' = pitch diameter of worm wheel.
T = throat diameter of worm wheel.
D = outside diameter of worm wheel.
F = face of worm wheel.
a = distance from center line to point of tooth.
b = length of side.
d' = pitch diameter of worm.
d = outside diameter of worm.
d'' = bottom diameter of worm.
e = radius at throat of worm wheel.
ϕ = angle of sides of face.
B = center distance.
R = number of revolutions of worm to one of wheel.
δ = angle of teeth in wheel with axis (used for gashing teeth).
π = 3.1416.
W = working depth.
W' = whole depth.
f = clearance.
t = thickness of tooth at pitch line.
t^n = normal thickness of tooth at pitch line.
p'^n = normal circular pitch.
s = addendum.
U = width of worm thread at top.
V = width of worm thread at bottom.
p = diametral pitch.

FORMULAS FOR WORM GEARS

$N = \frac{D' \pi}{p'}.$

$D' = N p' \, 0.3183.$

$T = (N + 2) p' \, 0.3183.$

$D = T + 2 (e - e \cos. \phi).$

$F = \frac{\left(\frac{d}{2} + 0.17 p'\right) \sin. \phi}{0.5}$, or $\frac{d + (0.34 p')}{2}$, when $\phi = 30$ degrees.

$a = F - (b \sin. \phi).$

$b = W' + (0.12 p').$

d' = as small as possible. (See discussion.)

$d = d' + 2 s.$

$d'' = d - 2 W'.$

$e = \frac{d'}{2} - s.$

$\phi = 30°$ to $35°$, or $\sin. \phi = \frac{F}{d + (0.34 p')}.$

$B = \frac{D' + d'}{2}.$

$p' = \frac{D'}{0.3183 N} = \frac{\pi}{p}$, or $p = \frac{\pi}{p'}.$

$L = p'n.$

$n = \frac{N}{R}.$

$Tan. \delta = \frac{L}{\pi d'}.$

$t^n = t \cos. \delta$, or $t = \frac{t^n}{\cos. \delta} = \frac{1.5708}{p}$ when $\delta = 14\frac{1}{2}$ degrees.

$f = 0.1 t.$ (See discussion.)

$U = 0.335 p'$, or $\frac{1.0536}{p}.$ (See discussion.)

$V = 0.31 p'$, or $\frac{0.9744}{p}.$ (See discussion.)

$p'^n = p \cos. \delta$, or $p' = \frac{p'^n}{\cos. \delta}.$

$W' = 0.6866 p'.$ (See discussion.)

Formulas for tooth parts as given for spur gears apply to worm gears.

DISCUSSION OF FORMULAS

N, *D'* and *p'*. The number of teeth, the pitch and the pitch diameter of worm gears are calculated in the same manner employed for spur gears.

T. The throat diameter of a worm gear or wheel corresponds to the outside diameter of a spur gear of the same number of teeth and the same pitch.

D. The extreme outside diameter of a worm-wheel can be found by the formula given, but ordinarily the measurement of a carefully drawn sketch is sufficiently accurate. Insufficient stock for a sharp tooth edge, depicted

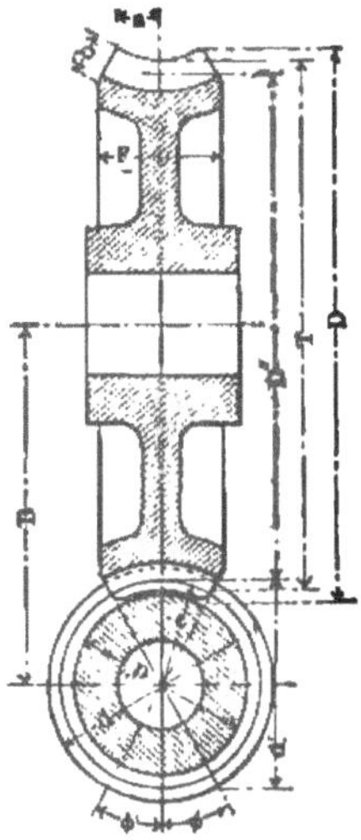

FIG. 110. DIAGRAM FOR WORM GEAR.

on Fig. 110, is preferable to enough stock for a perfect tooth as it is safer to handle a gear without the sharp-cornered teeth and also because such gear is of more pleasing appearance.

F. There is no gain in making ϕ greater than 30 degrees, so the second formula for *F* is the one usually employed.

d'. The angle of the worm, δ, governs in large part the efficiency of a worm drive, so the pitch diameter of the worm should be made as small as possible. When the lead is fixed, however, d' is also fixed for $\tan. \delta = \frac{L}{\pi d'}$.

ϕ. This angle is usually from 30 to 35 degrees, preferably 30 degrees. When made as great as 45 degrees and the face of the gear widened to correspond, the gear will wear out rapidly.

L. The lead is the advance of the worm thread in one complete revolution and is found by multiplying the circular pitch by the number of threads in the worm. For instance, a 1.5-inch pitch gear with double threads has a 3-inch lead.

n. The number of threads required for a worm is found by dividing the number of teeth in the worm-wheel by the velocity ratio required. For example: If a worm gear has 60 teeth and a velocity ratio of 30 to 1 is required, the worm should have two threads $\left(\frac{60}{30} = 2\right)$.

f, U, Y, and W'. The formulas given for these various dimensions apply to worms cut to 14½ degrees standard. Any deviation from this standard obviously calls for modifications of these formulas, but no fixed rule can be advanced.

The length of the worm (see Fig. 111) need be no more than three times the circular pitch, as seldom do more than two teeth come in contact with the

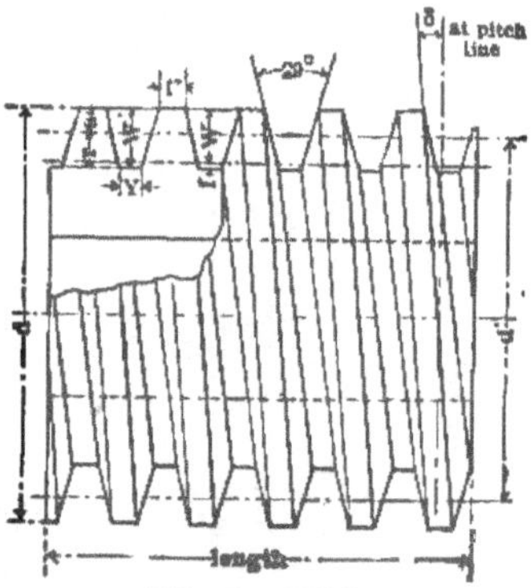

FIG. 111. WORM.

wheel teeth at the same time. It is good practice, however, to make the worm about six times as long as the circular pitch so that it may be shifted as it becomes worn, the worm nearly invariably wearing more rapidly than the wheel.

REVERSIBLE WORM AND GEAR

The surface of the worm thread constantly slips and slides over the surfaces of the wheel teeth, the direction of slippage following the plane of contact. The slippage plane being mutually tangent to the face of the thread and the face of the wheel teeth, the *gliding angle* must equal the sum of the lead angle and half the angle included between the faces of the worm thread. When this gliding angle equals 45 degrees it is evident that it is immaterial whether the worm or the gear is the driver, or which is the driven member—the worm and the gear will be perfectly reversible in this respect.

Expressed in the form of a simple equation, the conditions requisite for the worm and gear to be perfectly reversible are as follows:

$$Y = \frac{X}{2} = 45 \text{ degrees}$$

where

Y = lead angle,
X = angle included between the faces of the worm thread.

This equation is generally applicable, but there are reasonable limits to the lead angle that can be efficiently employed. The lower limit for the lead angle is about 25 degrees, making necessary an included angle of 40 degrees in order to make the gliding angle one of 45 degrees. A more acute lead angle would necessitate an included angle so obtuse as not to give adequate contact surface. The upper limit is, of course, 45 degrees. In this case there could be no included angle. This condition could exist only when the profile of the worm thread is perpendicular to the axis of the worm, when the worm thread is square or rectangular in cross-sections.

LEAD ANGLE, Y	INCLUDED ANGLE, X	LEAD ANGLE, Y	INCLUDED ANGLE, X
25	40	36	18
26	38	37	16
27	36	38	14
28	34	39	12
29	32	40	10
30	30	41	8
13	28	42	6
33	26	43	4
32	24	44	2
33	22	45	0
54	20		

THE HOB

The hob for cutting the teeth on the wheel must have an outside diameter equal to the outside diameter of the worm plus twice the clearance (see Fig.

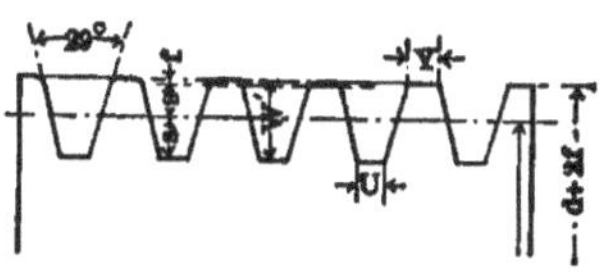

FIG. 112. HOB.

112), but this should advisably be increased by $0.03 \times p'$ to allow for wear.

The hob should always be a little longer than the section of the worm having any contact with the wheel (see Fig. 113) and, as this depends upon the diameter of the wheel to be cut, it should be proportioned to the diameter of the largest wheel liable to be handled. The correct hob length may be found from the following simple formula:

$$L = 2\sqrt{(D - W') - W'}$$

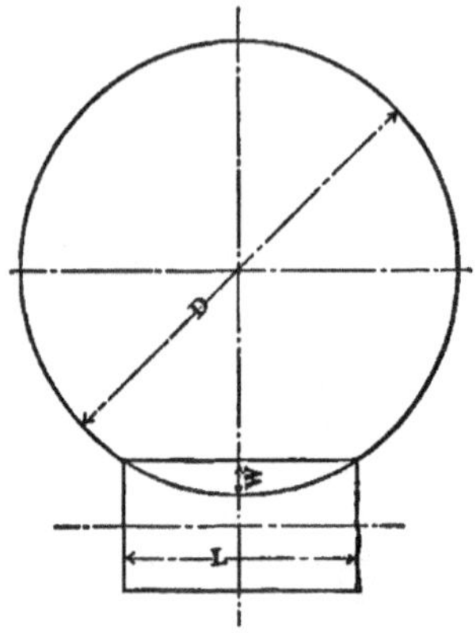

FIG. 113. LENGTH OF THE HOB.

where

L = length of hob,
D = outside diameter of largest gear to be cut,
W' = whole depth of tooth.

NUMBER OF FLUTES

The following article by Oscar J. Beale, which originally appeared in AMERICAN MACHINIST, June 22, 1899, very ably discusses the proper arrangement of teeth on a worm-gear hob.

"In the works of the Brown & Sharpe Manufacturing Company a pair of gears was wanted of the spiral or screw type, and it was thought better to

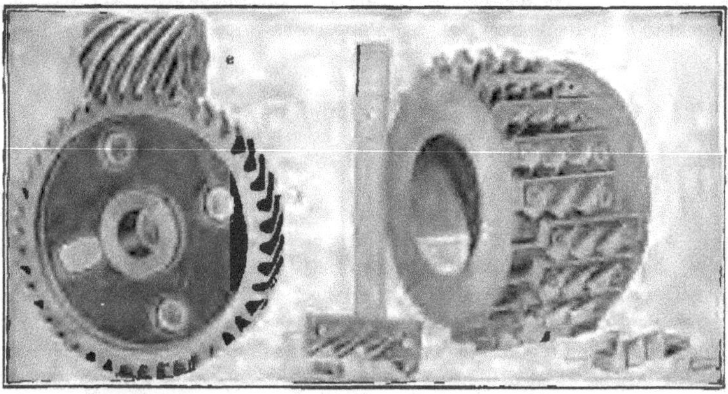

FIG. 114. WORM GEAR *B* AND WORM *A*.

FIG. 118. HOB USED FOR CUTTING WHEEL *B* OF FIG. 114.

make the large gear, or member, as a worm and the small member as a worm-wheel.

Fig. 114 shows the worm and wheel in mesh; *A* is the worm and *B* is the worm-wheel. The large member, or the worm *A*, has 43 threads; the lead of the worm is 60.3 inches, and the thread pitch, or the axial pitch, is 1.4 inches. The small member, or the worm-wheel *B*, has 7 teeth, and the circular pitch of the wheel is, of course, the same as the thread pitch of the worm, 1.4 inches.

Fig. 115 is an axial section of the worm threads. The threads incline 57 degrees from the plane perpendicular to the axis, which is so great that, while the axial thickness of the thread at the pitch line is 7/10 inch, the actual or the normal thickness is not quite 4/10 inch. In Fig. 116 the line *CD* shows the inclination of the threads; *CE* is the axial pitch, and *F G* the actual or normal pitch.

In cases where the inclination of the thread is more than 15 degrees, that is, in cases where the normal pitch is less than 0.96 of the axial pitch, it is well to have the depth and the addendum correspond to the normal pitch. Fig. 117 is a normal section of the thread, the depth being the same as a gear tooth of equal pitch, which makes the thread look shallow and thick when seen in the axial section, Fig. 115.

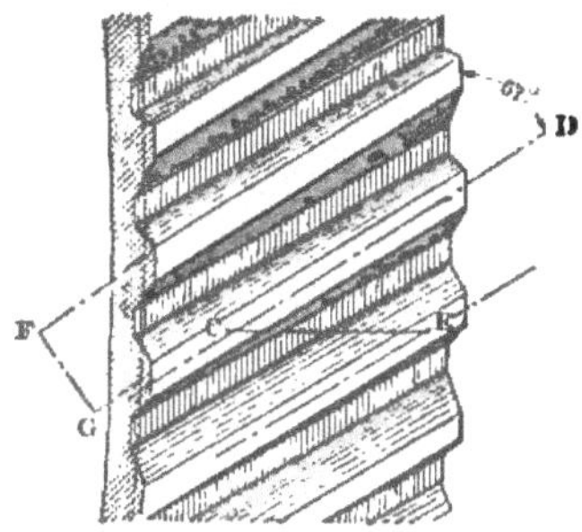

FIG. 116. AXIAL AND NORMAL PITCH.

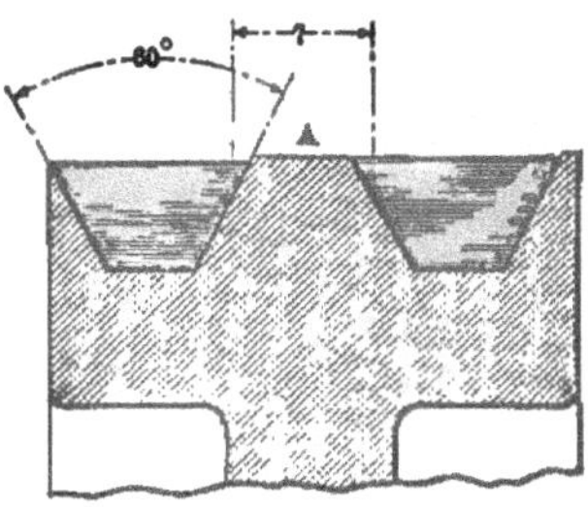

FIG. 115. AXIAL SECTION OF WORM THREADS.

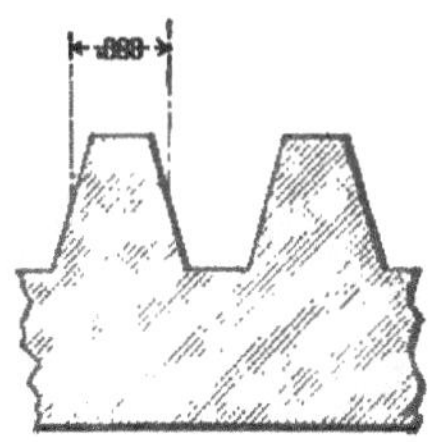
FIG. 117. NORMAL SECTION OF WORM THREADS.

The worm-wheel *B*, Fig. 114, was hobbed, or cut, with the hob shown in Fig. 118.

The worm has more than six times as many threads as the worm-wheel; the pitch diameter of the worm is four times that of the wheel; the wheel is the driver. The hob is made up of a cast-iron body, upon which are fastened lags that are arranged in steps in order to have the lags alike for convenience in manufacturing. Once a large hob was made that did not work, because the cutting edges of the hob teeth did not trim the tops of the worm-wheel teeth narrow enough to clear the backs of the hob teeth, which jammed so hard that the machine could not go. This jamming of the backs of the hob teeth upon the tops of the worm-wheel teeth was owing in part to incorrect spacing of the lags *HH*, Fig. 118, which will be explained.

A worm is a screw whose threads have the same outline, upon an axial section, as the teeth of a rack, the purpose of a worm being to mesh with a gear. A worm gear meshes with a worm. The action of a worm meshing with a worm gear is analogous to that of a rack with a spur gear, as stated by Professor Willis in his "Principles of Mechanism." In most worms the outlines of the threads upon the axial section have straight sides, as in Fig. 115, which corresponds to the sides of rack teeth in the involute system of gearing.

Fig. 118 is a hob made up of cast-iron body, into which are fastened lags *HH*. Two of these lags are shown detached. The lags were threaded in axial section like *A*, Fig. 115. The resulting teeth were trimmed and backed off, as in the detached lag on the left. The numbers in the scale are for inches.

I have spoken of the failure of a hob because the backs of the hob teeth jammed upon the tops of the wheel teeth. This interfering action can be explained in several ways; it is analogous to trying to thread a coarse screw in a lathe with a tool that does not lead or incline in the same direction that the thread inclines. A thread tool inclined for a right-hand thread would soon interfere in cutting a left-hand thread. Any grooving tool that has only one cutting edge or face must track in the same groove that it cuts. Sometimes a tool goes wrong and cuts a groove that bends the tool, which is occasionally noticed in cutting off a large piece in a lathe. In cutting a deep narrow groove a thin saw sometimes runs so much to one side that the saw is broken. In the case of the hob the interfering teeth would neither bend nor break, and so the machine had to stop. The teeth of a hob should be so arranged that there will be a cutting edge to take off an interfering point as it comes in the way. A worm-wheel can be cut with a tool that has only one cutting edge by bringing the tool into different positions in relation to the teeth of the wheel. In the AMERICAN MACHINIST for May 27, 1897, reference was made to the great number of cutting edges that a hob must have in order to cut a perfect wheel, and a description was given of a machine that cuts worm-wheels with a single tool acting in different positions. Such a machine was patented November 15, 1887, and another July 5, 1898.

In most hobs the cutting edges are straight, and in consequence the sides of the hobbed worm-wheel teeth are made up of straight lines in warped surfaces that meet in angles. These angles are often not noticed in worm-wheels of fine pitch and in wheels having a large number of teeth; but in wheels of coarse pitch and in wheels having few teeth the angles may be quite pronounced. Fig. 119 shows a worm-wheel that has teeth with hobbly sides on account of these angles. This wheel was cut with the hob shown in Fig. 120. The length and the diameter of the blank were great enough to extend beyond the teeth left by the hob that are available to work in connection with the worm. The available part of the teeth occupy about two-thirds the length of the wheel through the midpart, as between *I* and *J*. Though the teeth are available, yet their sides are so hobbly between *I* and *J* that they will need to have the angles finished off before the wheel can run smoothly with its worm.

Another kind of stepped action of the hob is seen as grooves near *KL*, Fig. 119, which are cut in consequence of the quick travel of the large part *KL*, in proportion to the narrow flats *MM*, Fig. 120, at the tops of the hob teeth. If the travel of *KL* had been slow enough or if the flats *MM* had been wide enough, there would have been no grooves.

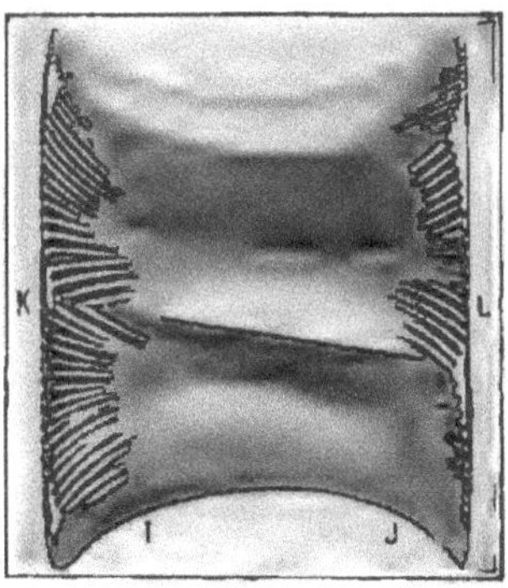

FIG. 119. A WORM GEAR OF FEW TEETH AND COARSE PITCH.

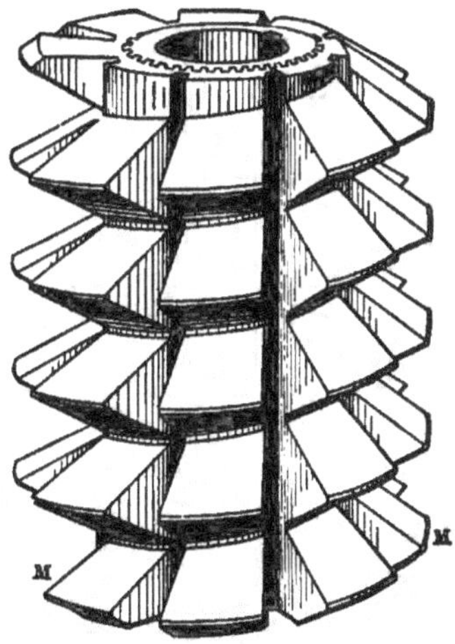

FIG. 120. HOB USED IN CUTTING WORM GEAR OF FIG. 119.

The circular pitch and the number of teeth of Fig. 119 are the same as in *B*, Fig. 114.

It is well known that the cutting edges of a hob must act upon the worm-wheel teeth in different positions, and that a tool with a single cutting edge must track in a groove cut with itself; but it was a surprise to learn that a hob of any number of cutting edges can be so made that it will absolutely refuse to cut a wheel that has only a few teeth like *B*, Fig. 114. When the workman told me that the hob jammed, I was incredulous, but a glance at the work proved that he was right. I could not believe that my previous experience had been such that I could have known how to make the hob, yet in a few minutes, when the solution came to me, I had the feeling that I must have known it well some time in the long past.

The things that affect this interference might be called variable; there are several of these variables. I am unable to give a rule that will indicate the conditions in which interference would be objectionable; yet, while limits may not be easily defined, an understanding of a few extremes may enable us to keep away outside these limits.

One way of explaining the interference is based upon the fact that, in a gear, any point outside the pitch circle moves through a greater distance, or faster, than a point in the pitch circle. Fig. 121 shows a single-threaded hob having only one row of cutting edges *OO*, the teeth, or the threads, extending nearly around the hob. Let the teeth in the wheel *P* be shaped as if they had been cut to the full depth with the cutting edges of the hob, and in a low-

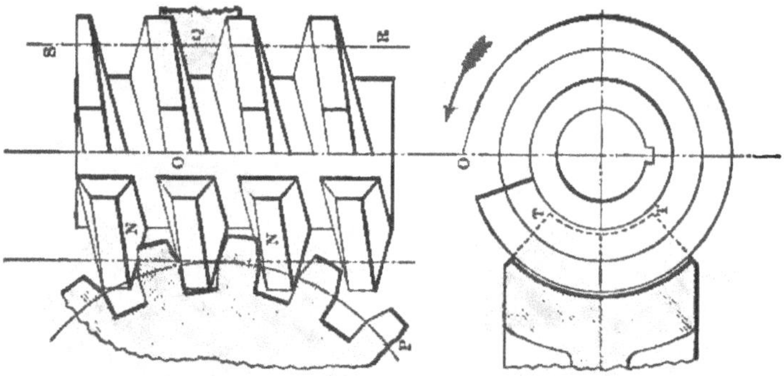

FIG. 121. A SINGLE-THREADED HOB WITH ONE ROW OF CUTTING EDGE.

numbered wheel we shall have tooth faces shaped as shown. Now, place this gear in mesh with the hob at the cutting edges, turn the hob in the direction of the arrow, and we shall soon find that the tooth faces of the wheel will interfere with the hob threads, as shown at *NN*, in consequence of the faces *NN* moving at a different speed from the pitch circle. There would also be interference upon the flanks of the gear, but it was thought that the cut would be quite as clear if the showing of this flank interference were not attempted. Only a slice section of the wheel is shown at *P*; in the real wheel we should have a still greater interference at the outer part of the teeth *TT*. In moving along a straight path, from *R* to *S*, a close-fitting tooth *Q* would not interfere; but interference would begin as soon as *Q* was moved in a curved path like that of a gear tooth.

From this consideration of Fig. 121 we should conclude that it is impractical to hob a wheel of few teeth with a hob having only one row of cutting edges, like the one shown. Even though we reduce the hob threads back of the cutting edges enough to clear the teeth of the wheel, so that it will be possible to hob the wheel, we shall not shape the teeth so that they will run correctly with the worm.

Another illustration of interference may be seen in Fig. 122. Let a small gear be cut, as shown, with a cutter that is shaped like a gear, as might be done in a Fellows gear shaper. Let every rotative movement of the gear, in order to take another cut, be through exactly one tooth, a cutter tooth always cutting on the line of centers, as shown. In this way cut to the full depth, moving the gear exactly one tooth at every setting. In our experimental cutting we can let the cutter rotate through one tooth at every movement of the gear, or we can let the cutter remain stationary, so far as rotation is concerned. When we have cut a few spaces to the full depth, we shall find that they are shaped as shown in Fig. 122, the spaces below the pitch line merely fitting a cutter tooth upon the line of centers without any enveloping or shaping of the gear teeth, as there would be in the ordinary working of the Fellows gear shaper. Now let us stop the cutter, leaving its cutting edges just above the side of the gear, and try to rotate the gear with the cutter in mesh, just as if they were a pair of gears, and we shall at once see that the teeth of the cutter interfere back of the cutting edges, as we should suppose from a mere inspection of Fig. 122.

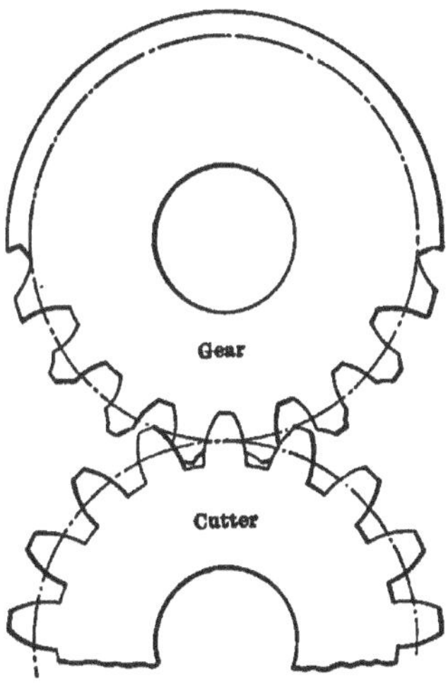

FIG. 122. ABSENCE OF ENVELOPING ACTION IN GEAR CUTTING.

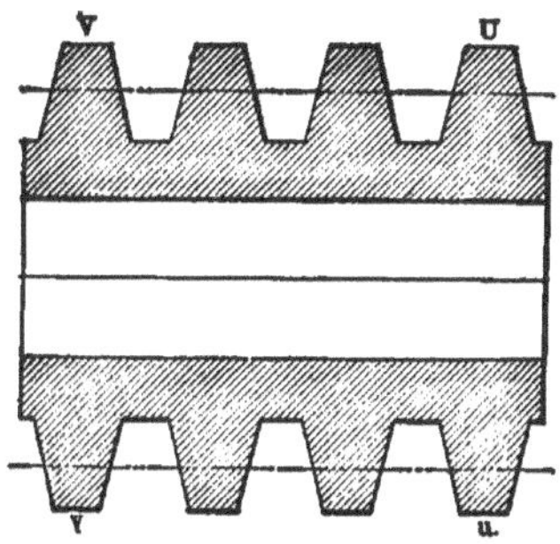

FIG. 123. A DOUBLE-THREAD HOB WITH TWO ROWS OF CUTTING EDGES.

The same kind of interference that we saw in a single-threaded hob, Fig. 121, will occur in a double-threaded hob that has only two rows of cutting edges if they are evenly spaced and are parallel to the axis. This can be understood from Fig. 123. Any tooth or thread *U* is exactly opposite another tooth *u*, because the thread is double, one thread starting at the end half way around from the other thread. One row of cutting edges *UV* will pass through the spaces cut by the other row *uv* in the same position as regards the worm-wheel teeth, and in consequence the backs of the teeth in both rows will interfere, as in Fig. 121.

A three-threaded hob with three evenly spaced rows of cutting edges will interfere, and so on.

From a careful consideration of the foregoing we arrive at the general principle—*The spacing of a hob must not be equal to the circumferential distance occupied by either one or to any whole number of threads.*

The more teeth there are in a worm-wheel the more teeth it is possible to have in contact with the worm threads at one time, in a worm that is long enough, and in consequence a long hob can possibly cut upon enough teeth at a time; or, what is the same thing, it can cut every tooth in enough positions so that even with only one row of cutting edges it can shape the teeth smooth and without interfering. In practice, however, it is never safe to trust to only one row.

My hob has 43 threads and 21 lags or cutting rows. It had spaced the lags 2/43 of the circumference apart, which gave just two thread spaces to each lag. Hence, so far as the shape of the worm-wheel teeth is concerned I was not doing any different with the 21 lags (shown in Fig. 118) than I could have done with only one lag.

Another body was made for the hob. The lags were spaced evenly, 21 in the circumference, which gave 2 1/21 thread spaces to each lag. This arrangement afforded twenty-one positions of lags. To accommodate these positions steps were provided, as seen at H. The hob was successful.

RELIEVING A SPIRAL FLUTED HOB WITHOUT SPECIAL FIXTURES*

Special fixtures are not necessary to relieve the teeth in a spiral fluted hob. This may be accomplished by indexing for a greater number of flutes than are actually contained in the hob.

Let L = lead of hob.
L_1 = lead of flute milled in hob.
C = pitch circumference.
I = distance gained by spiral flute in one revolution.
C' = circumferential length of each flute.
N = number of flutes to be added.

$$L_1 = \frac{C}{L}C.$$

$$I = \frac{C}{L_1}L.$$

$$N = \frac{I}{C'}.$$

If N turns out an inconvenient figure it may be changed to the nearest whole or fractional number and the lead of flute (L') changed to suit as follows:

$$L_1 = \frac{C}{NC'}L.$$

* R. J. Briney.

Example:

What will be the proper index for the relieving attachment for a hob 4 inches pitch diameter and 8-inch lead, number of flute cut in hob 5?

$$C = \pi\, 4 = 12.5664.$$

$$L' = \frac{C}{L} C = \frac{12.5664}{8} \times 12.5664 = 19.739 \text{ inches.}$$

$$I = \frac{C}{L_1} L = \frac{12.5664}{19.739} \times 8 = 5.088 \text{ inches.}$$

$$N = \frac{I}{C'} = \frac{5.088}{\frac{12.5664}{5}} = 2\,\frac{6}{25}.$$

Substituting $2\frac{6}{25}$ for 2 makes our index $5 + 2 = 7$, instead of 5.

Since the value N is changed from $2\frac{6}{25}$ to 2, we must change the lead of flutes to correspond.

$$L_1 = \frac{C}{NC'} L = \frac{12.5664}{2 \times \frac{12.5664}{5}} \times 8 = 20 \text{ inches.}$$

REDUCING THE DIAMETER OF WORM GEARS

Increasing the pitch diameter in order to avoid undercut is not good practice, as it tends to shorten the life of a gear, instead of lengthening it. By referring to Fig. 124 it is plain that the pitch of a worm gear at C is greater than at A and, since this pitch the worm can only be made to correspond with the pitch of the gear at one point, generally A, there must necessarily be a great amount of friction, with the necessary loss in efficiency at B and still more at C.

The efficiency and life of worm gears are greatly increased, therefore, by making the diameter and, therefore, the pitch of the gear to correspond with the pitch of the worm at point B, or the medium pitch diameter of gear. This will reduce the pitch diameter the following amount, it being assumed that angle of face ϕ is 30 degrees; or it can readily be found by a careful layout.

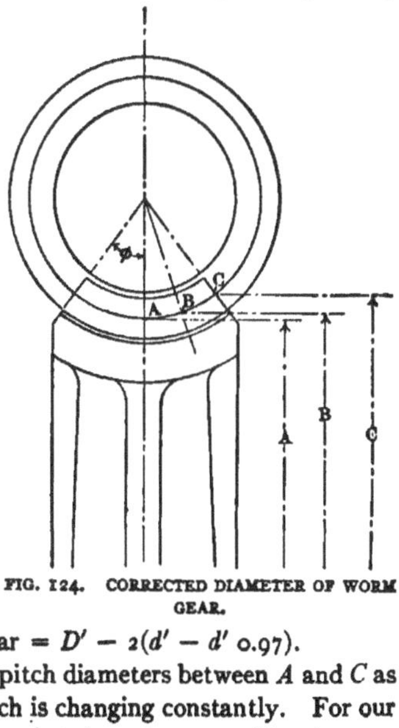

FIG. 124. CORRECTED DIAMETER OF WORM GEAR.

Corrected pitch diameter of worm gear $= D' - 2(d' - d'\,0.97)$.

There are in reality as many different pitch diameters between A and C as we would care to take sections, as the pitch is changing constantly. For our illustration, however, but the three main points have been considered.

GENERAL MANUFACTURING PROCESSES

The worm gear is first gashed out by a cutter, approximating the outside diameter of the hob for about two-thirds the full depth of the finished tooth, or else a taper hob of slightly smaller angle than the worm is used for roughing out the worm-wheel teeth. The finishing hob is then placed on the cutter spindle and dropped as far as possible into the gashed out tooth. The hob then completes the gear, driving it around and finishing the teeth at the same time. In starting the finishing out, care must be taken to prevent the teeth

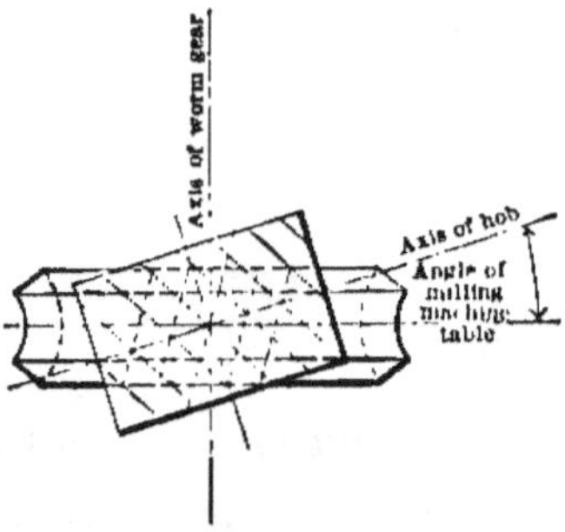

FIG. 125. CUTTING WORM GEAR WITH A HOB OF A DIFFERENT ANGLE FROM THE ENGAGING WORM.

of the hob locking on some sharp corner left by the gashing cutter. Gears which have been roughly cut with taper hobs avoid this danger to a great extent.

It is always advisable when hobbing the gear wheel teeth to take a hob that is similar to the worm to be employed, but when such a hob is not available one that is somewhat larger or smaller than the worm can frequently be used. See Fig. 125. By offsetting the axis of the hob, as shown in the figure, correct teeth can frequently be cut—in fact, it is sometimes possible to cut a right-hand wheel with a left-hand hob or *vice versa* by swinging the gear around until the angle of its thread corresponds with the angle of the hob.

STRAIGHT-CUT WORM GEARS

A modification of the ordinary type of worm and gear has been successfully employed in which the gear teeth are cut in a straight path, like a spur gear. See Fig. 126.

Advantages are claimed for this construction. It permits side adjustment which is impossible with the ordinary type of worm gear, and the contact is believed to be better as the pitch of the wheel corresponds with that of the worm the full width of the face. A disadvantage of the construction is that the helix angle of the worm that can be used with such a gear is limited to one of about 15 degrees.

This construction is often used for elevator service as it avoids a certain amount of the vibration that is practically unavoidable when employing

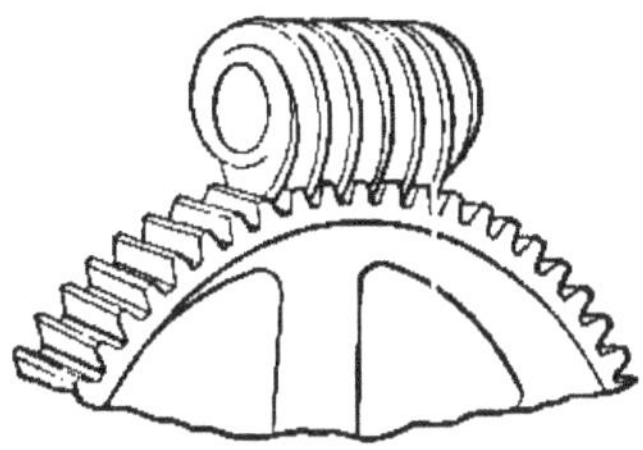

FIG. 126. STRAIGHT-CUT WORM GEAR.

gears of the hobbed type. Spacing errors are easier to avoid when cutting the plain straight teeth.

The teeth of the straight-cut worm gears are cut on milling machines, either of the plain or universal type. The table travels at right angles to

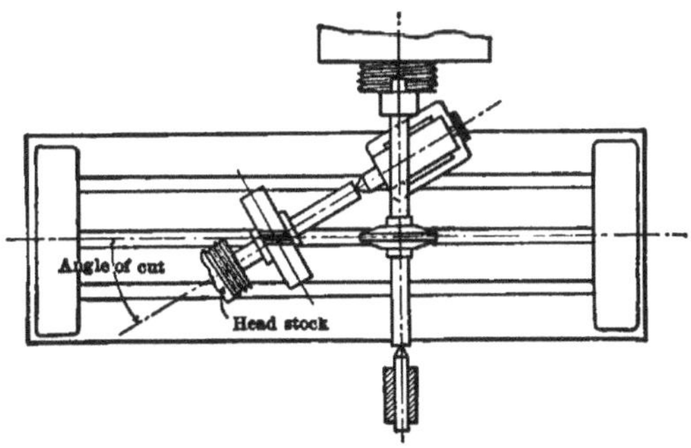

FIG. 127. CUTTING STRAIGHT-CUT WORM GEARS ON MILLING MACHINE.

the line of the cutter, the work being set up at the angle of the teeth in the worm gear. See Fig. 127.

Gears of this type should be laid out like spiral gears so that a standard spur-gear cutter may be employed to cut the teeth.

MATERIALS

The constant sliding between the thread surface of the worm and the face of the gear teeth limits and controls the materials or combinations of materials of which the worm and gear may be constructed. The worm being the active member as far as movement is concerned when slipping past the gear teeth, its thread is subject to constant and more destructive abrasion than are the teeth of the gear. The gear teeth though not subject to constant wear, coming into rubbing contact with the worm but occasionally, must nevertheless possess good wear-resisting qualities without being so hard as to increase unduly the wear on the worm thread. This relationship between the wearing qualities of the worm and of the gear is of the utmost importance.

It is generally recognized that the best materials from which a worm and a worm gear can be constructed, in order to realize good wearing qualities, high efficiency, etc., are case-hardened steel for the worm and phosphor-bronze for the gear. This combination gives excellent results.

Attempts have been made to substitute manganese-bronze as a material from which to make the gear teeth—usually only the teeth and rim of a worm gear are constructed of bronze, the hub, arms, etc., being made of cast iron or other cheaper material—but with disappointing results. The manganese-bronze was unsuitable on account of its hardness.

An excellent phosphor-bronze to employ for worm gears consists of: Copper, 80 parts; phosphorus, 1 part; tin, 10 parts.

The steel for the worm may be almost any low-carbon steel, which can be readily case-hardened and does not contain more than 3 or 3.5 per cent. of nickel nor more than 0.16 to 0.18 per cent. carbon. In case-hardening such a steel it should be carried to such a point that the scleroscope indicates a hardness of from 60 to 70.

The hardened steel worm should be carefully ground and trued up, but such operations are simple and evident for any standard type of worm.

POWER AND EFFICIENCY OF WORM GEARING*

In view of the good results now being obtained with worm gearing the old prejudice against that form of gearing, on account of its supposed low efficiency and short life, is dying out. These good results are the outcome of the application of principles which are by no means a late discovery, and it is expected that what follows will contain much that to some readers is not new. At the same time it is an undoubted fact that the best practice with worms is understood by but few, relatively speaking, and the corroboration of the theory by examples from practice which follow is believed to be new. No better illustration of the fact that good practice with worm gearing is not yet widely understood could be given than the statement in a recent and excellent work on gearing that "the diameter of the worm is commonly made equal to four or five times the circular pitch," the fact being that such proportions are distinctly bad if the worm is to do hard work.

It should be stated at the beginning that while what follows is not offered as a presentation of all the data necessary for assured success with worms under all conditions, it is hoped to make the general conditions of successful practice plain, and to present the "state of the art" as it exists to-day.

* F. A. Halsey, in the AMERICAN MACHINIST.

The essential change in practice which has improved the results obtained with worm gearing has been an increase in the pitch angle over what was formerly considered proper. There is no doubt whatever that this change has increased the efficiency of the gear, and, what is of more importance, has reduced the tendency to heat and wear. This is not only a fact, but it is a sound conclusion from theoretical considerations, which might have been predicted under proper examination.

THEORY OF WORM EFFICIENCY

The reason why an increase of pitch, other things being equal, or, in other words, an increase of the angle of the thread, gives these results, will be understood from Fig. 128. If *a b* be the axis of the worm and *c d* a line representing a thread, against which a tooth of the wheel bears, it will be seen that if the tooth bears upon the thread by a pressure *P*, that pressure may be resolved into two components, one of which, *e f*, is perpendicular, while the other, *e g*, is parallel to the thread surface. The perpendicular component produces friction between the tooth and the thread. The useful work done during a revolution of the thread is the product of the load *P* and the lead of the worm, while the work lost in friction is the product of the perpendicular pressure *e f*, the coefficient of friction and the distance traversed in a revolution, which is the length of one turn of the thread. Now, if the angle of the thread be doubled, as indicated, the load *P* remaining the same, the new perpendicular component *f h* of *P* will be slightly reduced from the old value *e f*, while the length of a turn of the thread will be slighty increased. Consequently their product and the lost work of friction per revolution will not be much changed. The useful work per revolution will, however, be doubled, because, the pitch being doubled, the distance traveled by *P* in one revolution will be doubled. For a given amount of useful work the amount of work lost is therefore reduced by the increase in the thread angle, and since the tendency to heat and wear is the immediate result of the lost work, it follows that that tendency is reduced. For small angles of thread the change is very rapid, and continues, though in diminishing degree, until the angle reaches a value not far from 45 degrees, when the conditions change and the lost work increases faster than the useful work, an increase of the angle of the thread beyond that point reducing the efficiency.

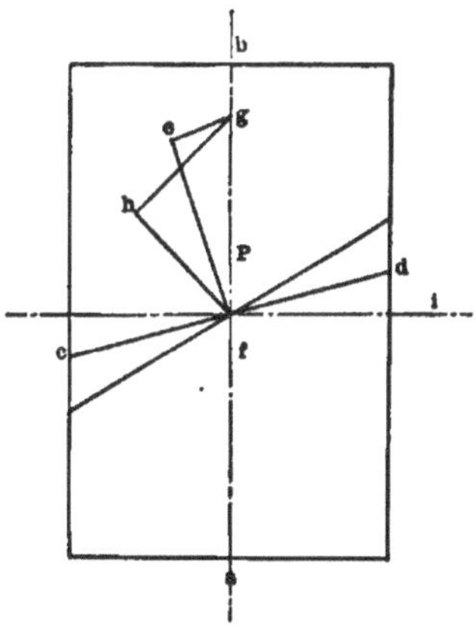

FIG. 128. THE PRINCIPLE OF WORM EFFICIENCY.

This general consideration of the subject shows the principles at the bottom of successful worm design, but a more exact examination is desirable. According to Professor Barr the efficiency of a worm gear, the friction of the step being neglected, is:

$$e = \frac{tan.\ a\ (1 - f\ tan.\ a)}{tan.\ a + f}$$

in which

e = efficiency,
a = angle of thread, being the angle $d\ f\ i$ of Fig. 128,
f = coefficient of friction.

To study the effect of the step, a convenient assumption is that the mean friction radius of the step is equal to that of the worm. This assumption would be realized only in cases where the step is a collar bearing outside the worm shaft, and the preceding and following formulas therefore represent extreme cases, one of a frictionless step, which would be approximated by a ball bearing, and the other of a step having about the extreme friction to be met with. Most actual cases would therefore fall between the two. Again, according to Professor Barr, the efficiency of a worm and step on the above assumption is:*

$$e = \frac{tan.\ a\ (1 - f\ tan.\ a)}{tan.\ a + 2f} \text{ (approximately).}$$

Notation as before.

These formulas give no clear indication of the manner in which the efficiency varies with the angle, and Chart 10 has been constructed to show this to the eye. The scale at the bottom gives the angles of the thread from 0 to 90 degrees, while the vertical scale gives the calculated efficiencies, the values of which have been obtained from the equations and plotted on the diagram. The upper curve is from the first equation, and gives the efficiencies of the worm thread only; while the lower curve, from the second equation, gives the combined efficiency of the worm and step. In the calculations for the diagram it is necessary to assume a value for f, and this has been taken at 0.05, which is probably a fair mean value. The experiments made by Mr. Wilfred Lewis for Wm. Sellers & Co. showed an increase of efficiency with the speed. The present diagram may be considered as confined to a single speed, and at the same time is not to be understood as showing the exact efficiency to be expected from worms, but rather to exhibit to the eye the general law connecting the angle of the thread with the efficiency.

The curves will be seen to rise to a maximum and then to drop. The exact values of the angle of thread to give maximum efficiency may be easily found by the methods of the calculus, the results being:

For worm thread alone the efficiency is at a maximum when

$$tan.\ a = \sqrt{1 + f^2} - f.$$

* In Professor Barr's formulas it is assumed that the worm thread is square in section. Thread profiles in common use affect the result but little.

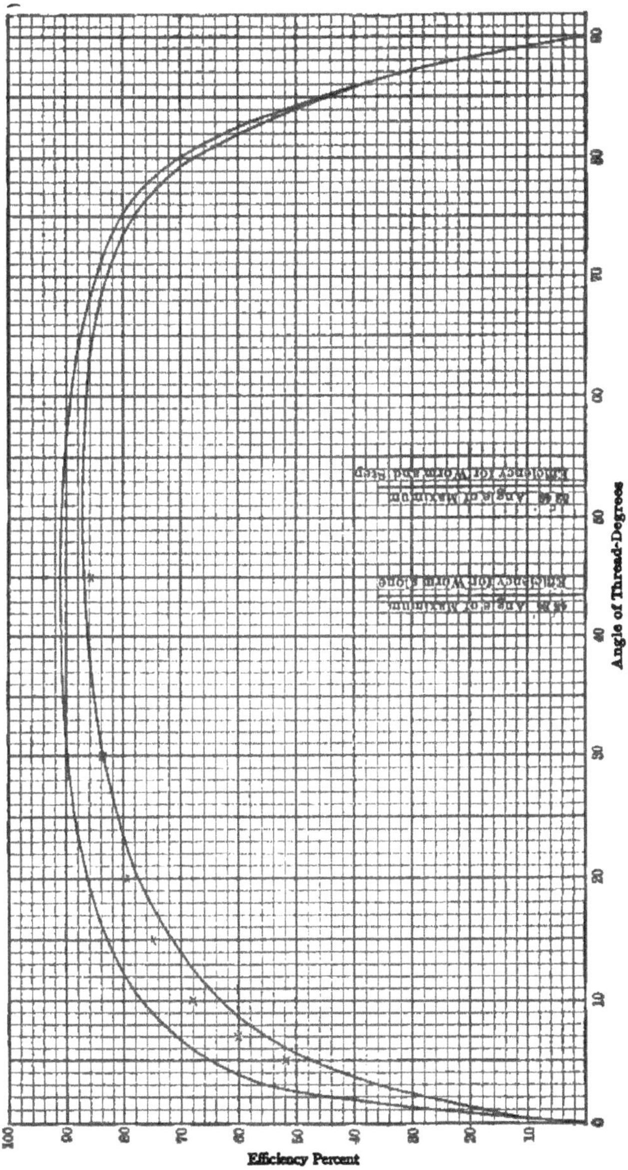

CHART 10. RELATION BETWEEN THREAD ANGLE AND EFFICIENCY.

Substituting the value of f (0.05) used in calculating the diagram, this becomes *tan.* α for maximum efficiency = 0.9512, and by referring to a table of natural tangents we find that α for maximum efficiency = 43° 34°. Similarly for the worm and step the result is *tan.* α for maximum efficiency =

$$\sqrt{2 + f4^2} - 2f, \text{ which for } f = 0.05 = 1.318,$$

and a table of tangents tells us again that α for maximum efficiency = 52°49′.

Of more importance than the angle of maximum efficiency is the general character of the curves, of which the most pronounced peculiarity is the extreme flatness, showing that for a wide range of angles the efficiency varies but little. Thus, for the upper curve there is scarcely any choice between 30 and 60 degrees of angle, and but little drop at 20 degrees.

At first sight the lower curve might be thought the more useful of the two, as it includes the effect of the step, but a little consideration will show that this is not the case. For most cases in which worms are used the efficiency of the transmission, as such, is of very little account. What the designer concerns himself with is the question of durability and satisfactory working, and the results to be expected in this respect are best shown by the upper curve, in which high efficiency means a durable worm. Throughout this discussion, in fact, the chief significance of efficiency lies in the fact that low efficiency means rapid wear, and *vice versa*.

EXPERIMENTAL CORROBORATION OF THE THEORY

The experiments of Wm. Sellers & Co., before referred to, go far to confirm the soundness of the above views. From the present standpoint it is unfortunate that those experiments did not cover a wider range of worm-thread angles—those actually used being 5 degrees, 7 degrees, and 10 degrees. Other experiments were, however, made on spiral pinions of higher angles, spiral pinions being understood by Mr. Lewis to mean those pinions having the mating gear a true spur, the pinion shaft being at a suitable angle with the gear shaft to bring the pinion in proper mesh—a construction which is exemplified in the well-known Sellers planer drive. Mr. Lewis gives a formula by which the efficiencies of worms can be calculated from those for spiral pinions, and in the absence of direct experiments on worms of high angles, his results for spiral pinions have been modified by this formula to read for worms. The results for the two forms of gearing differ by less than 5 per cent. for the extreme case of his experiments To compare the results obtained by Mr. Lewis with Professor Barr's formula, a speed has been selected from the experiments giving the nearest coefficient of friction to that used in obtaining the curves of Chart 11. The results have been plotted in Chart 11, where they appear as small crosses, and will be seen to have a very satisfactory agreement with the lower curve, with which they should be compared, as the steps of the worms used by Mr. Lewis were of the usual pattern without balls.

The variation of the coefficient of friction with the speed lends an interest to Chart 11, which is a series of curves obtained from the results published by Mr. Lewis in the same manner as the crosses of Chart 10, the curve for 20 feet velocity being in fact the same as that appearing as crosses on Chart 11. The other curves of Chart 11 are obtained from those of Mr. Lewis, and cover a range of velocities from 3 to 200 feet per minute at the pitch line, as noted at

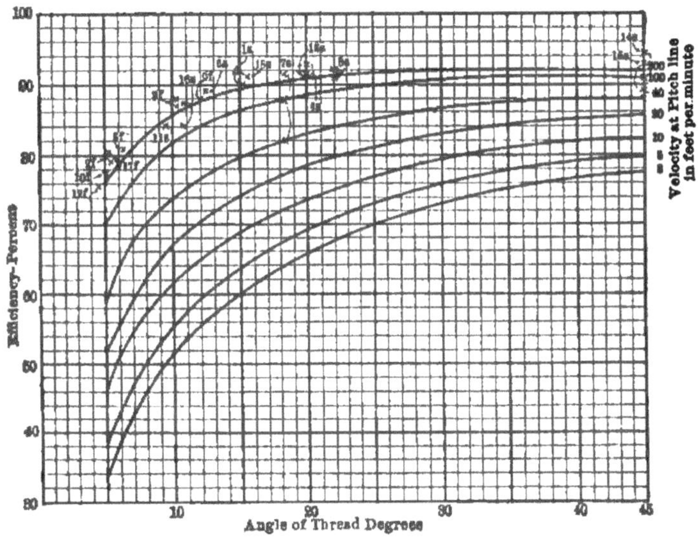

CHART 11. RELATION BETWEEN THREAD ANGLE, SPEED AND EFFICIENCY WITH CASES FROM PRACTICE.

the right. In this diagram the results obtained by Mr. Lewis on worms are plotted direct, but the experiments on spiral pinions have been modified as explained above. Inspection of the curves shows that while there is a progressive increase of efficiency with the speed, there is, nevertheless, not much probability, or indeed room, for further improvement beyond the speed of 200 feet per minute. It will furthermore be seen that the efficiency drops off much less for low angles of thread at high speeds than at low.

In interpreting this diagram, it should be remembered that the durability of a worm depends upon the amount of power lost in wear, and not upon the percentage so lost. The ability of a given worm to absorb and carry off the heat due to friction is fixed, and does not vary with the speed. That is, a given worm running at 100 revolutions under a given pressure can carry off as much friction heat as the same worm at 200 revolutions, while it, under the same pressure, would transmit but one-half the power in the former case that it would in the latter. In other words, the percentage of lost work might be twice as much at the lower speed as at the higher without increasing the tendency to heat.

The increase of efficiency with the speed is a valuable property of worms, and enables them to do much more work than they otherwise would. Thus the 20-degree worm at 20 feet per minute lost 21½ per cent. of the work in friction. Increasing the speed to 40 feet doubled the work applied, and, had the efficiency remained constant, would have doubled the friction heat to be dissipated. In point of fact, this increase of speed diminished the percentage of loss to 17, and the amount of loss and heat, instead of being doubled, was only increased in the ratio of 160 to 100. It is plain from the diagram, however, that this action does not continue much beyond a velocity of 200 feet per minute, beyond which the amount of loss must be more nearly proportional to the speed, and this doubtless has some connection with the fact observed by Mr. Lewis that 300 feet per minute is the limit of speed when the gears are loaded to their working strength, and that the best conditions are obtained at about 200 feet per minute. It is proper to add, however, that in the cases from practice given later there are three which have been made repeatedly, and which are conspicuously successful, in which the velocity exceeds 600 feet, and one in which it exceeds 800 feet. No doubt, in all such cases, if the pressure on the teeth could be known it would be found to be light.

It will be seen that an increase of speed for any worm under constant pressure leads to an increase of friction work, and the limit is reached when the worm is no longer able to carry off the heat generated fast enough to prevent undue rise in temperature. Furthermore, this limiting speed depends upon the pressure, it being higher for low pressures than for high. A worm having an angle which might be successful at low speed may fail at high speed; but it would seem that any worm which is successful at high speed should also be successful at low, which is in accordance with mechanical instinct.

There are, it will be observed, two methods of increasing the pitch angle. The diameter may be kept constant and the pitch be increased, or the pitch may be kept constant and the diameter be reduced. From a mathematical standpoint, these two methods are identical; that is, at a given pitch-line velocity a worm of a given angle should have the same efficiency, regardless of the diameter; but in a mechanical sense the methods are not identical. The worm of the larger diameter would naturally have a gear of wider face, and the pair, having greater area of tooth surface in contact, would carry a larger load.

EXAMPLES FROM PRACTICE

It is impossible to say who was the first to recognize the significance of the pitch angle as a factor in the satisfactory performance of worm gearing, but it may be mentioned as a matter of interest that the exhibit of the Hewes & Phillips Iron Works at the Newark Industrial Exhibition of 1873 included several worm-driven planers, in which the worms were double-threaded and had a pitch angle of 15° 15′, a pitch diameter of 3½ inches, a lead of 3 inches, and a speed cutting of 256 and backing of 640 r.p.m., which give pitch-line velocities of 237 and 590 feet. This worm was successful, and was many times repeated; but later on Hewes & Phillips were struck by the high-belt speed idea, and in order to increase the belt speed they changed the worm to 6.16 p. d., 1¾-inches pitch, single thread; speed cutting, 446, and backing 1,110 r. p. m., giving a pitch angle of 5° 15′ and pitch-line velocities of 720 and 1,780 feet. This worm was a failure, and was soon changed to 6.16 p. d., 3½-inches lead, double-thread, speed cutting, 281, and backing 700 r. p. m., giving an angle of 10° 15′ and pitch-line velocities of 452 and 1,130 feet. This worm did better than the last, but not so well as the first. By this time the lesson was learned, and Hewes & Phillips set out to use a worm of 30 degrees pitch angle. Structural considerations, however, prevented the use of so high an angle and they compromised on 20 degrees, the final worm resulting from this experience having a pitch diameter of 2.63 inches, with 3 inches lead, quadruple thread, the speed cutting being 300 and backing 700 r. p. m., giving pitch-line velocities of 205 and 480 feet, and this remained the standard angle as long as these planers were manufactured.

The writer has seen one of these 20-degree worm gears, opened up after twelve years' use, and the wear disclosed was very slight—no shoulder being in existence. As a result of the experience outlined above, this house adopted the standard practice of the worms as small as possible in diameter, and giving the threads in all cases a pitch angle of 20 degrees. The form of tooth used was the epicycloidal, while the materials used were hard cast iron for the gear and case-hardened open-hearth steel for the worms.

These Hewes & Phillips worms are plotted in Chart 11 as crosses 1, 2, 3, 4, of which 1 is the 15° 15′, 2 the 5° 15′, 3 the 10° 15′, and 4 the 20°, the first and last being successes, and the second and third failures.

In plotting these worms, and all others having pitch-line velocities above 200 feet, the crosses are placed near and above the 200 feet curve. It is unfortunate that we have no curves for higher speeds, but Mr. Lewis recommends the use of the 200 feet line for all higher speeds. Leaders connecting different crosses indicate the same worm at different speeds in all cases. The letters *s* and *f* on the diagram mean success or failure in all cases.

Fig. 129 is a drawing of a worm 3 (failure) and Fig. 130 shows worm 4 (success), and no more instructive pair of drawings could be imagined than these. The pitches are not far different, and what difference there is is in favor of the larger worm. The duty is the same, the gears are of about the same diameter, and the revolutions per minute are nearly the same. The essential change is in the increase of the pitch angle by a reduction of the diameter, and this changed failure to success.

The Newton Machine Tool Works use worm gearing in many of their machines, notably their cold saw cutting-off machines. In the earlier machines of this class the worm had a pitch diameter of 2⅞ inches, with a pitch of 1 inch, single thread, the revolutions per minute being 765. These figures give a pitch angle of 6° 20′, and a pitch-line velocity of 572 feet. This machine could be operated, but not with satisfaction on account of the heating and short life of the worm. The worm was then increased in lead by making it double-threaded, giving a pitch angle of 12° 30′, the speed being reduced to 500 revolutions per minute, giving a pitch-line velocity of 375 feet.

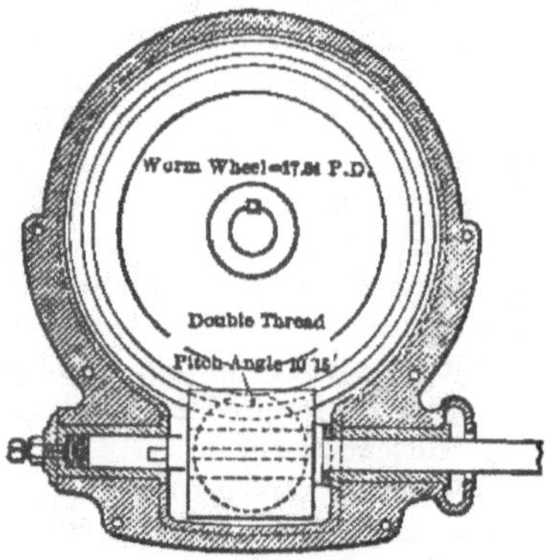

FIG. 129. HEWES & PHILLIPS UNSUCCESSFUL WORM.

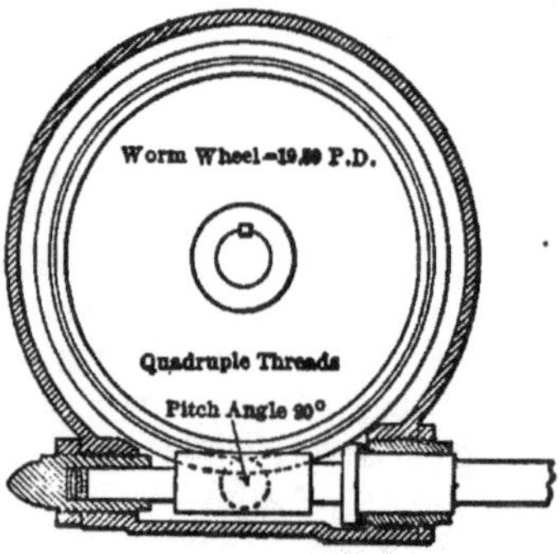

FIG. 130. HEWES & PHILLIPS SUCCESSFUL WORM.

The change proved to be a great improvement, heavier work than was before possible being done after the change without distress or difficulty, and this worm has since been applied to a large number of machines with entire success. A still later worm used on these machines has a pitch diameter of 3⅞ inches and a lead of 4 inches, triple threads, giving a pitch angle of 18° 15′, and this is found to be a still further improvement. This last worm is used on a wide variety of mchines and at a variety of speeds from 40 to 680 r. p. m., giving pitch-line velocities of from 40 to 685 feet, and with uniformly good results. In many cases it is used without an oil cellar, though for comparatively light work. The form of thread used is the involute, and the material is hardened steel for the worm and bronze for the wheel. These Newton worms appear in Chart 11 as 5, 6, 7, of which 5 is nearly a failure, while 6 and 7 are entirely successful. The second Newton worm—the one appearing in Chart 11 as 6—is shown in Fig. 131.

Another habitual user of worms is John Bertram & Sons, of Dundas, Ontario, Canada, who employ them in all their planers, and use largely a worm of 3.18-inches pitch diameter, 4 inches lead, quadruple threads, the speed cutting being 186 and reversing 744 r. p. m. These figures give a pitch angle of 22 degrees, and pitch-line velocities of 155 and 620 feet. This worm appears in Chart 11 as 8, the vertical position for the higher speed being again uncertain. These worms are highly successful, as the writer knows from repeated observation. Both worm and wheel are of cast iron, the thread being Brown & Sharpe standard. The Bertram worm is shown in Fig. 132. In reading this drawing it should be remembered that the conventional representation of a worm, with the threads shown by straight lines, shows a larger apparent pitch angle than the true one, as shown by a true projection.

Another case of failure was a worm drive applied to a large boring machine, the worm being 12-inches pitch diameter, 8-inches lead, quadruple thread, speed 80 r. p. m. and above, worm of forged steel, wheel of bronze, oil cellar

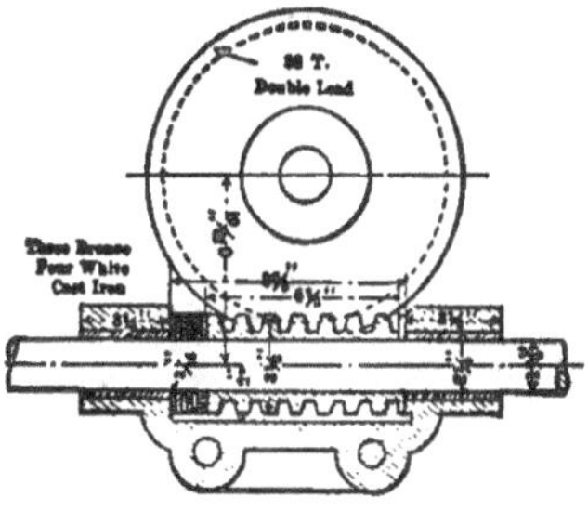

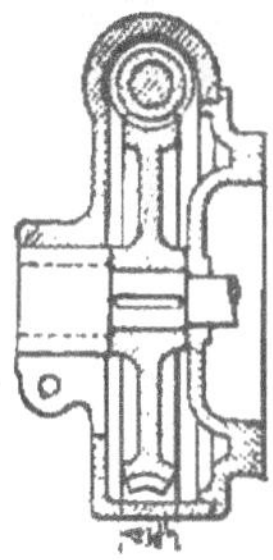

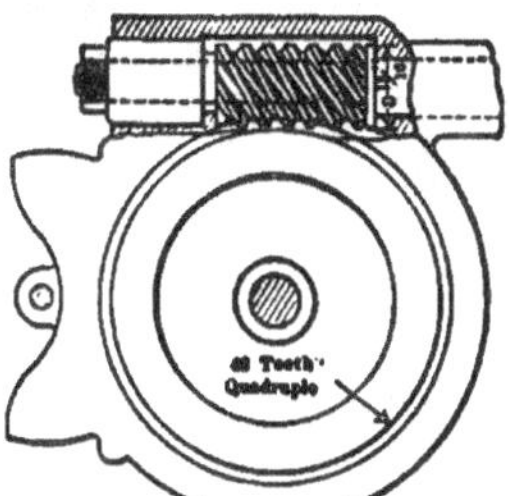

FIG. 131. THE NEWTON WORM AND STEP. FIG. 132. THE BERTRAM WORM.

lubrication. These figures give a pitch angle of 12 degrees and a pitch-line velocity of 250 feet. This worm is located on Chart 11 as 9.

Still other cases of change from failure to success are supplied by Mr. Jas. Christie, of the Pencoyd Iron Works. The first of these relates to a boring machine, which was, by the makers, supplied with a worm drive having a worm of 5½-inches pitch diameter, 1½-inches pitch, single thread, steel worm and cast-iron wheel, average speed 150 r. p. m. These figures give a pitch angle of 5 degrees and a pitch-line velocity of 215 feet. This was a failure, but was successfully replaced by a worm of 4⅞-inches pitch diameter, 2½ inches lead, and the same number of revolutions, which figures give a pitch angle of 9° 15′ and a pitch-line speed of 190 feet. These two worms appear as 10 and 11. This successful worm lies in the region of unsuccessful ones, but the influence of the increased lead angle is unmistakable. The fact of its success is probably due to the pressure on the teeth being well below the working strength, or to the speed being moderate, or both.

The second case, of which the data were supplied by Mr. Christie, relates to two heavy milling machines, in which the cutter spindles were driven by worms 6-inches pitch diameter by 1½-inches pitch, single thread. It was found that the cutters could be run much faster than was originally contemplated, and the worms were consequently speeded up to about 500 r. p. m. In these machines cast-iron worm-wheels were speedily destroyed, while hardened steel worms and bronze wheels would last about a year. Later two more machines were built having steel worms and bronze wheels, the worms being 4½-inches pitch diameter by 5-inches lead, quadruple threads, speed 280 r. p. m. These worms have been in use six years, and are described as being "as good as new." The data given for the first worm give a pitch of 4° 30′ and a pitch-line velocity of 785 feet. It appears in Chart 11 as 12. The pitch angle of the second worm is 19° 30′, and its pitch-line velocity is 328 feet. It appears in Chart 11 as 13.

Mr. Christie has made many successful changes, of which these are typical, and he now uses worms with great freedom and success. His general conclusion is that good worms begin with those having the pitch about equal to the diameter, giving a pitch angle of 17° 15′.

Another equally striking case of success accompanying an increase of the pitch angle is supplied by Mr. W. P. Hunt, of Moline, Ill., who says:

"In building a special double-spindle lathe I wished to use a worm drive, and having a single thread ¾-inch pitch hob, 2¾-inch outside diameter, I decided to work to that, and made my gear with 26 teeth, giving a speed reduction of 26 to 1. The worm was to run at 460 revolutions per minute, but upon starting the machine I found it impossible to keep the worm and gear cool, and the belts would not pull the cut.

"Accordingly I decided to make a new worm and hob having the same outside diameter as the one first tried, but with double thread and 1-inch pitch, 2-inch lead and a new gear having 48 teeth, giving me a speed reduction of 24 to 1, or less than at first.

"Upon starting the machine with the new worm and gear, not only did it run perfectly cool, but the belts have ample power. We use graphite and oil on the worm, and it is not enclosed."

Mr. Hunt does not give the pitch diameter of his worms, but assuming the threads to have been in accordance with the Acme standard, the pitch diameters are 2⅜ and 2¼ inches respectively, the thread angles being 5° 44′ and 15° 48′, and the pitch-line speeds 286 and 271 feet per minute. Mr. Hunt's worms are plotted in Chart 11 as 17 and 18.

Three other cases of successful worms under heavy duty are found in milling machines which have been repeated many times. The first two worms would ordinarily be described as spiral gears. The shafts are at right angles.

The first of these, which appears as 14 in the diagram, has a pitch diameter of 2¼ inches, a pitch angle of 45°, and a speed varying between 180 and 945 r.p.m., giving pitch-line velocities of 106 to 555 feet per minute. Both gears are of cast iron. The second, 15 in the diagram, is of the same style, and has the same pitch diameter, with speeds varying between 90 and 472 *R.P.M.*, giving pitch-line velocities of 53 to 277 feet per minute. The third, 16 in the diagram, is a true worm, 2¼-inches pitch diameter, lead 1.333, triple thread, speed 200 to 1,442 r. p. m., bronze wheel and hardened steel worm. These figures give a pitch angle of 10° 45′, and a pitch line velocity of 118 to 845 feet per minute. While this worm is entirely successful, it was at first a failure.

LIMITING SPEEDS AND PRESSURES

A very important point connected with worm design, and one on which data are very scarce, is the limiting pressures for various speeds at which cutting begins. The paper by Mr. Lewis contains some information on this subject, and the accompanying table supplied by Mr. Christie, from experiments made by him, supplies the most definite additional data on the subject known to the writer. In all cases the worms were of hardened steel and the worm-wheels of cast iron. Lubrication by an oil bath.

	SINGLE-THREAD WORM 1″ PITCH 2⅞ PITCH DIAMETER				DOUBLE-THREAD WORM 2″ LEAD 2⅞ PITCH DIAMETER			DOUBLE-THREAD WORM 2½″ LEAD 4½ PITCH DIAMETER		
Revolutions per minute....	128	201	272	425	128	201	272	201	272	425
Velocity at pitch line in feet per minute............	96	150	205	320	96	150	205	235	319	498
Limiting pressure in pounds	1,700	1,300	1,100	700	1,100	1,100	1,100	1,100	700	400

LIMITING SPEEDS AND PRESSURES OF WORM GEARING.

There is real need of a comprehensive series of experiments on this subject. It is obvious enough that a worm, otherwise well designed, might fail from having too high a speed for its load. Were such data at hand it would seem that with existing knowledge of the influence of the angle of the thread, worm design might be made a matter of comparative certainty. Especially should the behavior of worms at speeds above 200 feet per minute be subjected to further experiment, as it is frequently necessary to use speeds above that figure, and there can be no doubt that higher speeds are entirely feasible if suitable pressures accompany them. The speed as a factor should be kept in mind equally with the pitch angle. A worm may fail because of too high a pitch-line velocity as well as because of too low a pitch angle.

The number of cases cited is too few for certainty in drawing general conclusions, but the testimony is unmistakable in its confirmation of the theory of the influence of the angle of the thread. It will be seen that every case having an angle above 12° 30′ was successful, and every case below 9° unsuccessful, the overlapping of the successful and unsuccessful worms in the intervening region being what is to be expected in the border region between good and bad practice. This band of uncertain results is in fact narrower than we would have any right to expect from a collection of data from miscellaneous sources, and could the inquiry be widened in scope the width of this band would doubtless be increased. As throwing light on these cases, it should be remembered that case 16 is known to have been made successful only by careful attention to the material used, the first worms made having been failures, and that 3, which is near 16 and was a failure, had an excessive speed, while 11, at a lower angle and a success, had a very moderate speed. At a higher speed 11 would probably have failed, and at a lower speed 3 would probably have been a success. It is believed that Chart 11 points out clearly the nature of the worm problem and the conditions of success in its solution.

EFFICIENCY AND TEMPERATURE

The unavoidable sliding action between the threads of the worm and the teeth of the gear represents lost energy which must be measured by the amount of friction heat generated. The elevation in the temperature of the lubricating oil about a worm drive should then approximate the efficiency of the construction.

The relationship between the efficiency of a journal bearing and its rise in temperature is similar and is generally recognized as a reliable indication of the efficiency of the bearing. Unfortunately, a number of experiments that have been conducted with a view of discovering such relationship in connection with worm drives have been rather misleading. This has been partly due to inadequate lubrication, but more probably to the fact that both liquid and solid friction have been present.

Comprehensive experiments to ascertain the relationship between efficiency and temperature of worm drives at the engineering laboratory of the Royal Technical High School, Stuttgart, Germany, however, have conclusively shown that some relationship does exist. The conclusions arrived at by the experimenters may be summarized as follows:

1. The difference in oil temperature is approximately proportionate to the tooth pressure when the speed of the worm remains constant.

2. The tooth pressure decreases according to a definite law with any increase in worm speed when the temperature remains constant. (A curve depicting such decrease in tooth pressure is of distinctly hyperbolic character.)

As confirmation of the above deductions, the conclusions arrived at by the firm of Henry Wallwork & Co., Ltd., Manchester, England, which has conducted many individual tests to ascertain the laws governing worm-gear efficiency, are of particular interest. A. V. Wallwork summarized these conclusions, in a letter that appeared in AMERICAN MACHINIST, Dec. 5, 1912, as follows:

"That the efficiency of a correctly designed gear increases under light load and decreases under heavy load, and that the temperature rise of the oil gives an exact measure of the power loss in the gear."

EFFICIENCY OF LANCHESTER WORM GEAR

The Lanchester worm, which is very similar to the Hindley worm, being of the hour-glass variety, has been subject to exceedingly exhaustive experiments at the National Physical Laboratory of Teddington, England, which were briefly described in American Machinist, June 12, 1913.

The engineers making these tests arrived at the conclusions that the efficiency of the worm gear itself depends entirely on the condition of the oil film between the worm and the wheel and that the efficiency would remain practically constant as long as this oil film is perfect. The fall in efficiency of the gear at slow speed, which was quite apparent in all tests, was believed to be due in part to the reduction in the quantity of oil carried around by the worm.

In conclusion, the report submitted to the Daimler Co. of England, for which the tests were made, says:

"The conclusion to be drawn from these tests is that the efficiency of the gear lies between 93 and 97 per cent. under all circumstances, and, taking the normal running speed of the worm at 1, 000 r. p. m., the efficiency of the gears lies between 95 and 97 per cent., only falling slightly below the lower figure when the temperature approaches 100 degrees Centigrade."

AUTOMOBILE WORM DRIVES

The prejudice that appears to exist in this country against the worm drive for automobiles is not so pronounced in Europe. There does not seem to be any reasonable ground upon which to base the objections made by many manufacturers in regard to worm-gear transmissions. English manufacturers in particular have secured very gratifying success with worm drives. David Brown & Sons, Ltd., Lockwood, Huddersfield, England, claim excellent results for their worm-gear auto drives—stating that 95 per cent. efficiency is easily realized. E. G. Wrigley & Co., Ltd., Birmingham, England, claim for their worm drives an efficiency as high as that of a bevel-gear drive at normal speed, equal or greater efficient life and practical silence in operation.

Ralph H. Rosenberg presented a paper before the Society of Automobile Engineers, excerpts of which were printed in American Machinist, Feb. 29, 1902, in which he summarized his conclusions on the worm drive for heavy power vehicles in part as follows:

"My contention is—and I have proved it empirically—that, first, the tooth angle and angle of lead or advance must coincide, and second, that the advance angle fixes the diameter of the worm. It is determined by extending the lines describing the flanks of the teeth to points where the intervening distance is equal to the lineal pitch times the number of leads; the distance from these points to the pitch circle of the gear is the diameter of the worm.

"I have adopted the following method for determining the width of the gear and face of the teeth. They are described by diverging lines from the center of the worm, including an angle of 120 degrees. Gears made according to this formula will permit a reasonable amount of variation between the pitch circle of the worm and the pitch circle of the gear, the surfaces remaining complementary. This is not permissible with any other form and allows a certain latitude in manufacture. It is absolutely necessary, however, to maintain proper relations of the axes.

COST

"I have heard it asserted by those conceding the utility and desirability of the worm gear that it was an expensive device. My endeavors to ascertain upon what ground this assumption was made and what particular item entered into the consideration of cost were usually met by general statements. So I conclude that the cost-of-manufacture information while not as vague as that relative to designing is, nevertheless, indefinite. I take the liberty of quoting from E. E. Whitney's paper of June, 1911, relative to the cost of worm-gear construction, wherein he states the worm-gear drive is not a cheap device and that the indicated efficiency and durability results cannot be expected unless the gears are properly designed, constructed of the best materials and adequately mounted in high-grade anti-friction bearings.

"I concede this statement to cover the essential facts generally, but on the question of cost take issue, believing the worm gear to be the cheapest form of final drive. It is admittedly true that proper design is essential to the success of any mechanism, but it does not follow that proper design will entail any expense over and above improper design, so far as it relates to the cost of manufacture. In my experience I have found that the materials used in the worm and gear are not more expensive than those employed in the bevel-gear drive or the side-chain drive, where double reductions are used. Furthermore, a distinction should be made between experimental work and work of actual production where the facilities are provided for executing large quantities. In substantiation of my statement I give the following data, taken from records covering the cost of producing a worm and gear for a 5-ton truck:

Bronze ring gear blank	$18.00	
Steel for worm	9.00	
Time to machine worm		2 hours
Labor on worm, rough-turning		2 hours
Labor on worm, molding		2 hours
Labor on worm, grinding		3 hours
Ring gear, turning		3 hours
Ring gear, cutting		1 hour and 10 minutes

"The parts are then in condition for assembling. Concerning the desirability of the worm gear, I am in accord with Mr. Whitney, and can say that

I have inspected gears after they have run 120,000 miles and found them in excellent condition. Granting that the expense of production is higher, it is offset by the greater life of the gear."

There is no question that properly proportioned and constructed worm-gear drives can be made with high guaranteed efficiency; that they are compact and may be constructed for considerably higher ratios without requiring undue space for their accommodation; that their wearing qualities can be as great as those of a bevel-gear drive; and that they possess other advantages of silence, smooth operation, etc. These numerous advantages, which are now lessened by no serious drawbacks, should appeal strongly to the automobile manufacturer, particularly in the construction of auto trucks requiring large reduction in speed between the motor and the driving wheels.

THE HINDLEY WORM GEAR

An interesting modification of the standard worm gear is frequently encountered in elevator service, and is known as the Hindley worm gear or, from its form, the *globoid gear*. The worm differs from the ordinary straight worm because, instead of being cylindrical in shape, it is formed somewhat

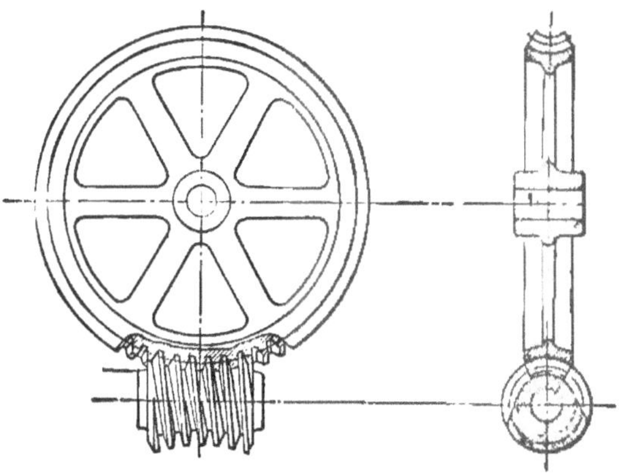

FIG. 144. THE HINDLEY WORM GEAR.

like an hour-glass or spool. The pitch diameter of the worm varies so as to coincide with that of the gear for the full length of the worm. The thread of the worm is in mesh with several gear teeth at the same time, the worm enveloping a section of the gear wheel. See Fig. 144. There is no bottom clearance in this gear and the length of tooth and depth of thread are somewhat greater than in the common gear and worm, thus increasing contact surface and therefore decreasing wear.

The smallest diameter threads, at the center of the worm, evidently engage the gear teeth as do the threads in an ordinary worm gear, the contact of which is on the pitch line. The threads toward the ends of the worm are of greater diameter, however, and as they do not revolve on axes tangent to the pitch diameter of the gear, but about the axis of the worm, contact is not simply on the pitch line, but on the entire surface of the engaging gear teeth. That is, the center teeth are in line contact, while the end teeth are subject to surface contact.

An article by John L. Wood, which appeared in AMERICAN MACHINIST, June 23, 1914, takes up in detail the method of calculating the strength of Hindley worm gears and the methods of calculating them. Excerpts of this discussion follow:

METHOD OF CALCULATING STRENGTH

"The following example is given to show a method of calculating the strength of this construction:

"In a carriage the theoretical load to be supported by means of two worms in the direction of their axes was 25,000 pounds.

"For the construction the following dimensions for the worm and rack were used:

WORM	DIMENSIONS
Smallest diameter at root of teeth	1.072 inches.
Corresponding pitch diameter	1.348 inches.
Corresponding diameter at top of thread	1.587 inches.
Obliquity of teeth	15 degrees.
Number of teeth in contact with rack	7
Pitch = 0.375 inch.	
Lead = 0.75 inch.	

RACK	DIMENSIONS
Thickness = length of tooth = face of tooth	0.95 inch.
Pitch diameter	12.652 inches.
Number of teeth in complete circle	106
Load for each worm and rack	12,500 pounds.

"Assuming that all of the seven teeth are effective and that each takes its proportionate load, we have,

$$\textit{Load for each tooth} = W = 1{,}785 \text{ pounds.}$$

"Using the Lewis formula for the rack and assuming, as stated before, that f = face of tooth = thickness of the rack at the root of the rack teeth, we get,

$$W = Spfy,$$

or

$$S = \frac{W}{pfy} = \frac{1{,}785}{0.375 \times 0.95 \times 0.118} = 42{,}450 \textit{ pounds per square inch.}$$

"The rack was made of steel with an elastic limit of approximately 53,000 pounds per square inch, and no trouble was experienced with this rack and no change was ever necessary.

"The worm was made of bronze with a tensile strength of 52,000 pounds per square inch. Since its smallest diameter was larger than the face of the rack, it might have been supposed that the teeth would have been sufficiently strong. This proved not to be the case, and the material was changed to steel with an elastic limit of 53,000 pounds per square inch, and no more difficulties were experienced.

"In many other places where this Hindley worm is used it has been found that the rack can be calculated by using the Lewis formula and that a safe rule for calculating the strength of the worm teeth is:

"Divide the total load to be supported by the number of teeth in contact. Assume this load to be applied at the top of the tooth. Consider the tooth a cantilever, of which the base is a line drawn tangent to the root of the tooth of the smallest diameter of the worm, the length of this base being the distance between the points of intersection of this line and the pitch line of the tooth, and the width being the thickness of the tooth at the root.

"Since this method of calculating the worm is not convenient, the following rules which have shown satisfactory results are given:

"Instead of considering as the base of the tooth the chord of the pitch diameter, which touches the root of the tooth, assume this base to be twice the chord subtended by half of this angle.

"Call this distance f. Then $f = 2\sqrt{Dm}$, in which D is the pitch diameter of the worm, and m is the depth of space below pitch line. Let p = circular pitch of worm and rack, then $m = 0.3683p$. And

$$f = 2\sqrt{D \times 0.3683p} = 1.2\sqrt{pD}.$$

"Now substitute for f in Lewis formula, and we have for the worm $W = 1.2Spy\sqrt{pD}$, and for the rack $W = Spfy$, in which S, p, y have values the same as for any other gears, while f is the width of the rack, and D the pitch diameter of the worm; y is the same for the worm as for the rack.

"It is evident that a gear of this kind must be made most accurate. This is especially so since the teeth in the rack are cut with a hob of as many teeth as there are in the worm itself. If any inequalities are found in the shape of the teeth in the hob, it will be seen that the rack teeth being cut to suit the largest section of the hob tooth and the worm being exactly like the hob, all the load might be taken on one tooth only.

"The following method is the manner of manufacturing these gears at the Rock Island Arsenal:

OPERATIONS FOR MAKING WORM

"First: Cut off stock ⅛ inch longer that the dimension required for the length of the worm and then center the part. Second: Between centers of engine lathe rough-turn all diameters 1/8 inch large and all length dimensions 1/16 inch long. Note: There should be 1 inch of stock left on the length of threaded part to be cut off after thread has been cut. This is to insure against any error created by the spring of thread tool on entering and leaving the cut and to make sure that the thread tool is cutting on both sides of the thread when at the proper length of worm. This rule is important and if not followed the worm will have an error in the lead both at beginning and ending of thread. Third: Between the centers of the lathe, with special fixtures fastened to the carriage, finish the radius. Use the micrometer with

FIG. 145. SPECIAL FIXTURE CUTTING WORMS OR HOBS.

double ball points to measure the diameter at the center of the radii or smallest diameter. Fourth: Strike a fine line around the piece at the smallest diameter. Note: Care must be taken to get this line accurately located, as all horizontal measurements are to be taken from this line. A special pointed tool is used for this operation. Fifth: Rough out the thread with special roughing tool. Sixth: Finish thread with the special thread tool. Seventh: Face off the ends of the threaded parts to a proper distance from the center line and finish-turn both bearings complete. Fig. 145 shows the lathe set up with special fixture-cutting worms or hobs.

MAKING THE HOB FOR CUTTING HINDLEY WORM-WHEELS

"First Operation: Between centers of engine lathe rough out the blank of the hob, leaving the stock at each end of the part to be threaded equal to not less than one-half the pitch. This is important as the end teeth of the hob must be equidistant from the center line and have full cutting surfaces.

Second: Between centers in the lathe, with the special fixture, similar to that shown in Figs. 1 and 2, fastened on the carriage, cut the radius to finish, using the double half-point micrometer for measuring the diameter at the smallest diameter. Third: Strike a line around the piece at the smallest diameter. Note: Care must be taken to get this line accurately located as all horizontal measurements are to be taken from it. Also, it is used in cutting the worm-wheel in proper relation to the hob when a located tooth is required. Fourth: Rough-out thread with special tool. Fifth: Finish thread with special thread tool. Note: The hob should be larger in diameter than the worm by one-tenth of the thickness of the tooth at the pitch line. Use the gage or template for depth and width of thread, also, micrometer with ball point, sleeve. Sixth: Strike a line parallel with the axis, crossing the line described in operation No. 3, in the exact center of the top of the thread. This line is essential when a located tooth in the worm-wheel is required. Seventh: Mill the flutes deep enough to establish a cutting edge at the bottom of the thread, as the hob must finish the face of the worm-wheel, also, care must be taken to have the cross lines come in the center of the tooth. Mill the flutes square with the helical angle of the thread. Eighth: Cut the hob to same length as the worm and remove any teeth on each end of hob which would be liable to break off while the hob is cutting. Care must be taken to have the same number of teeth on each end of the hob, counting from the one with the center line or cross-line. Ninth: Face the ends of the threaded part, measuring from the center line to get the faces an equal distance from the center of the radii. Tenth: Back off teeth. Note: Care must be taken in this operation to get an equal amount of clearance on each side of the tooth. Eleventh: Turn the bearing surfaces at each end of the threaded part. Allow about 0.015 inch for grinding after hardening the hob. Twelfth: Harden the hob. Note: Temper this hob only at the point of the teeth. Care must be taken that the bottom of the teeth is hardened as this part of the hob must form the diameter of the wheel. Thirteenth: After tempering, set the hob in the centers of the lathe and get the teeth running true. Then recenter each end and turn to finish.

MAKING THE WORM-WHEEL

"First Operation: Turn up the worm-wheel, leaving the outside diameter about 0.02 inch large, to be finished with the hob. Second: Mill the teeth in the wheel. Note: The exact centers of the hob must be in the center plane of the worm-wheel when teeth are required to be in a fixed relation to some other part of the wheel or segment. Turn the hob around in the machine until the center of the tooth which has the center line is exactly on top, and then set the work in the machine, locating from the center of the cross-lines. To get the proper depth of the tooth, measure with a double ball-point micrometer from the center of the bore to the center of the radii on top of the tooth. If the hob is accurately made and located, this measurement will be found reliable. Note: The worm-wheel must be driven at the proper lead by an independent set of gears.

"The illustration, Fig. 146, shows a machine in operation with the gear

FIG. 146. MACHINE IN OPERATION.

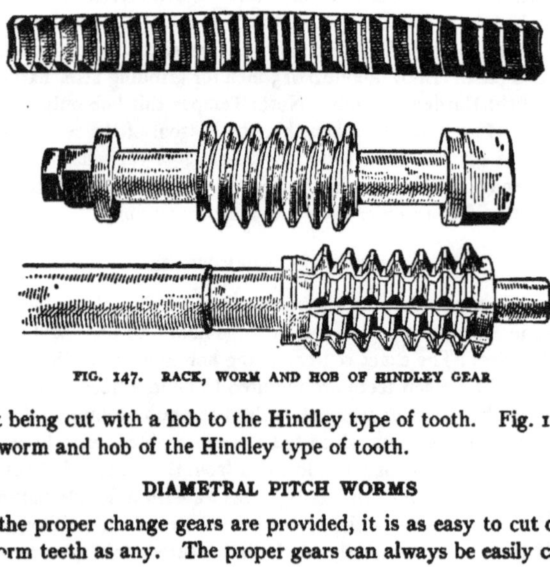

FIG. 147. RACK, WORM AND HOB OF HINDLEY GEAR

segment being cut with a hob to the Hindley type of tooth. Fig. 147 shows a rack, worm and hob of the Hindley type of tooth.

DIAMETRAL PITCH WORMS

"If the proper change gears are provided, it is as easy to cut diametral pitch worm teeth as any. The proper gears can always be easily calculated by the rule that the screw gear is to the stud gear as 22 times the pitch of the lead screw of the lathe is to seven times the diametral pitch of the worm to be cut. For example, it is required to cut a worm of 12 diametral pitch, on a lathe having a leading screw cut six to the inch. We have:

$$\frac{\text{Screw gear}}{\text{Stud gear}} = \frac{22 \times 6}{7 \times 12} = \frac{11}{7};$$

and any change gears in the proportion of 11 and 7 will answer the purpose with an error of 1/10,000 of an inch to the thread of the worm. If 22 and 7 give inconvenient numbers of teeth, the numbers 69 and 22 can be used with sufficient accuracy, and 47 and 15, or even 25 and 8, may do in some cases."*

Care should be taken when using these calculations that the same change gears are used to chase both the hob and the worm, as a slight difference in the lead of one tooth may prove a serious matter in a worm engaging a large gear where several teeth will be in contact. This same precaution applies to worms and hobs of a fractional circular pitch.

It should also be remembered that 4 pitch does not mean 4 threads per inch measured on the axis of worm, but 4 threads per inch of diameter of the engaging worm gear. The corresponding circular pitch is 0.7854 inch, not 0.25 inch.

* George B. Grant's Treatise on Gearing, Section 120.

SECTION VII

Helical and Herringbone Gears*

The exacting demands of smoothness and silence in operation, long life and high efficiency for high-speed gear transmission, such as those imposed upon the reduction gears for steam turbines, are simply met by helical or herringbone gears only. Such gears are superior to the common spur gear in all the requirements made by this trying service. The obliquity of the teeth keeps one set in mesh until the following set of teeth is well engaged so that at no time is there a sudden transference of load from one tooth to the next, as occurs in ordinary spur gearing. The load is gradually put on a tooth and as gradually taken off so that the strain on the teeth is kept practically constant and the sudden shock of impact, common to spur gearing, is avoided.

In the ordinary spur gear the teeth come in contact over their entire length at one time and the whole load is first thrown on the end of the tooth, producing the maximum leverage strain as soon as contact takes place. This leverage strain is subsequently reduced, but it takes place suddenly and, therefore, is much more serious than if led up to gradually. In gears with oblique teeth, the load is put on each tooth gradually and as gradually removed so that no severe leverage strain is created at any time.

Helical gears, examples of which are diagrammatically depicted in Figs. 148–153, possess the drawback of exerting a more or less serious axial or side thrust on account of the action of the teeth. This thrust varies with the angle of the spiral, so that the highest efficiency is obtained when the angle is only great enough to assure the accurate and gradual meshing of a set of teeth during the equally gradual releasing of the preceding set of teeth. The angle of the spiral need be such only that the end of one tooth will just overlap the end of the adjoining tooth. It follows, therefore, that the wider the face of a helical gear, the less the angle of spiral need be and the less the side thrust produced.

When employing helical gears it is always desirable to use two gears of opposite pitch or spiral in the same shaft, so that the axial thrust of the two gears will balance. Ordinarily this is accomplished by the use of duplicate sets of gears similarly mounted on common shafts. Where space is limited, gears of opposite spiral may be mounted side by side to form virtually one gear, such gear being actually a herringbone gear. The exact balancing of the helical gear thrust may not always be feasible, nor may it be advisable to employ herringbone gears, and in such instances partial balance of thrust

* See Section XVI for rolled Helical and Herringbone Gears.

may often be realized. Fig. 150 illustrates such a case; the thrust on shaft *B* is balanced, and that on shafts *A* and *C* is reduced to a minimum.

Two circular pitches are employed for helical gearing: The "normal circular pitch," which is the shortest distance between the center of con-

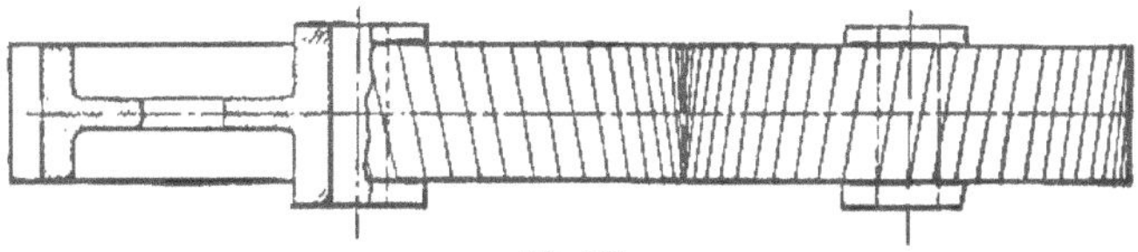

Fig. 148

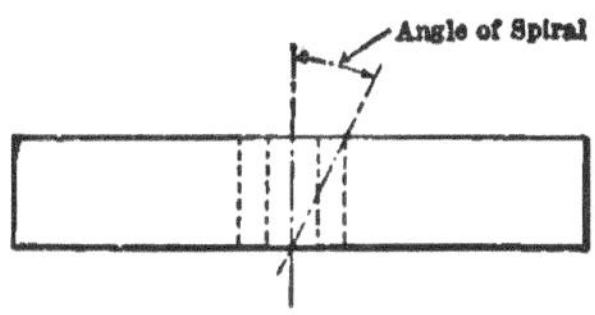

Fig. 149

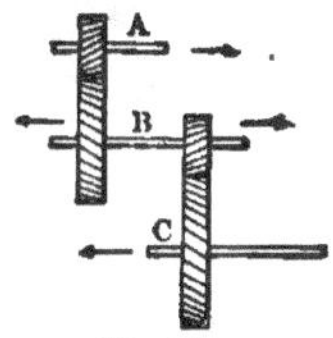

Fig. 150

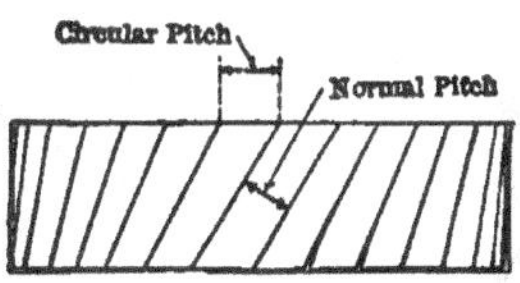

Fig. 151

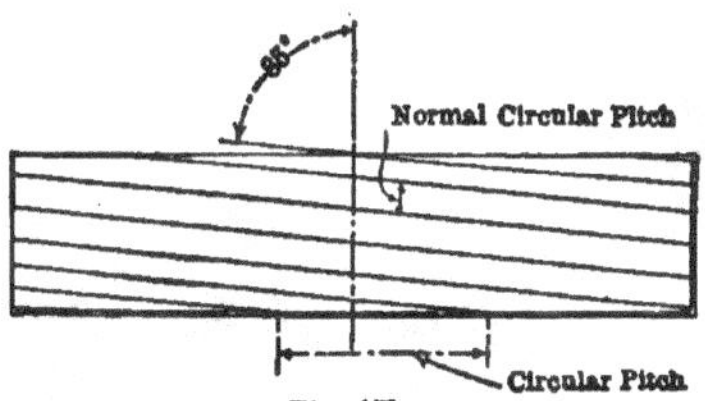

Fig. 152

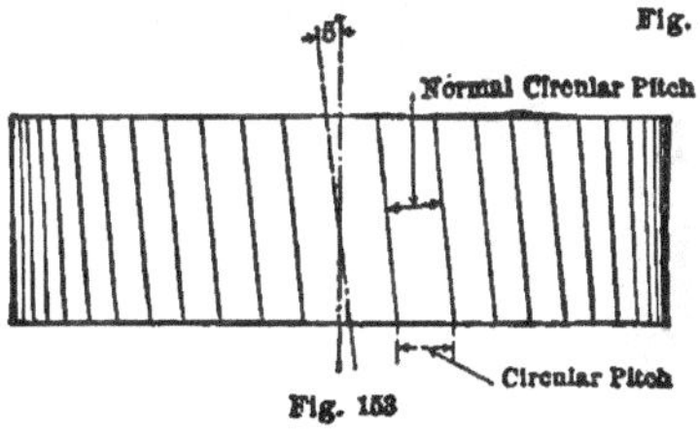

Fig. 153

THE DESIGN OF HELICAL GEARS.

secutive teeth and is measured on an imaginary pitch cylinder, and the "circular pitch," which is the distance between the center of two teeth in the plane of the gear and which is measured on the pitch circle as for spur gears. See Figs. 151, 152 and 153.

The pitch diameters must be in proportion to the number of teeth and both the circular and normal pitches must be the same in both gears of a pair. The angle of spiral must also be the same but of opposite hand.

The form of tooth employed for helical and herringbone gears is the involute or a close approximation of that form.

NOTATION FOR HELICAL AND HERRINGBONE GEARS

D' = pitch diameter.
D = outside diameter.
N = number of teeth.
p = diametral pitch.
p' = circular pitch.
p'^n = normal circular pitch.
E = angle of spiral.
s = addendum.
f = clearance.
W' = whole depth of tooth.
t = thickness of tooth.
L = center distance } pair of gears.
$\frac{V}{v}$ = speed ratio } pair of gears.

FORMULAS FOR HELICAL AND HERRINGBONE GEARS

$$D' = \frac{p'N}{3.1416}, \qquad (1)$$

$$= 2\left(\frac{B}{\frac{V}{v}+1} \times \frac{V}{v}\right) \text{ for gear,} \qquad (1\text{-}a)$$

$$= 2\left(\frac{B}{\frac{V}{v}+1} \times 1\right) \text{ for pinion.} \qquad (1\text{-}b)$$

$$N = \frac{3.1416D'}{p'}. \qquad (2)$$

$$D = D' + 0.6366p'^n. \qquad (3)$$

$$Cos.\,E = \frac{p'n}{p'}. \qquad (4)$$

$$p'^n = p'\,cos.\,E = \frac{3.1416}{p}. \qquad (5)$$

$$p' = \frac{p'^n}{cos.\,E}, \qquad (6)$$

$$= \frac{3.1416D'}{N}. \qquad (6\text{-}a)$$

$$p = \frac{3.1416}{p'^n}. \qquad (7)$$

$$s = 0.3183p'^n. \qquad (8)$$

$$W' = 0.6866p'^n. \qquad (9)$$

$$t = \frac{p'^n}{2}. \qquad (10)$$

$$L = 3.1416D'\,cot.\,E. \qquad (11)$$

The corresponding spur cutter $= \frac{N}{(cos.\,E)^3}$ (12)

DISCUSSION OF FORMULAS

Only the pitch diameter and the number of teeth are figured from the circular pitch.

The outside diameter and all tooth parts are figured from the normal pitch, the relationship existing to the normal pitch being similar to the relationships of spur gears to their circular pitches.

The angle of the spiral, in practice, is usually selected with a view to the spur cutter available. The normal pitch divided by the circular pitch gives the cosine of the angle of the spiral. This angle should be kept as low as possible, ordinarily not exceeding 20 degrees, in order to avoid undue axial thrust. When the angle is fixed as well as the pitch of the cutter, the diameter necessary to give the proper combination may be found by first determining the circular pitch. A change in the angle to accommodate an even number of teeth makes no difference in helical gears.

The angle of the spiral must be accurately adhered to, for a slight deviation will interfere with the proper contact of the teeth.

The normal pitch should conform as nearly as possible with some standard pitch, so it is customary to assume the normal pitch and proportion the angle of spiral accordingly. If a variation between the pitch of the cutter and the normal pitch cannot be avoided, it is better to have the cutter pitch finer rather than coarser than the normal pitch.

The circular pitch, found by dividing the normal pitch by the cosine of angle of spiral, must always correspond to an even number of teeth—*i.e.*, the product of the circular pitch and the number of teeth must represent the pitch circumference of the gear in question.

After settling the pitch of the cutter, the number of teeth for which the cutter should be made is found by dividing the actual number of teeth in the gear by the third power of the cosine of the spiral angle.

The spiral lead is the distance traveled by the thread in one complete revolution of the pitch circle. As the angle of spiral becomes smaller and the form of the tooth approaches that of a spur gear, the lead becomes longer until, when the spiral angle is zero, it lengthens to infinity.

EXAMPLES IN THE DESIGN OF HELICAL GEARS

1. REQUIRED:

Pair of helical gears cut with a 4-pitch cutter, speed ratio 4 to 1 and center distance 12½ inches.

4-pitch cutter = 0.7854 inch circular pitch (Table I),
= normal pitch for helical gears.

SOLUTION:

DIMENSION	FORMULA	GEAR	PINION
Pitch diameter	(1–a and b)	20.000″	5.000″
Number of teeth	(2)	76	19
Normal pitch	Table I	0.7854″	0.7854″
Circular pitch	(6–a)	0.8267″	0.8267″
Angle	Cosine from (4)	18° 11′	18° 11′
Addendum	(8)	0.250″	0.250″
Whole depth of tooth	(9)	0.539″	0.539″
Thickness of tooth	(10)	0.392″	0.392″
Outside diameter	(3)	20.500″	5.500″
Lead	(11)	191.292″	47.823″
Cutter used		No. 2–4p.	No. 5½–4p.
Hand		R. H.	L. H.

2. REQUIRED:

Pair of helical gears to replace two spur gears of 60 and 15 teeth respectively, 3 pitch, 12½-inch centers. Speed ratio to remain the same and the helical gears to be cut with the same pitch cutter but to have fewer teeth in order not to change the center distance.

3-pitch cutter = 1.0472 circular pitch (Table I),
= normal pitch for helical gears.

SOLUTION:

DIMENSION	FORMULA	GEAR	PINION
Pitch diameter	(1–a and b)	20.000″	5.000″
Number of teeth	Assumed	56	14
Normal pitch	Table I	1.0472″	1.0472″
Circular pitch	(6–a)	1.1220″	1.1220″
Angle	Cosine from (4)	21° 3′	21° 3′
Addendum	(8)	0.3333″	0.3333″
Whole depth of tooth	(9)	0.9075″	0.9075″
Thickness of tooth	(10)	0.5236″	0.5236′
Outside diameter	(3)	20.666″	5.666‴
Lead	(11)	163.456″	40.864″
Cutter used		No. 2–3p.	No. 7–3p.
Hand		R. H.	L. H.

When space permits or when helical gears can be arranged in pairs so as to overcome the axial thrust that constitutes the chief drawback of these highly efficient gears, they are usually to be preferred to herringbone gears. They are easier to machine and possess the always desirable feature of greater simplicity.

A well-balanced helical gear drive is shown in Fig. 154, which illustrates the arrangement of gears on a heavy rack cutter. Before the installation of these gears, ordinary spur gears had been employed with discouraging results. The spurs had broken, worn out rapidly and had proved entirely inadequate for the service demanded. The helical gears, on the other hand, have proved eminently satisfactory. After several years of operation they show no appreciable wear, run smoothly and with little noise, and permit the rack cutter to be driven at far higher speed than was formerly possible.

Helical speed-reduction gears have also proved very successful and some remarkably high efficiencies are reported for such apparatus. The De

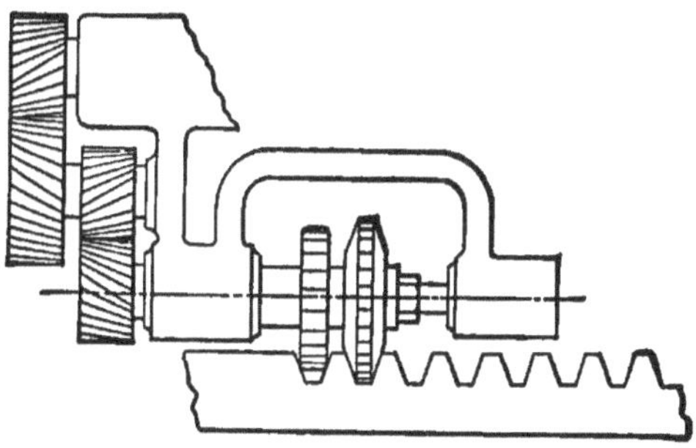

FIG. 154. HELICAL GEARS FOR CUTTER DRIVE.

Laval Steam Turbine Co., which has adopted such mechanism in connection with its steam turbines, claims efficiencies as high as 99 per cent. and states that an efficiency of 98½ per cent. is a conservative figure. This enviable record being only possible through excellent design, the use of proper materials and high-grade workmanship, a brief description of this gear will be of interest.

DE LAVAL SPEED REDUCTION GEAR

Two helical gears with their pinions are used to overcome the axial thrust and to permit a somewhat greater spiral angle to be employed than is customary. The angle of spiral is such that several teeth are in contact at the same time. The form of tooth used is the involute to secure true line contact as long as the tooth is in mesh.

The gears consist of a rigid cast-iron center or spider upon which steel bands of a special grade of steel are shrunk. The gear blanks are carefully mounted and trued up before the teeth are cut in order to insure accuracy.

The pinion is cut directly on the pinion shaft, which is a special nickel-steel forging that is oil-tempered to the desired degree—the pinion having to be considerably harder than the gear bands to assure uniform wear and long life.

After the gear and pinion have been cut, they are carefully polished to remove any tool marks and unevenness. This is accomplished by running the gear and pinion with similar and as accurately cut "dummy" gears. This polishing process naturally adds to the cost of manufacture, as the "dummy" gears rapidly deteriorate from wear and have to be discarded. The added cost of polishing is well warranted, however, as it greatly increases the life and efficiency of the finished gear.

The bearings, lubrication and details of the gear case for such speed reducers are not dissimilar to those of any other high-grade speed-reduction gear.

Special flexible couplings for connecting the pinion shaft to the steam turbine and the gear shaft to the machine to be driven are necessary, of course, for any mis-alignment would be suicidal to the high efficiency demanded.

Speed reduction gears built with this care have been opened up after more than 10 years of constant and exacting service and have shown no wear.

HERRINGBONE GEARS

A double-helical or herringbone gear avoids the axial thrust of the single-helical gear which is not properly balanced, consisting as it does of virtually

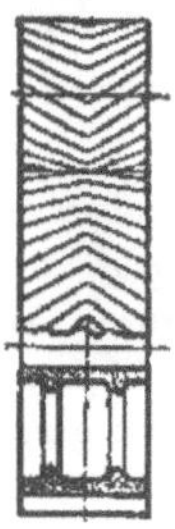

FIG. 155.

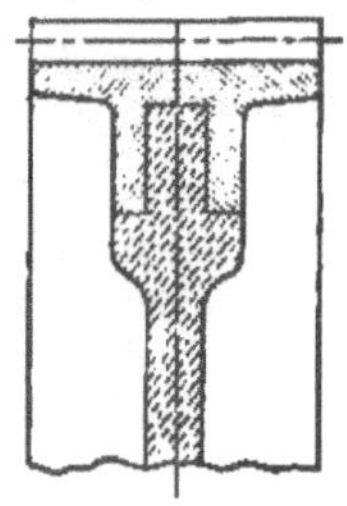

FIG. 156.

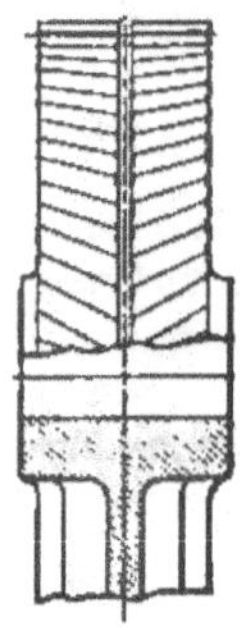

FIG. 157.

THE DESIGN OF HERRINGBONE GEARS.

two gears exerting axial thrusts in opposite directions and thereby nullifying any unbalanced side pressure.

The formulas for herringbone gears are the same as for helical gears, but the angle of spiral can be made considerably more obtuse. In fact, the angle of the teeth with the axis of the gear is only limited by constructional difficulties. Angles of 30 degrees are quite common and angles as great as 45 degrees are quite frequently used.

Fig. 155 shows a common type of herringbone gear, and Figs. 156 and 157, modifications that are frequently resorted to to facilitate manufacture. The gear shown in Fig. 155 really consists of two helical gears fastened together. The gears are cut separately and not connected until after machining. This simplifies the cutting of the teeth but also adds to the cost of manufacture. The type shown in Fig. 156 differs only in that the rim is made in two pieces and separate from the spider. The same ease in cutting the teeth is thus secured as in the other type and the added advantage of simplifying replacements should the gear teeth be damaged or wear out.

The grooving of the type shown in Fig. 157 allows the gear to be made in one piece, as the cutter can be run out at the groove. In this type of construction, the teeth may be staggered to somewhat lessen the width of the central groove.

The great difficulty of accurately cutting herringbone gears so that they may be interchangeable has led to the adoption of many makeshifts. When the gear is made in one piece, the pinion is sometimes made in two pieces to

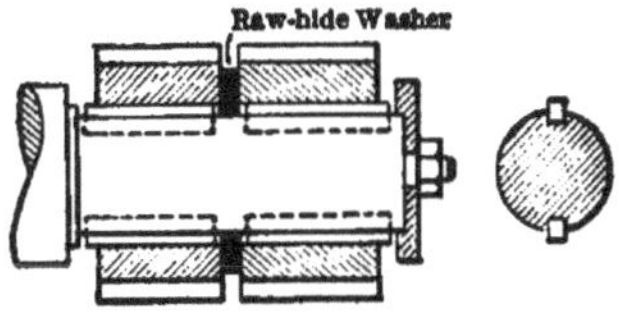

FIG. 158. ARRANGEMENT OF PINION.

facilitate the proper engagement of the teeth. The pinion is run with the gear and not keyseated until after the teeth have accurately adjusted themselves to the position of the teeth in the gear.

Another plan that was suggested by "Attic" in the AMERICAN MACHINIST consists of placing a washer of some elastic material between the two halves of the pinion, as shown in Fig. 158, so that they can adjust themselves by a slight axial movement, a movement along the shaft having the same result as if the half gears were slightly turned about the shaft.

INTERCHANGEABLE SYSTEM FOR HERRINGBONE GEARS

Percy C. Day presented an interchangeable standard for herringbone gears before the American Society of Mechanical Engineers.

The proposed standard, which has been adopted by at least two important manufacturers of herringbone gears, is as follows:

Tooth shape	Involute.
Pressure angle	20 degrees.
Spiral angle	23 degrees.

$$\text{Pitch diameter (20 teeth and over)} = \frac{\text{number of teeth}}{\text{diametral pitch}}.$$

$$\text{Blank diameter (20 teeth and over)} = \frac{\text{number of teeth} + 1.6}{\text{diametral pitch}}.$$

$$\text{Pitch diameter (under 20 teeth)} = \frac{0.95 \times \text{number of teeth} + 1}{\text{diametral pitch}}.$$

$$\text{Blank diameter (under 20 teeth)} = \frac{0.95 \times \text{number of teeth} + 2.6}{\text{diametral pitch}}.$$

Addendum.. $\frac{0.8}{\text{D.P.}}$.

Dedendum.. $\frac{1.0}{\text{D.P.}}$.

Full tooth depth.................................. $\frac{1.8}{\text{D.P.}}$.

Working tooth depth............................... $\frac{1.6}{\text{D.P.}}$.

Standard face width for gears with pinions of not less than 25 teeth, six times the circular pitch.

Face widths for high-ratio gears with small pinions, six to twelve times the circular pitch.

When a pinion of less than 20 teeth is used with a standard gear, the center distance must be slightly increased to suit the enlargement of the pinion. If it is desired to keep the center distance to the standard dimensions, the gear diameter may be reduced by the amount of the enlargement given to the pinion.

For example: If a pinion of 10 teeth, 5 diametral pitch (D.P.) is to mesh with a gear of 90 teeth at 10-inch centers:

$$\text{Pitch diameter of pinion } \frac{(0.95 \times 10) + 1}{5} = 2.1 \text{ inches.}$$

Enlargement over standard pinion = 0.1 inch.

$$\text{Pitch diameter of standard gear} = \frac{90}{5} = 18 \text{ inches.}$$

Reduced pitch diameter of gear = 18 − 0.1 = 17.9 inches.

$$\text{Center distance } \frac{17.9 + 2.1}{2} = 10 \text{ inches.}$$

Strictly speaking, there can be no enlargement for reduction of the pitch diameter in a pinion or gear of given pitch and number of teeth. It is convenient to assume such enlargement or reduction, however, when using teeth of long and short addenda but standard depth.

MACHINING HERRINGBONE GEARS

Three general methods of cutting herringbone gear teeth are in vogue: (1) A double-hobbing machine of special design is employed by which the right- and left-hand teeth are cut simultaneously. In such machines the cutting profile of the hob teeth, cutting their way into the gear blank on a diagonal line conforming to the obliquity of the teeth, are modified from the true involute form in order that the space cut may closely approximate the correct involute profile in the direction of rotation. For extreme accuracy, therefore, a uniform obliquity of tooth is necessary, or else a special hob for each angularity of tooth. (2) The method second in importance is that of planing the teeth. Planers are employed which use a cutting tool having the same section and form as the finished tooth space, or machines with templets for guiding simple planing tools are used. (3) Herringbone gears of the smaller sizes are also cut with single rotary cutters.

HOBBING PROCESS FOR CUTTING HERRINGBONE GEARS

Machines of two general types are employed in the hobbing process. The working principle of the type shown diagrammatically in Fig. 159–*a* is as follows:

The gear blank *a* is mounted on the vertical mandrel *c*, which is rotated by the face plate *b* through the spindle *d* actuated by the driving worm gear *e*.

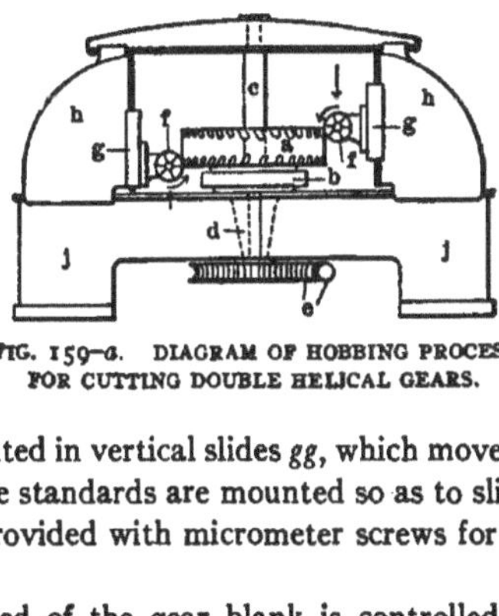

FIG. 159–*a*. DIAGRAM OF HOBBING PROCESS FOR CUTTING DOUBLE HELICAL GEARS.

The hobs *ff* are mounted in vertical slides *gg*, which move up and down on the standards *hh*. These standards are mounted so as to slide on the bed of the machine *j* and are provided with micrometer screws for adjusting the depth of cut.

The rotating speed of the gear blank is controlled by a train of wheels which operate a differential gear so as to give the required spiral lead to the teeth. The process is entirely automatic and evades the necessity of inclining the hob axes, the lead being governed by the speed at which the gear blank is revolved. This allows the hobs to be set at right angles to the axis of the gear blank and the same hobs to be used for gears of different obliquity of tooth, provided extreme accuracy is not essential.

Fig. 159–*b* shows the other type of hobbing machine for cutting herringbone gear teeth. This single-headed machine has two hobs, a right- and a left-hand one, carried on a single saddle. The hobs rotate in opposite directions and are fed downward, cutting both halves of the gear at the same time. The cutting pressures are opposed and neutralized through the

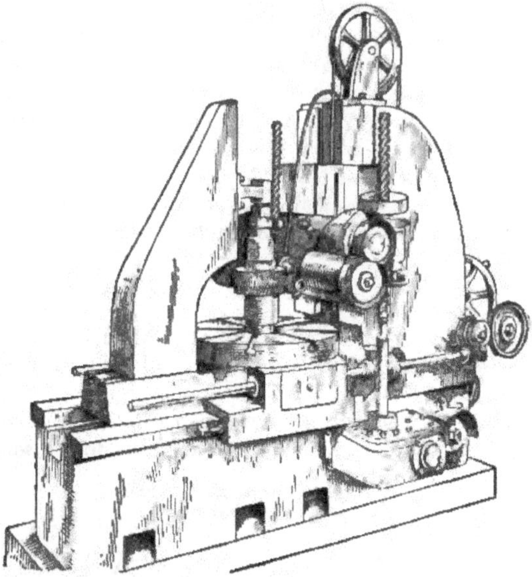

FIG. 159-b. HOBBING MACHINE FOR CUTTING DOUBLE HELICAL GEARS.

FIG. 159-c. MACHINE FOR PLANING DOUBLE HELICAL GEARS.

section of the gear being cut, thus relieving the machine of any such unbalanced stresses.

The lower hob is carried on a spindle, while the upper hob is carried on a slide which is vertically adjustable. This allows the hobs to be set for cutting gears of various widths, and the adjustable features of the hobs permit them to be set so as to cut teeth with apexes on the center line of the gear or teeth of the staggered variety.

The desired depth of cut is fixed by horizontal adjustment of the work carriage, after which the cutting operations require little attention, the retardation and acceleration of the hobs, the rotation of the work table or carriage, etc., being entirely automatic.

PLANING PROCESSES FOR CUTTING HERRINGBONE GEARS

Two general schemes of planing herringbone gear teeth are employed: First, that in which a tool of the exact section and form of the space to be cut is used; and second, the one in which templets are employed to guide the simple cutting tools employed—the "former" method. The first method is only employed for cutting comparatively small gears with circular pitches usually under 1 inch, the second for machining larger gears of coarser pitch.

A large herringbone gear planer in which the tooth profiles are shaped by formers is shown in Fig. 159–*c*. The two tool saddles move toward each other on the cutting stroke, both halves of the gear being cut at the same time by tools advancing from the outer edges of the gear to the center apex line of the tooth. The cutting pressures of the tools are thus resisted by the gear itself and neutralized, instead of being taken up by some part of the machine. The main shaft carries two worm-wheels, the larger for indexing and the smaller for rotating the gear blank. Two sets of change gears, one for the angle of the tooth and the other for the number of teeth, together with a reversing drive constitute the special features of this particular planer.

MILLING PROCESS FOR CUTTING HERRINGBONE GEARS

The following description of the milling process for cutting herringbone gear teeth is taken from an article in AMERICAN MACHINIST by Percy C. Day, as is also the test under the three following sub-titles.

"In the milling process the teeth are sometimes cut by means of end mills formed to the tooth shape on the normal section. The working principle of the machines usually employed is shown in the diagram, Fig. 160. The end mill *a* is supported by the saddle *b*, which traverses the bed *c*. The mill is driven by the bevel gears *d* from the splined shaft *e* and driving cone *f*. The feed and differential motions are driven from *e* through speed cones or gears *g* and clutch *h*. The traverse of the saddle *b* is actuated by the feed screw *j*. Motion is also transmitted from *j* through change wheels *k*, reversing gears *l*, dividing change wheels *m*, worm *o*, dividing wheel *w*, and work spindle *p* to the blank *q*.

"While the end mill traverses from one edge of the blank to its center, the blank is rotated through an angle which gives the requisite spiral form to the tooth. The saddle then operates a stop which is in connection with the reversing gear *l*, and the rotation of the blank is reversed until the end of the cut. A quick-return mechanism, not shown in the diagram, comes into action at the end of the cut, and the mill is returned to the starting position. The dividing mechanism *m* is then operated by hand, and the cutting process is repeated on another tooth.

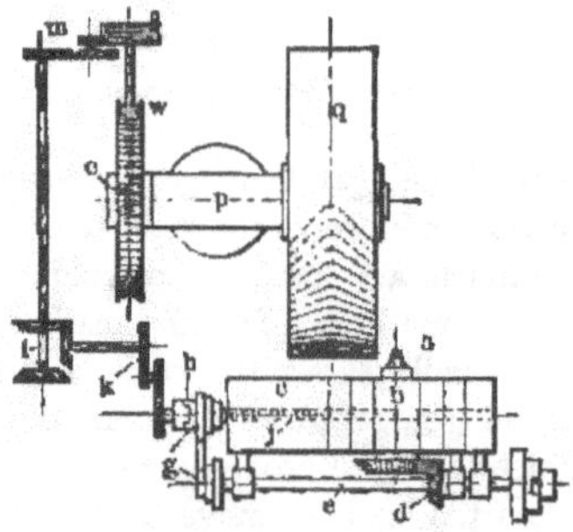

FIG. 160. DIAGRAM OF MILLING PROCESS OF CUTTING DOUBLE HELICAL GEARS.

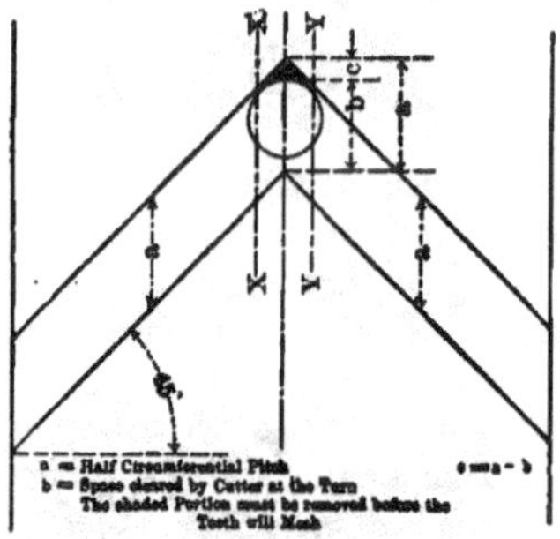

FIG. 161. DIAGRAM OF SPACE NOT CUT BY MILLING PROCESS OF CUTTING DOUBLE HELICAL GEARS.

"The end-milling process can be readily adapted for cutting double helical bevel gears.

"The disadvantages of the process are principally of a practical nature. End mills are small tools, and are liable to rapid wear. Since the teeth are cut singly, any wear on the mill causes a change of tooth shape and thickness. The reversal of the angular motion of the blank while cutting proceeds allows the inevitable backlash in the mechanism to take effect in a manner which is not conducive to accurate work. The cutter must be formed to the normal tooth section, and has not the circumferential shape of the teeth which it cuts. The width of the tooth space at the apex corresponds to the normal instead of to the circumferential pitch, hence the space must be cleared out by hand or in a separate operation (see Fig. 161).

"The tendency to wear is greater when the end mills are small, and wheels on this system are generally made of coarser pitch than is really necessary or even desirable from the user's point of view, in order to minimize the manufacturing difficulties by the use of large mills.

ADVANTAGES OF THE DOUBLE HELICAL SYSTEM

"The adoption of the double helical principle in gearing, if properly applied, reduces noise to a minimum and practically eliminates vibration without any necessity for departure from sound mechanical principles. In this type of gear, pinions may be chosen of sufficient hardness to wear evenly with the wheels, and soft materials do not enter the proposition. This is due to the absolute continuity of engagement which is characteristic of double helical gears when accurately cut and correctly designed to suit the working conditions. The effect of vibration is not by any means confined to the gears themselves, but acts injuriously on the shafts and machinery connected therewith. Many failures of haulage and other gear-driven shafts have been directly traced to this cause.

"Consider a pair of wheels transmitting 100 horse-power with an efficiency of 96 per cent. If we assume only one-tenth of the lost energy to be dissipated in vibration which is absorbed in the wheel shaft, the result is somewhat surprising. Under such conditions the shaft is called upon to absorb energy at the rate of nearly eight million foot pounds during each working day of 10 hours' duration. The result is finally expressed in crystallization of the shaft material. . . .

"Another interesting application of machine-cut, double helical gears is the reduction of speed from high-power steam turbines. No other type of gear can be used for this class of work, because absolute smoothness of action is essential. The essence of this problem is to avoid excessive velocity by keeping the pinion diameter small, but at the same time it is undesirable to reduce the number of teeth below a certain point because absolute continuity of engagement must be maintained. The result of these conditions is that the gears must be of extremely fine pitch and great relative width. For example, a set of gears recently constructed for a 500-horse-power steam turbine, to reduce from 3,000 to 300 revolutions per minute, were 4 diametral pitch with face width 10 inches and pinion of 19 teeth.

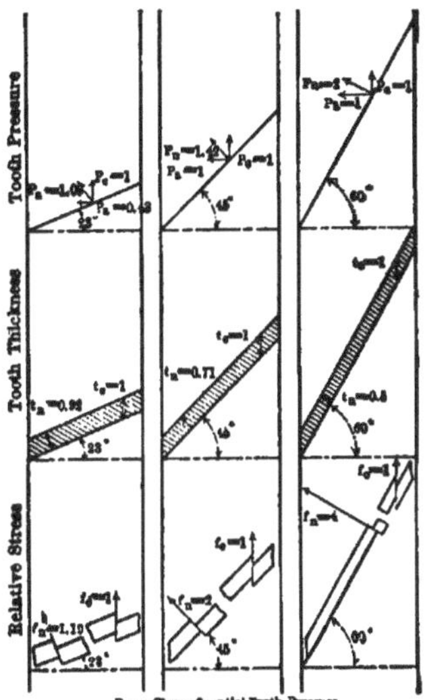

P_c = Circumferential Tooth Pressure
P_n = Normal Pressure
P_a = Axial Pressure
t_c = Circumferential Tooth Thickness
t_n = Normal Thickness
f_c = Circumferential Stress
f_n = Normal Stress

FIG. 162. COMPARISON OF TOOTH PRESSURE, THICKNESS, AND RELATIVE STRESS FOR THE TOOTH ANGLES OF DOUBLE HELICAL GEARS.

"There is probably no field of application for double helical gears which offers such substantial advantages as for driving machine tools. In most modern machine shops there is a tendency to dispense with shafting as far as possible, and to drive the tools individually from separate motors. This method allows a more economical distribution of machines over the available floor space, and leaves the space overhead clear for rapid handling. On the other hand, motor-driven tools require far more gearing than when the drive is effected by belts, and it has been found difficult to obtain uniformity of motion under the new conditions. If machine-cut, double helical gears are used for this purpose, the quality of the work turned out is much improved and, by reason of reduced vibration, higher speeds and coarser feeds can be employed.

"The diagram, Fig. 162, shows the relative normal tooth pressures, pitch line sections and stresses for angles of 23, 45, and 60 degrees.

"One of the greatest advantages of machine-cut, double helical wheels is to be found in their adaptability for high ratios of reduction. The number of teeth which can be used with success in the smallest pinion lies far below the practical limit for straight spurs, and pinions of four or five teeth are by no means uncommon for special purposes. Since, however, the pitch can be made very fine, it is rarely necessary to reduce the number of teeth so far, and most high-ratio gears are made with pinions of 11 to 20 teeth. Pinions for high ratios are generally cut solid on their shafts, in order that the diameters may be kept low to bring the wheels within reasonable proportions.

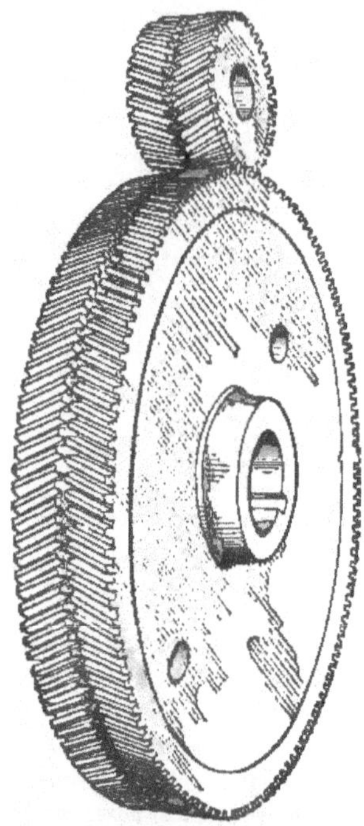

FIG. 163. TYPICAL HIGH-RATIO GEARS WITH STAGGERED TEETH.

"Single wheels and pinions will transmit heavy powers with ratios between 10 and 20 to 1, so that they can be used in place of worm gears or double trains of ordinary spurs. As against worm gears the gain lies in the direction of increased efficiency and life. A set of double helical gears with 20 to 1 ratio has an efficiency of about 95 per cent. against a maximum of about 80 per cent. for a worm gear of equal ratio.

AN INTERESTING APPLICATION OF DOUBLE HELICAL GEARING

"An interesting example of this difference came under my notice a short time ago. A worm gear of first-class manufacture and modern design had been in use for some 2½ years, driving a deep-well pump from a 50-horse-power motor with reduction 480 to 22 revolutions per minute. This gear was replaced by a double train of machine-cut, double helical wheels, the ratios being 480 to 60, and 60 to 22. The records of power consumption and pump duty were regularly kept, and after the new gear had been running for a year the figures showed a net saving of over 17 per cent. in its favor as against the average for the whole life of the worm gear. It was also shown that the efficiency of the double helical gear had actually improved after a year's daily work, while the worm gear had steadily deteriorated in this respect from the day it was started.

"Fig. 163 shows a set of double helical gears that are representatives of their design and construction.

IMPORTANT POINTS IN APPLYING DOUBLE HELICAL GEARS

"In conclusion it is desirable to add a word of caution to those who are about to adopt this class of gear for the first time. It must not be forgotten that there are three fundamental points of difference between machine-cut, double helical wheels and ordinary spur gearing:

"(*a*) The pitch is finer.
"(*b*) The face width is greater.
"(*c*) The tooth pressures are generally higher.

"To insure satisfactory working it is necessary that the shafts shall be parallel, true, and rigidly supported. The center distance must also be adjusted with great care on account of the fine pitch and small clearances allowed. Motor pinions of high-ratio gears should be mounted on extended shafts with an outer bearing. Anything in the nature of an overhung drive should be avoided wherever possible.

"To avoid undue wear from magnetically controlled end-thrust in motors, the pinions should be mounted on two parallel feathers set at 180 degrees and carefully bedded to the keyways (see Fig. 164). The pinions should be a good tight fit on the motor shafts, but there should be just sufficient freedom to allow them to move along under the influence of continued side pressure, so that the motor armature can reach a neutral position where the pressure ceases. It is unnecessary to allow the pinions to slide freely on the shafts, and if this is done there may be trouble from excessive wear of the keys and keybeds."

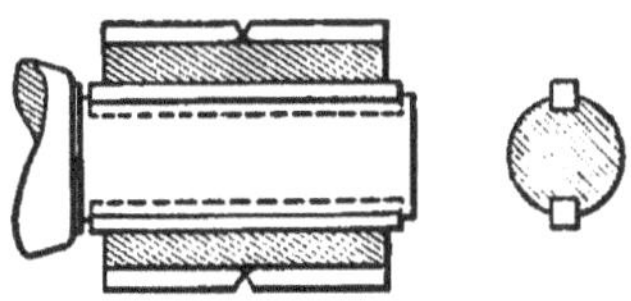

FIG. 164. AN IMPROVED METHOD OF KEYING FOR GEARS.

DETERMINING LEAD AND ANGLE FROM SAMPLE

To produce a herringbone gear to operate with a sample, the calculations for which are unknown, is generally a matter of cutting and trying until a satisfactory gear is produced, as for herringbone or helical gears the angle and lead must be exceptionally accurate, the teeth having contact their entire length, and a slight error is noticeable. There is more or less leeway for spiral gears, but the method as described can be applied to them as well.

Cover the points of the teeth in sample with an application of lampblack, or anything that will make a clear impression on a piece of clean white paper. Roll the gear thus treated on the surface of the paper, being careful not to allow it to slip, until a sharp impression of the points of the teeth is made, as illustrated in Fig. 165. This will represent a development of the teeth at the outside circumference.

The angle of the teeth at the outside circumference may then be measured with a protractor by extending the lines of the tooth as developed on the paper.

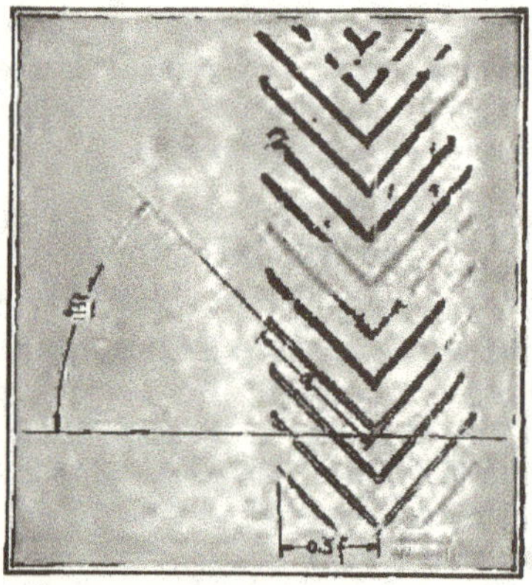

FIG. 165. IMPRESSION MADE BY ROLLING SAMPLE HERRINGBONE GEAR.

f = Face of herringbone gear.
α = Length of the tooth from center of face.
β_1 = Angle of spiral at outside diameter.
β = Angle of spiral at pitch diameter.
L = Lead of spiral.
C_1 = Outside circumference.
C = Pitch circumference.

$$\cos.\ \beta_1 = \frac{0.5f}{\alpha}.$$

For helical gears this formula would be:

$$\cos.\ \beta_1 = \frac{f}{\alpha}.$$

The next step is to find the lead:

$$L = \frac{C_1}{\tan.\ B_1}.$$

As the lead is necessarily the same at the outside diameter as it is at the pitch diameter of a helical or spiral gear when cut with a rotary cutter, the angle of spiral at the pitch line may be found by formula 4.

$$Tan.\ \beta = \frac{C}{L}.$$

The fact that the lead is the same at all points when cutting a spiral, helical, or herringbone gear cutter, using a single rotary cutter, makes the solution of this problem a simple matter. Fig. 166 is self-explanatory.

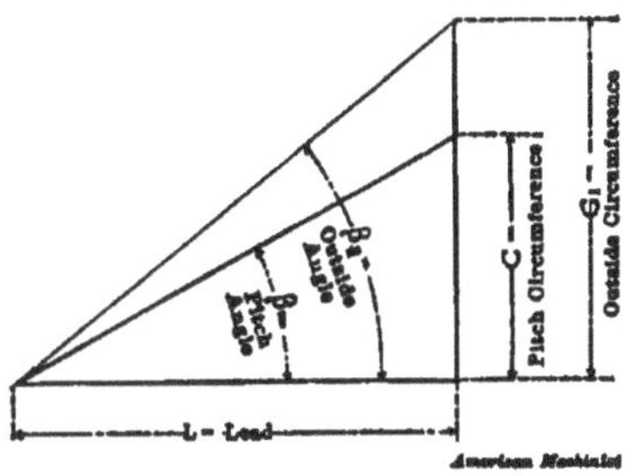

FIG. 166. DIAGRAM OF ANGLES OF HERRINGBONE GEAR.

This being the case, it is apparent that such a cutter cannot reproduce its own shape in the gear blank, as to do this the angle and lead must be proportional to all parts of the tooth. When the teeth are generated this condition is fulfilled and the angle at the pitch line will be proportional to the pitch and outside circumferences, or:

$$\beta = \frac{\beta_1 C}{C_1}.$$

EFFICIENCY AND STRENGTH OF HERRINGBONE GEARS

The efficiency of accurately cut herringbone gears for ratios up to 10 to 1 is about 98 per cent. For greater speed ratios, their efficiency shows a slight falling off, but if a single reduction is not abnormal and the gear and pinion well mated, the efficiency should not be less than 96 per cent. Even better records have been realized by gears which have been particularly accurately cut, polished and run in an oil bath. Frequently an efficiency of 99 per cent. or even slightly higher has been obtained by such gears with speed ratios not exceeding 10 to 1.

The life of herringbone gears is far greater than that of even the most carefully cut spur gears, and though accurate data are hard to find on this subject, it may be safely stated that the usual life of a carefully cut herringbone gear is at least three or four times that of a similar spur gear.

W. C. Bates, Mechanical Engineer, Fawcus Machine Co., prepared for AMERICAN MACHINIST a comprehensive article on the design and strength of herringbone gears, from which the following important points are abstracted.

Due to the advantages of the herringbone construction, compared to that of spur gears, the load is always distributed over more than one tooth, the transference of load from tooth to tooth is without shock, the bearing pressure angularly placed on the tooth diminishes the strain on the root of the tooth, the tooth is shorter and therefore more sturdy, and the spacing of the teeth and tooth form are more accurate on account of the hobbing process employed in cutting the teeth, so that the indeterminate tooth stresses are at a minimum when running at high speeds.

Such advantages naturally allow considerable modification of the well-known Lewis formula for the strength of gears, as such empirical formula pertains to the strength of herringbone gears.

It then becomes:

$$S = s' \frac{1,200}{1,200 + v}$$

where

S = allowable stress per square inch at a speed of v feet per minute.
s' = static stress = 8,000 pounds per square inch for cast iron,
= 20,000 pounds per square inch for steel,
v = speed in feet per minute.

MODIFIED HERRINGBONE GEARS

An efficient modification of the standard herringbone gear may be constructed with a form of tooth that relies only upon its rolling action and eliminates all sliding contact. A tooth of the regular involute form on the pitch line is employed with the surplus metal above and below this contact line cut away. Such a gear will run as smoothly as the standard herringbone with full tooth section as long as no appreciable wear takes place. This naturally limits the practical value of the type, but it is of interest as representing the ideal in gearing action.

A radical departure from any ordinary type of gear has recently been put on the market by the R. D. Nuttall Co., Pittsburg, Pa., which has been designated as "The Circular Herringbone." A description of this gear appeared in AMERICAN MACHINIST, Oct. 16, 1913, from which the following excerpts are taken.

"It has a continuous tooth curved across the gear face, the curve being a circular arc. This approximates the shape of a herringbone tooth, hence the name "The Circular Herringbone." The tooth profiles at the middle of the face are true involutes; other profiles vary slightly from this, but are close approximations to the involute form. Though any tooth proportions can be cut, those standardized are: A pressure angle of 20 degrees, addendum 0.25, dedendum 0.25, clearance 0.05, working depth 0.5, and whole depth 0.55 of circular pitch."

These gears are generated, the cutter and blank rolling together during the process of cutting in such a manner that the line of tangency is along the pitch line of the cutter and the pitch surface of the blank.

HERRINGBONE BEVEL GEARS

The commercial development of a practical system of rolling gears from metal blanks heated to a semi-plastic condition—Anderson process—has made possible for the first time the production of herringbone bevel gears possessing all the merits of the efficient herringbone spur gear and in addition the capacity of absorbing in themselves the axial thrust common to other types of bevel gears. This noteworthy achievement in gear development gives promise of greatly extending the field for bevel gearing, particularly as herringbone bevels are as readily and cheaply made as ordinary bevel gears—see Section XVI.

SECTION VIII

Spiral Gears

Before going into the matter of calculations it may be well to direct the readers to a careful consideration of the accompanying perspective sketches originally published in American Machinist, October 11, 1906, by H. B.

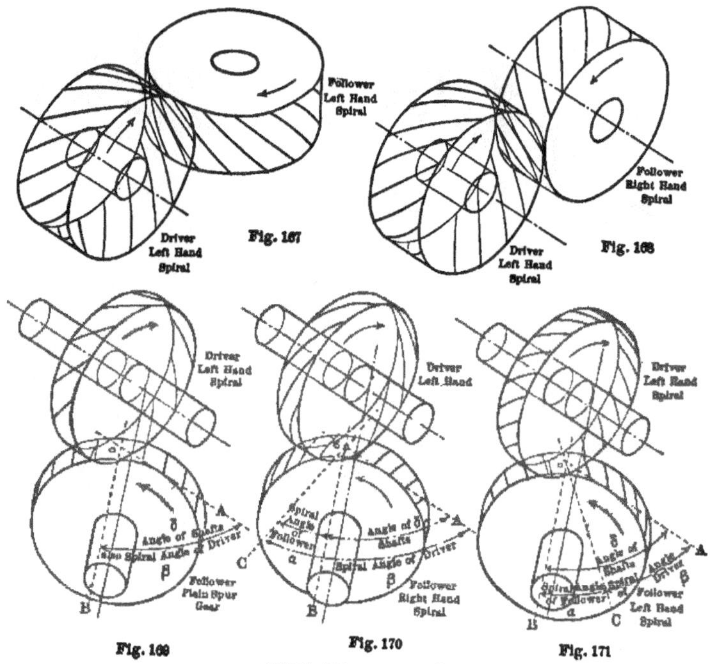

SPIRAL-GEAR DIAGRAMS.

McCabe. "In Figs. 167 to 171, inclusive, the driving gear of each pair is shown as if transparent, the teeth being represented by lines. Fig. 167 shows a pair of gears on shafts at right angles and Fig. 168 a pair on parallel shafts. Note that in Fig. 167 both spirals are left hand, while in Fig. 168 one is left and the other right hand; that is, in the first case they are the same hand and in the second case they are opposite hands. (The word hand as here used has the same significance as in the case of threads.) It is evident that these two are fixed conditions for shafts respectively at 90 degrees and parallel.

"Now when the shafts are at any angle between 90 degrees and 0 degrees either of these conditions may exist; that is, the spirals may be both the same hand or they may be opposite hands. This may be made plain by observing carefully Figs. 169, 170, and 171, in which the shafts are at an acute angle, all conditions in the three views being exactly alike except that the teeth are at different spiral angles in each. Note that in Fig. 169 the spiral angle of the driver is the same as the angle of the shafts which makes the follower a plain spur gear. Also note that the spiral of the driver is left hand. Now letting the spiral of the driver remain left hand, but increasing its angle a little we have the condition of Fig. 170. By decreasing it a little we have the condition in Fig. 171, making in the first case the spirals opposite hands and in the second case the spirals the same hand.

The lines *OA* and *OB* in these figures are drawn parallel to the shafts and the line *OC* is drawn tangent to the spiral of the teeth and makes with *OA* and *OB* respectively the spiral angles of the driver and of the follower. Note that in Fig. 171 the angle of the shafts *AOB* equals the spiral angles $AOC + BOC$, and in Fig. 170 the same angle *AOB* equals $AOC - BOC$

RELATION OF SHAFT AND SPIRAL ANGLES

"The following general rules are now evident:

"1. When the spirals are the same hand the angle of the shafts is the sum of the spiral angles.

"2. When the spirals are opposite hands the angle of the shafts is the difference of the spiral angles.

"3. When the spiral angle of one gear is the same as the angle of the shaft the spiral angle of the other will be zero, making it a plain spur gear.

"4. When the shafts are at right angles the spirals must both be the same hand.

"5. When the shafts are parallel the spirals must be opposite hands. (Helical gears.)

"6. When the shafts are at any acute angle the spirals may be either the same hand or opposite hands."

The following is an extract from an article on spiral gearing originally published in AMERICAN MACHINIST by F. A. Halsey:

"Spiral gears are not to blame for the undoubted fact that they are somewhat troublesome to lay out, the difficulties of the problem being due to the limitations of workshop facilities and not to the geometrical nature of the gears themselves. It is easy to understand and explain the action of an existing pair of spiral gears. More than this, it is easy to lay out a pair of such gears which shall exactly meet all the conditions of the case except one; they cannot, except through rare good luck, be made with the appliances at hand. To be more specific, the circumference cannot usually be divided into an exact whole number of teeth by any stock cutter, and the real problem becomes the readjusting of the diameters of the gears and the angle of the teeth, so that stock cutters shall make an exact whole number of teeth.'

"With spur gears it is only necessary to multiply the (circular) pitch of the cutter by the number of teeth to be cut to obtain the circumference of the gears. With spiral gears this operation gives the length of a portion of a spiral or, more properly, helix, wound upon the pitch surface. We do not know the angle of this helix, the diameter of the pitch cylinder upon which it is wrapped, or even what part of a complete turn the known portion comprises. The length is known for each gear and nothing more, and it becomes a matter of trial to find the diameters of the gears and the helix angle to suit

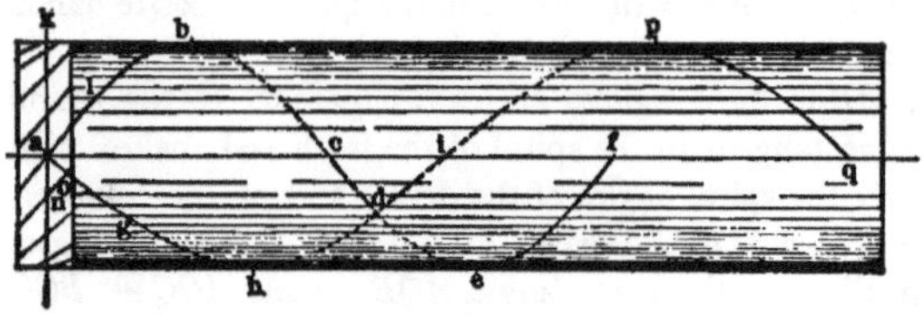

FIG. 172.

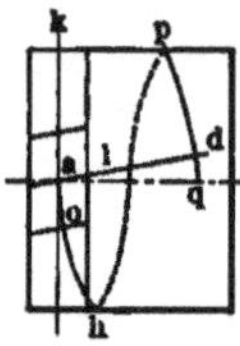

FIG. 173.

this portion of the helix and at the same time to fill the required center distance.

"Fig. 172 is a conventional representation of the pitch surface of a spiral gear, the surface being extended beyond the limits of the gear in order that the two helixes with which we are concerned may be shown. The first of these, *abcdef*, is the tooth helix and the second, *aghdip*, is the normal helix. The tooth helix is of importance because it defines the angle of the teeth. Given the diameter of the pitch surface, the helix may be defined by the angle *kal* or by the length *af*, in which it makes a complete turn—that is, by its pitch. For the determination of the speed ratio of a pair of gears the former method is the more convenient, but the tables supplied with universal milling machines which are used in setting up the machine employ the latter method.

"In all spiral gear problems we have two pitches to deal with—the pitch of the tooth helix and the pitch of the teeth. The latter may be measured in several ways. First is the value *an* measured on the circumference or the *circular pitch*, which is analogous to the pitch of spur gears; second is the value *ao* measured on the normal helix or the *normal pitch*, for which the cutters must be selected; third is the value *ar* measured parallel with the axis or the *axial pitch*. Since the cutters must be selected with reference to the normal pitch, the length of the normal helix is naturally of importance in conection with the number of teeth in the gear. The normal pitch multiplied by the number of teeth must naturally equal the length *aghd* of this helix measured between its intersections *a* and *d* with the helix of a single tooth. Note that the length of the normal helix to be considered is the length *aghd* between its intersections with the tooth, and not the length *aghipq* of a complete turn around the cylinder. That this is true may be seen by reference to Fig. 173, in which the angle *kal* is nearly a right angle. It is apparent from this illustration that the length of the normal helix from *a* to *d* takes in all the teeth, and that *ao*, multiplied by the number of teeth, must equal *ahpd* and not *ahpq*. This length *ahpd* is always less than *ahpq*, and usually much less. Fig. 174, *A*, is a development of Fig. 173 on a reduced scale, *ad* being the developed length of the normal helix. Fig. 174, *B*, and Fig. 174, *C*, show

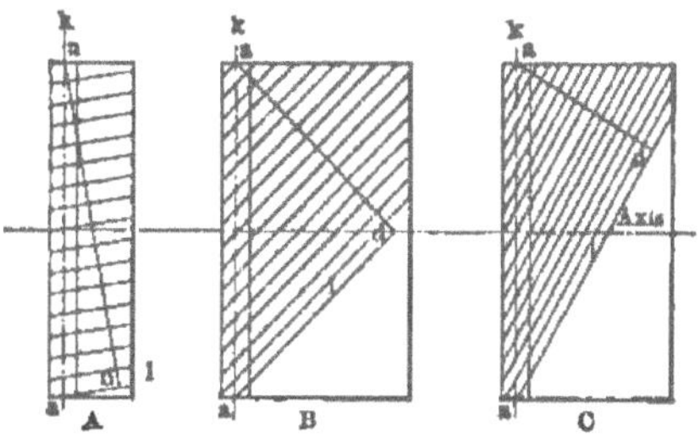

FIG. 174.

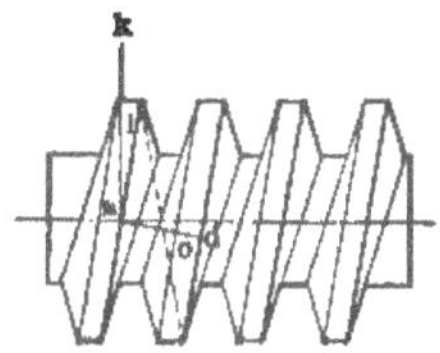

FIG. 175.

how with the same circumferential pitch and the same number of teeth but a reduced value of the angle *kal*, the length of the normal helix which cuts all the teeth grows shorter until it may make but a small part of a complete turn around the cylinder. It is clear that in all cases the line *ad* cuts all the teeth precisely as does the circumference *aa*, which goes completely around the cylinder. It is also clear that if the normal pitch is decided upon at the start, a diameter of cylinder and a helix angle must be found such that the normal pitch, multiplied by the number of teeth, shall equal the length of the normal helix between two intersections with the tooth helix.

"It is natural to ask, Why not employ the circumferential pitch and so deal directly with the circumference instead of the normal helix? Because we do not know what it is. The normal pitch is determined by the cutter used, while the circumferential pitch depends also upon the helix angle, and until this angle is known the circumferential pitch is not known.

"In the extreme case of a spiral gear in which the helix angle is so small that the gear becomes a single thread worm, as in Fig. 175, points *o* and *d* coincide and the length of the helix between *a* and *d* becomes the normal pitch. It is, however, true as before that the normal pitch, multiplied by the number of teeth, which is now one, is still equal to the length of the normal helix between two intersections with the tooth helix.

"A glance at Fig. 174 will show that in gears of the same diameter the length of the normal helix* grows shorter as the angle *kal* grows less, and

* "Length of normal helix" is to be understood as meaning the length of that helix between two intersections with the same tooth helix.

hence that it and its gear will contain successively fewer and fewer teeth of the same normal pitch. That is to say, the number of teeth in a gear varies with the helix angle as well as with the diameter, and *the number of teeth in two gears of the same normal pitch is not necessarily proportional to the diameters.* In fact, it is never so proportional, except when the angle *kal* is equal to 45 degrees. *The diametral pitch of the cutters and the diameter of the gear thus do not determine the number of teeth.*

"The two facts thus developed are fundamental and will bear restating:

"First, *the number of teeth is equal to the length of the normal helix divided by the normal pitch.*

"Second, *the numbers of teeth in a pair of gears are not proportional to the diameters, except when the angle of the tooth helix is 45 degrees.*

THE SPEED RATIO

"Fig. 176 illustrates the simplest possible case of a pair of spiral gears. The gears are of equal size and the tooth helix has an angle of 45 degrees. Such a pair of gears will obviously run at the same speed—that is, have a

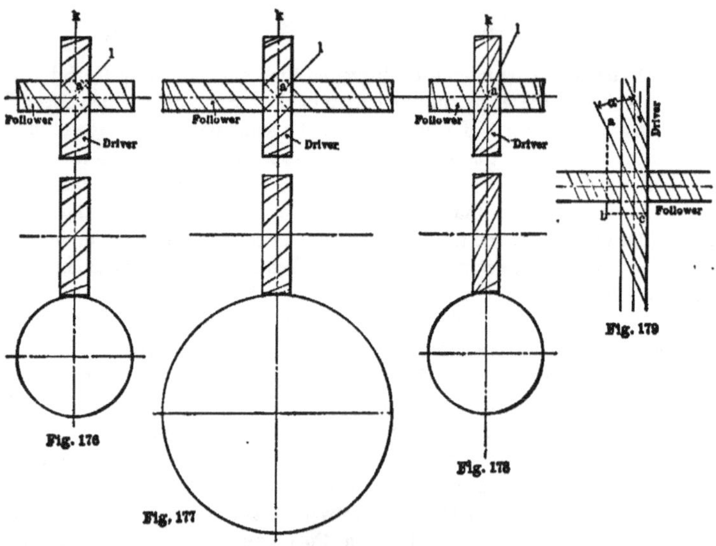

THE SPEED RATIO.

speed ratio of 1—and as obviously both will have the same number of teeth. Now, unlike spur gears, there are two ways in which the speed ratio of such a pair of spiral gears may be varied. First, the diameters of the gears may be changed, as with spur gears, the angle of the tooth helix remaining unchanged, as in Fig. 177; and second, the angle of the helix may be changed, the diameters of the gears remaining unchanged, as in Fig. 178. These methods act in very different ways. The first method is analogous to the procedure with spur gears. As with spur gears, the circumferential or pitch-line speed of the two gears remains, as before the change, equal, but the length of the circumference of the two gears is unequal and the largest one thus has a less number of revolutions than the smaller one. The second method is entirely unlike anything seen in connection with spur gears. By it the pitch-line speeds of the two gears are made unequal, and hence, while their diameters are equal, the lower one revolves the more slowly. This points out another fundamental difference between spiral and spur gears: With spiral gears, unless the helix angle is 45 degrees, *the pitch-line speeds of two mating gears are not the same.*

"The two methods of changing the speed ratio shown in Figs. 177 and 178 may be combined. That is, part of the desired change in speed may be obtained by changing the diameters of the gears and the remainder by changing the angle of the helix. Given the speed ratio and the diameter of one of the gears, we may assume a helix angle and find a diameter for the second gear to go with it which shall give the desired speed ratio, and, having done this, a second angle may be assumed and a second diameter be found. There are thus an indefinite number of combinations of angles and diameters which will give the required speed ratio. Note, however, that with the diameter of one gear fixed, every change in the diameter of the other changes the distance between centers, that not every angle of helix can be obtained by the gears which are furnished with universal milling machines, and that if ready-made cutters are to be used the lengths of both normal helixes must be exact multiple of the normal pitch of the teeth.

"The limitation of the helix angle is not, however, as serious as is usually supposed. The tables for spirals which have heretofore been supplied with universal milling machines give but a few of the spirals which can be obtained with the change gears which are regularly supplied with the machines. For universal milling machines, about two thousand spirals can be cut with these gears.

"Geometrically speaking, there is a wide range of choice in the helix angle. As regards the desirability of different angles from the standpoint of durability, the conditions are essentially the same as in worm gearing. Reference to Charts 10 and 11 under worm gears will show that the most favorable angle for durability is at about 45 degrees. There is, however, but a trifling increase in wear down to 30 degrees, no serious increase down to 20 degrees, and no destructive increase down to about 12 degrees. As the angle of worm is the complement of the angle of the driving spiral gear, the angle selected from Charts 10 and 11, for worm gears, should be the angle of the follower α, which is measured from the axis. Where gears are to transmit considerable power the best results should attend the use of angles between 30 and 45 degrees, while angles as low as 20 degrees may be used without hesitation, and as low as 12 degrees if the gears are to run in an oil bath or do light work only. The angle may also be increased about 45 degrees by similar amounts and with similar results.

"Fig. 179 is a development of the gears of Fig. 178, the angle α of Fig. 179 being equal to *kal* of Fig. 178, but in reversed position, because in Fig. 178 the upper side of the driver is seen, while in Fig. 179 the direction of the teeth is that of the lower side of the driver."

NOTATION FOR SPIRAL GEARS

The angle as given for spiral gears is from the axis, which is the opposite or complement of the angle for a worm, therefore the angle governing the efficiency of spiral gears should be determined from tables on worm gears as the angle of the follower (α).

The greatest angle must always be the driver, except where the angle is 45 degrees, when either gear may drive.

All of the tooth parts are derived from the normal pitch. The pitch diameters are derived from the circular pitch, which is never the same in both gears of a pair, except where the angle of both gears is 45 degrees.

As the diameter of the spiral gear is no indication of its speed ratio, the terms gear and pinion are liable to be confusing, therefore follower and driver are used.

N_2 = number of teeth in follower.
N_1 = number of teeth in driver.
d_2 = pitch diameter of follower.
d_1 = pitch diameter of driver.
α = angle of follower.
β = angle of driver.
p'_2 = circular pitch of follower.
p'_1 = circular pitch of driver.
p'^n = normal circular pitch (the same in both gears of a pair).
P = normal diametral pitch (the same in both gears of a pair).
L_2 = lead of follower (length of tooth helix).
L_1 = lead of driver (length of tooth helix).
D_2 = outside diameter of follower.
D_1 = outside diameter of driver.
s^n = addendum of normal pitch.
r_2 = revolutions of follower.
r_1 = revolutions of driver.
δ = angle of shafts.
C = center distance.

EXAMPLES

Specifications for a pair of spiral gears are sometimes given in this manner: Required a pair of spiral gears; ratio 3 to 1, to operate on 5-inch centers.

	DRIVER		FOLLOWER		REMARKS
	TO FIND	FORMULA	TO FIND	FORMULA	
1	β	$Tan\ \beta = \frac{d_1 r_1}{d_2 r_2}$	α	$90° - \beta$	Axes at right angles only.
2	β	$Tan\ \beta = \frac{p'_1}{p'_2}$	α	$90° - \beta$	Axes at right angles only.
3	β	$Cos\ \beta = \frac{p'^n}{p'_1}$	α	$\delta - \beta$	
4	β	$Tan\ \beta = \frac{d_1 \pi}{L_1}$	α	$\delta - \beta$	
5	p'^n	$\frac{d_1 \pi}{N_1} cos\ \beta$	p'^n	$\frac{d_2 \pi}{N_2} cos\ \alpha$	Same in both gears.
6	p'^n	$p'_1 cos\ \beta$	p'^n	$p'_2 cos\ \alpha$	Same in both gears.
7	p'_1	$\frac{p'^n}{cos\ \beta}$	p'_2	$\frac{p'^n}{cos\ \alpha}$	
8	p'_1	$\frac{d_1 \pi}{N_1}$	p'_2	$\frac{d_2 \pi}{N_2}$	
9	L_1	$p'_2 N_1$	L_2	$p'_1 N_2$	Axes at right angles only.
10	L_1	$d_1 \pi\ tan\ \alpha$	L_2	$d_2 \pi\ tan\ \beta$	
11	N_1	$d_1 P\ cos\ \beta$	N_2	$d_2 P\ cos\ \alpha$	
12	N_1	$\frac{d_1 \pi}{p'_1}$	N_2	$\frac{d_2 \pi}{p'_2}$	
13	d_1	$\frac{2C}{\left(\frac{r_1}{r_2} tan\ \alpha\right) + 1}$	d_2	$\frac{2C}{\left(\frac{r_1}{r_2} tan\ \beta\right) + 1}$	Axes at right angles only.
14	d_1	$\frac{2C}{\left(\frac{r_1}{r_2} \frac{cos\ \beta}{cos\ \alpha}\right) + 1}$	d_2	$2C - d_1$	
15	d_1	$N_1 p'_1 0.3183$	d_2	$N_2 p'_2 0.3183$	
16	d_1	$\frac{N_1}{P\ cos\ \beta}$	d_2	$\frac{N_2}{P\ cos\ \alpha}$	
17	D_1	$d_1 + 2s^n$	D_2	$d_2 + 2s^n$	
18	D_1	$d_1 + \frac{2}{P}$	D_2	$d_2 + \frac{2}{P}$	14½° standard only.
19	Cutter See chart 14	$\frac{N_1}{cos^3\ \beta}$	Cutter	$\frac{N_2}{cos^3\ \alpha}$	
20		$C = \frac{N_1}{2P\ cos\ \beta} + \frac{N_2}{2P\ cos\ \alpha}$			

FORMULAS FOR SPIRAL GEARS.

The outside diameter of the driven gear must not exceed 7 inches; to be in the neighborhood of 6 diametral pitch.

As the most efficient spiral angle is in the neighborhood of 45 degrees, the follower should be made as large as possible, as to obtain this angle the diameter of both gears must be in proportion to their number of teeth, as for spur gears. As the pitch mentioned in connection with spiral gears is always the normal pitch, to obtain a trial pitch diameter for the follower twice the addendum of the normal pitch subtracted from the outside diameter will give the pitch diameter, according to formula 18:

$$d_2 = D_2 - \frac{2}{P} = 7 - \frac{2}{6} = 6\tfrac{2}{3} \text{ inches.}$$

and,

$$d_1 = 5 \times 2 - 6\tfrac{2}{3} = 3\tfrac{1}{3} \text{ inches.}$$

The next step is to find the angle of driver by formula 1.

$$Tan.\, \beta = \frac{d_1 r_1}{d_2 r_2} = \frac{3\tfrac{1}{3} \times 3}{6\tfrac{2}{3} \times 1} = 1.5, \text{ or } 56° \; 19'.$$

The angle of follower $= 90° - 56°19' = 33° \; 41'$.

Find the provisional number of teeth by formula 11.

$$N_1 = d_1 P \cos. \beta = 3\tfrac{1}{3} \times 6 \times 0.5546 = 11.092.$$
$$N_2 = d_2 P \cos. \alpha = 6\tfrac{2}{3} \times 6 \times 0.8321 = 33.284.$$

Naturally the number of teeth must be whole numbers, so it will be necessary to change either the center distance, or to make numerous calculations and shift the diameters. Practically, however, it is possible to have quite an error in the normal pitch; the normal pitch, or the pitch of the cutter, preferably being under size rather than over. The teeth are thus cut enough deeper than standard to secure the proper thickness of tooth at the pitch line.

This difference may be 0.02 of the circular pitch in some cases.

If the cutter is heavier than the normal pitch it will be impossible to secure enough clearance at the bottom of the tooth as the proper thickness of tooth will be reached before getting the depth of tooth required.

If the center distance can be changed the pitch diameters may be shifted by the method explained on page 66 of Mr. F. A. Halsey's book—"Worm and Spiral Gearing," as follows:

$$\frac{\text{final diameter}}{\text{provisional diameter}} = \frac{11}{11.094}$$

or

$$\text{final diameter} = \text{provisional diameter} \times \frac{11}{11.094}.$$

That is:

$$\text{final } d_1 = 3\tfrac{1}{3} \times \frac{11}{11.094} = 3.305;$$

and,

$$\text{final } d_2 = 6\tfrac{2}{3} \times \frac{33}{33.282} = 6.610;$$

and

$d_1 + d_2 = 3.305 + 6.610 = 9.915 =$ twice the corrected center distance.

In the present example the normal circular pitch for the nearest even number of teeth, 11 and 33, by formula 5 would be:

$$p_1^n = \frac{d_1\pi}{N_1} \cos.\beta = \frac{3\tfrac{1}{3} \times 3.1416}{11} + 0.5546 = 0.5280.$$

As the pitch of the cutter is 0.5236, this error will not prevent a first-class job being turned out if proper precautions are taken, and no change will be required in the center distance.

These points being settled, the remaining calculations are simple. Before making any calculations, the requirements should be put in the form of a table to avoid confusion, as follows:

DIMENSIONS	DRIVER	FOLLOWER
Pitch diameters	3⅓	6⅔
Revolutions	3	1
Angles	56° 19′	33° 41′
Number of teeth	11	33
Circular pitch	0.9520″	0.6345″
Normal pitch	0.5280″	0.5280″
Cutter used	No. 2–6p	No. 3–6p
Lead, exact	6.9795″	31.4160″
Lead, approximate	6.9670″	31.5000″
Addendum	0.1680″	0.1680″
Outside diameter	3.6690″	7.0030″
Whole depth of tooth	0.3630″	0.3630″
Thickness of tooth	0.2640″	0.2640″
Gear on worm	86	72
First gear on stud	48	40
Second gear on stud	28	56
Gear on screw	72	32

An error of 0.5 inch in a lead of 50 inches would not ordinarily be prohibitive, but the angle must be changed to suit any alteration of the lead or the cutter will drag. If too much alteration is made in the lead and angle, the teeth must be cut a little deeper than standard to allow the gears to assemble on the proper shaft angles.

The amount of adjustment that can be made depends, of course, upon the accuracy required, and should be done by some one accustomed to the work. This is not possible when cutting helical or herringbone gears, as the tooth has contact the entire length of face and a slight error is noticeable. The accuracy of the final calculations may be checked by the angles, obtained from the circular pitch by Formula 3.

Another way of presenting this problem is as follows:

Required, a pair of spiral gears; ratio 4 to 1; about 8 diametral pitch (0.3927-inch circular pitch). Angle of spiral for driver, β, to be about 55 degrees. $\alpha = 90 - \beta = 35$ degrees.

Find the diameter of driver by formula 13.

$$d_1 = \frac{2C}{\left(\frac{r_1}{r_2}\,tan.\,\alpha\right) + 1} = \frac{2 \times 6}{\left(\frac{4}{1} \times 0.7002\right) + 1} = 3.1572 \text{ inches.}$$

The diameter of follower $d_2 = 2C - d_1 = 12 - 3.1572 = 8.8428$ inches.

The remaining dimensions are found as in the first example.

Still another example:

The ratio of a spiral gear drive is 4 to 1. The diameter of the driver cannot be less than 8 inches, on account of the size of the shaft. The distance between centers to be 5½ inches. No pitch mentioned.

Assumed diameter of driver $d_1 = 8$ inches.

Diameter of follower $d_2 = (5\frac{1}{2} \times 2) - 8 = 3$ inches.

According to formula 1:

$$Tan.\,\beta = \frac{d_1 r_1}{d_2 r_2} = \frac{8 \times 4}{3 \times 1} = 10.66 \text{ or } 86^\circ\,25'.$$

Try 7 diametral pitch:

According to formula 11:

$$N_1 = d_1 P\,cos.\,\beta = 8 \times 10 \times 0.0625 = 5 \text{ teeth;}$$
$$N_2 = d_2 P\,cos.\,\alpha = 2 \times 10 \times 0.9980 = 19.96, \text{ say 20 teeth;}$$

which just happens to come out even.

If the center distance is not specified, the best plan is to assume number of teeth, angles, and pitch of cutter, P or p'^n and find the corresponding center distance by formula 20.

Example:

What center distance will be required for a pair of spiral gears 11 and 33 teeth, 6 diametral pitch, the angle of the 11 tooth drive being 56° 19′ and the angle of the follower 33° 41′.

According to formula 20:

$$C = \frac{N_1}{2P\,cos.\,\beta} + \frac{N_2}{2P\,cos.\,\beta} = \frac{11}{2 \times 6 \times 0.5546} + \frac{33}{2 \times 6 \times 0.8321} = 1.6529$$
$$+ 3.3049 = 4.9578 \text{ inches center distance;}$$

1.6529 being the pitch radius of the pinion, and 3.3049 the pitch radius of the gear.

When the center distance is approximate, this is the simplest solution of the problem, the speed ratio being used in place of a trial number of teeth, and the number of teeth made to suit the desired center distance.

A CHART FOR LAYING OUT SPIRAL GEARS

Chart 13 with the following explanation of its deviation and use will be an aid in solving spiral gear problems once the provisional number of teeth are obtained. This diagram and explanation were originally published in AMERICAN MACHINIST, February 27, 1902, by J. N. Le Conte.

"The provisional numbers of teeth will not in general be whole numbers,

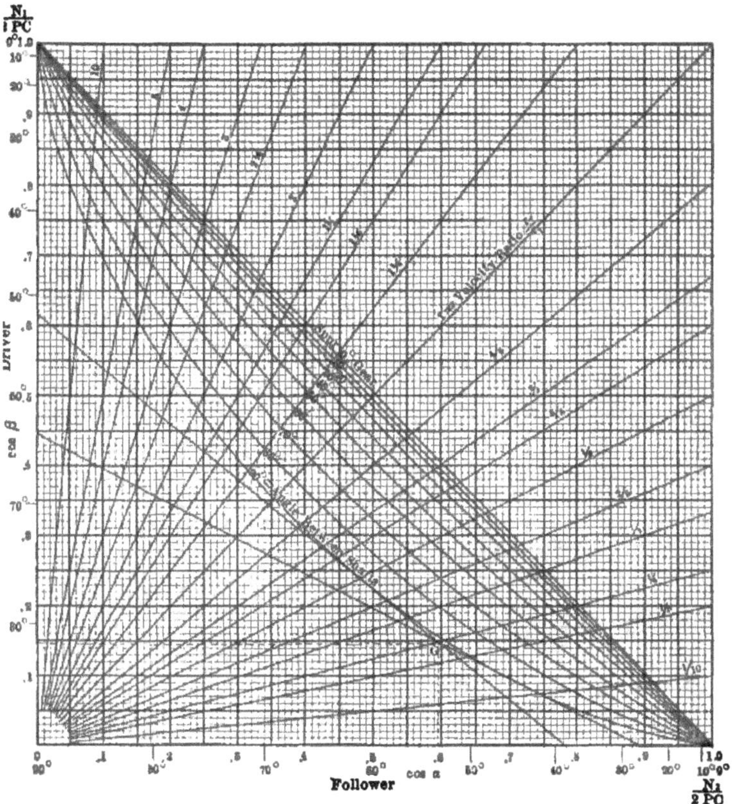

CHART 13. DIAGRAM FOR LAYING OUT SPIRAL GEARS.

but we must choose the nearest whole numbers to the ones obtained, and recalculate the angles and radii to fit the new case, as has been previously shown. The direct solution of this depends upon the solution of the equation:

$$\frac{N_2}{cos.\alpha} + \frac{N_1}{cos.(\alpha - \delta)} = 2PC$$

"In which N_1 and N_2 are the nearest whole numbers of teeth to the calculated ones, and δ is the shaft angle or $\delta = \alpha + \beta$. As is well known, this equation cannot be solved by any simple means, for it is of the fourth degree, and, though possible of solution, such solution is not practical. Furthermore there are four real values of α which will satisfy it. Graphic methods of solution, or continued approximations, must then be resorted to. Chart 13 gives a method by which the angle can be read off directly. Having obtained the nearest whole number of teeth on the gears, find on the diagram the point G whose co-ordinates are $\frac{N_1}{2PC}$ and $\frac{N_2}{2PC}$ on the inner scales. Through this point draw a line or merely lay a straight-edge tangent to the curve representing the shaft angle. The outer scales on the bottom and left will give roughly the angles α and β respectively, and the inner scales the values of cos α and cos β quite accurately. The radial lines of velocity ratio will facilitate the location of the desired point, for if the ratio be one of those given, the point must lie on its line. It will also be noticed that for shafts crossing at 90° the position of the line gives the desired angles at its two extremities.

"It is interesting to note that two lines can be drawn through a given point tangent to the curves, as shown. As a matter of fact, four such lines could be drawn provided the whole of the curves were laid in, but that portion shown is the only portion giving positive angles, *i.e.*, angles within the angle δ. But there will be two separative positive values of the angle α, which, with a given velocity ratio, number of teeth and shaft distance, will work correctly together, giving of course different values of the radii. Which of these is the one required can always be told as lying nearest to the first approximation of the angle. If the point G lies on one of the curves, the two positions coincide (a limiting case), and if it lies on the concave side the solution is impossible within the angle δ.

"As an example of the use of the diagram, take the oft-quoted case of Mr. De Leeuw. Here angular velocity

$$y_1 = \frac{\gamma^2}{\gamma^1} = \tfrac{1}{4},\ N_1 = 8,\ N_2 = 32,\ C = 4.468'',\ \delta = 90°,\ \text{and}\ P = 6.$$

Then

$$\frac{N_1}{2PC} = 0.149,\ \frac{N}{2PC} = 0.596.$$

"These co-ordinates give the point G on the diagram. A line through G drawn tangent to the lower part of the 90° curve gives quite accurately *cos.* $\alpha = 0.894$, and *cos.* $\beta = 0.447$, or:

$$\alpha = 26°\,35'\ \text{and}\ \beta = 63°\,25',$$

agreeing quite closely with the result derived analytically. If the second line be drawn through G tangent to the upper portion of the curve, it gives: *cos.* $\alpha = 0.787$ and *cos.* $\beta = 0.617$, or:

$$\alpha = 38°\,61'\ \text{and}\ \beta = 51°\,54'.$$

These fulfill the requirements."

SPIRAL GEAR TABLE

While it is better in every case to understand the principles involved before using a table, as this tends to prevent errors, they can be used with good results by simply following the directions carefully. The subject of spiral

	To obtain the circular pitch for one tooth divide by the required diametral pitch.	To obtain the pitch diameter, divide by the required diametral pitch and multiply the quotient by the required number of teeth.	To obtain the lead of spiral, divide by the required diametral pitch and multiply the quotient by required number of teeth.		To obtain the pitch diameter, divide by the required diametral pitch and multiply the quotient by the required number of teeth.	To obtain the circular pitch for one tooth divide by the required diametral pitch.	
ANGLE OF SPIRAL DEGREES	CIRCULAR PITCH	ONE TOOTH OR ADDENDUM	LEAD OF SPIRALS		ONE TOOTH OR ADDENDUM	CIRCULAR PITCH	ANGLE OF SPIRAL DEGREES
Small Wheel.	Small Wheel.	Small Wheel.	Small Wheel.	Large Wheel.	Large Wheel.	Large Wheel.	Large Wheel.
1	3.1419	1.0001	180.05	3.1420	57.298	180.01	89
2	3.1435	1.0006	90.020	3.1435	28.653	90.016	88
3	3.1457	1.0013	60.032	3.1458	19.107	60.026	87
4	3.1491	1.0024	45.038	3.1492	14.335	45.035	86
5	3.1535	1.0038	37.077	3.1527	11.473	36.044	85
6	3.1589	1.0055	30.056	3.1589	9.5667	30.055	84
7	3.1652	1.0075	25.728	3.1651	8.2055	25.778	83
8	3.1724	1.0098	22.573	3.1724	7.1852	22.573	82
9	3.1806	1.0124	20.082	3.1807	6.3924	20.082	81
10	3.1900	1.0154	18.092	3.1901	5.7587	18.092	80
11	3.2003	1.0187	16.464	3.2003	5.2408	16.464	79
12	3.2145	1.0232	15.076	3.2105	4.8097	15.104	78
13	3.2242	1.0263	13.966	3.2294	4.4454	13.988	77
14	3.2377	1.0306	12.986	3.2378	4.1335	12.986	76
15	3.2522	1.0352	12.138	3.2524	3.8637	12.138	75
16	3.2679	1.0402	11.393	3.2678	3.6279	11.397	74
17	3.2848	1.0456	10.417	3.2821	3.4203	10.745	73
18	3.3116	1.0514	10.192	3.3032	3.2360	10.166	72
19	3.3225	1.0576	9.6494	3.3225	3.0715	9.6494	71
20	3.3430	1.0641	9.1848	3.3433	2.9238	9.1854	70
21	3.3650	1.0711	8.7662	3.3652	2.7904	8.7663	69
22	3.3882	1.0785	8.3862	3.3833	2.6694	8.3862	68
23	3.4127	1.0863	8.0399	3.4129	2.5593	8.0403	67
24	3.4451	1.0946	7.7379	3.4391	2.4585	7.7242	66
25	3.4661	1.1033	7.4332	3.4663	2.3662	7.4336	65
26	3.4953	1.1126	7.1664	3.4952	2.2811	7.1663	64
27	3.5258	1.1223	6.9198	3.5257	2.2026	6.9197	63
28	3.5579	1.1325	6.6912	3.5575	2.1300	6.6916	62
29	3.5918	1.1433	6.4799	3.5919	2.0626	6.4799	61
30	3.6276	1.1547	6.2778	3.6277	2.0000	6.2832	60
31	3.6650	1.1666	6.0979	3.6652	1.9416	6.0997	59
32	3.7043	1.1791	5.9282	3.7044	1.8870	5.9282	58
33	3.7457	1.1923	5.7710	3.7459	1.8360	5.7680	57
34	3.7894	1.2062	5.6181	3.7826	1.7882	5.6178	56
35	3.8349	1.2207	5.4754	3.8351	1.7434	5.4770	55
36	3.8830	1.2360	5.3431	3.8834	1.7013	5.3448	54
37	3.9336	1.2521	5.2201	3.9261	1.6616	5.2200	53
38	3.9867	1.2690	5.1028	3.9921	1.6242	5.1026	52
39	4.0482	1.2867	4.9866	4.0416	1.5890	4.9920	51
40	4.1010	1.3054	4.8873	4.1012	1.5557	4.8874	50
41	4.1626	1.3250	4.7885	4.1540	1.5242	4.7884	49
42	4.2273	1.3456	4.6949	4.2272	1.4944	4.6948	48
43	4.2956	1.3673	4.6065	4.2956	1.4662	4.6062	47
44	4.3671	1.3901	4.5223	4.3675	1.4395	4.5225	46
45	4.4428	1.4142	4.4428	4.4428	1.4142	4.4428	45

TABLE 22—SPIRAL GEAR TABLE SHAFT ANGLES 90 DEGREES

For one Diametral Pitch.

gears is so much more complicated than other gears that many will prefer to depend entirely on tables.

This table gives the circular pitch and addendum or diametral pitch and lead of spirals for one diametral pitch and with teeth having angles from 1 to 89 degrees to 45 and 45 degrees. For other pitches divide the addendum given and the spiral number by the required pitch, and multiply the results by the required number of teeth. This will give the pitch diameter and lead of spiral for each gear. For the outside diameter add twice the addendum of the normal pitch, as in spur gearing.

Suppose we want a pair of spiral gears with 10 and 80 degree angles, 8 diametral pitch cutter, with 16 teeth in the small gear, having 10-degree angle and 10 teeth in the large gear with its 80-degree angle.

Find the 10-degree angle of spiral and in the third column find 1.0154. Divide by pitch, 8, which is 0.1269. Multiply this by the number of teeth; $0.1269 \times 16 = 2.030 =$ pitch diameter. Add two addendums or $\frac{2}{4} = 0.25$ inch. Outside diameter $= 2.030 + 0.25 = 2.28$ inches.

The lead of spiral for 10 degrees, for small gear, is 18.092. Divide by pitch $= \frac{18.092}{8} = 2.2615$. Multiply by number of teeth, $2.2615 \times 16 = 36.18$, or lead of spiral, which means that the tooth helix makes one turn in 36.18 inches.

PITCH OF CUTTER	CORRESPONDING CIRCULAR PITCH	CORRESPONDING DIAMETRAL PITCH	ADDENDUM
P	*p'*	*p*	*s*°
2	2.2214	1.4142	.70710
2¼	1.9745	1.5909	.62853
2½	1.7771	1.7677	.56568
2¾	1.6156	1.9445	.51426
3	1.4809	2.1213	.47140
3½	1.2694	2.4748	.40406
4	1.1107	2.8284	.35355
5	.8885	3.5355	.28284
6	.7404	4.2426	.23570
7	.6347	4.9497	.20203
8	.5553	5.6568	.17677
9	.4936	6.3638	.15713
10	.4443	7.0710	.14142
12	.3702	8.4853	.11785
14	.3173	9.8994	.10101
16	.2776	11.3137	08838
18	.2468	12.7279	.07856
20	.2221	14.1421	.07071
22	.2019	15.5563	.06428
24	.1851	16.9705	.05892
26	.1708	18.3847	.05439
28	.1586	19.7990	.05050
30	.1481	21.2116	.04714
32	.1388	22.6274	.04419
36	.1234	25.4558	.03928
40	.1111	28.2842	.03535
48	.0925	33.9411	.02988

TABLE 23—SPIRAL GEARS OF 45 DEGREES

For determining the pitch diameters of spiral gears when the pitch of cutter is assumed and angle of spiral is 45 degrees: Multiply the addendum of the normal pitch found in fourth column by number of teeth.

For the other gear with its 80-degree angle, find the addendum, 5.7587. Divide by pitch, 8 = 0.7198. Multiply by number of teeth, 10 = 7.198. Add two addendums, or 0.25, gives 7.448 as outside diameter.

The lead of spiral is 3.1901. Dividing by pitch, 8 = 0.3988. Multiply by number of teeth = 3.988 the lead of spiral.

When racks are to mesh with spiral gears, divide the number in the circular pitch columns for the given angle by the required diametral pitch to find the corresponding circular pitch.

If a rack is required to mesh with 40-degree spiral gear of 8 pitch, look for circular pitch opposite 40 and find 4.101. Dividing by 8 gives 0.512 as the circular pitch for this angle. The greater the angle the greater the circular or linear pitch, as can be seen by trying an 80-degree angle. Here the circular pitch is 2.261 inches.

Without the aid of the table, even such a relatively simple problem in spiral gear design would entail a complexity of computations which would not only be tedious but would be liable to introduce errors which might lead to costly shop time wastes and ruined material. For all practical purposes, the table will be found extremely valuable and even when odd spiral angles have to be employed, necessitating the more laborious calculations, it will prove of great assistance in checking the results of the computations.

DIRECTION OF ROTATION AND THRUST OF SPIRAL GEARS

The use of spiral gears generally causes some study on the part of the designer to determine the proper direction of the teeth, having given the direction of rotation of the two shafts which are to be connected. Another point about spiral gears which also causes some study after the direction of the teeth has been determined is the direction of the axial thrust, that is, the direction in which the spiral gears tend to move along their axes when transmitting motion. The proper direction of thrusts is very important to locate correctly the ball-thrust bearings of other suitable anti-friction devices.

It sometimes occurs that a newly designed machine, when started for the first time, has a shaft which is driven by spiral gears running in the opposite direction to that which was intended, or that the anti-friction washers have been located on the wrong side of the helical gear. To obviate this and reduce the chance for mistakes in directions of rotation and strains in spiral gears, four diagrams, 180 to 183, are arranged to readily illustrate every possible combination; giving the direction of the teeth, their rotation, and the direction of the lateral strains when they are transmitting motion in the directions indicated. These diagrams eliminate the necessity of consulting a gear model, nor is it necessary to go through a series of hand manipulations describing the rotations in the air.

In the diagrams, Figs. 180 and 181 represent a pair of right-hand helical gears with the direction of rotation of the drivers reversed.

The diagrams Figs. 182 and 183 each show a pair of left-hand helical gears, also with the directions of their drivers reversed. It should be noted that reversing the direction of rotation of the drivers reverses the directions

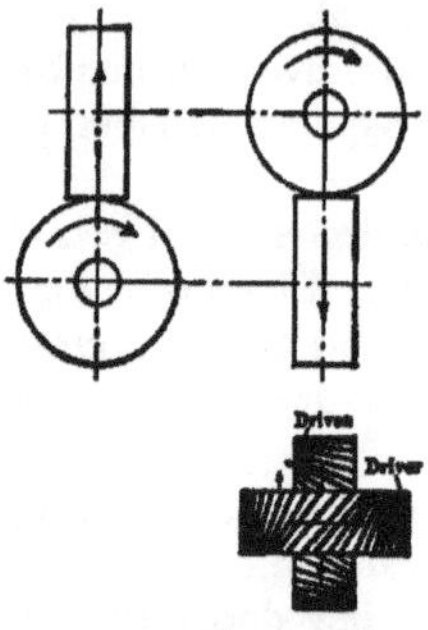

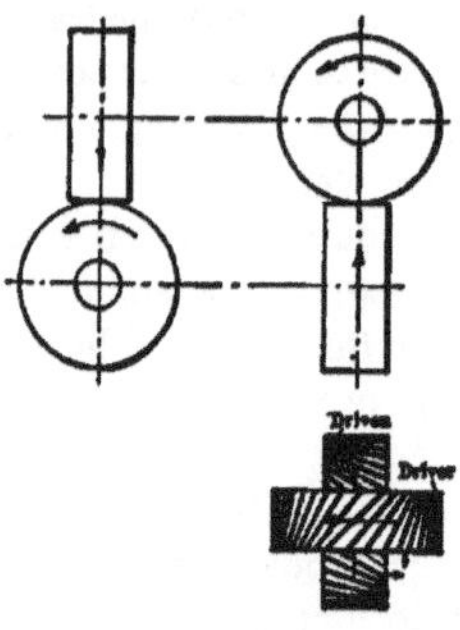

FIG. 180. FIG. 181.

RIGHT-HAND SPIRAL GEARS.

of their axial thrusts. Also, if the driven gears are made the drivers and rotating in the same direction, as shown, the lateral strains are also reversed; that is, if in Fig. 180 the driven gear is made the driver and rotates as indi-

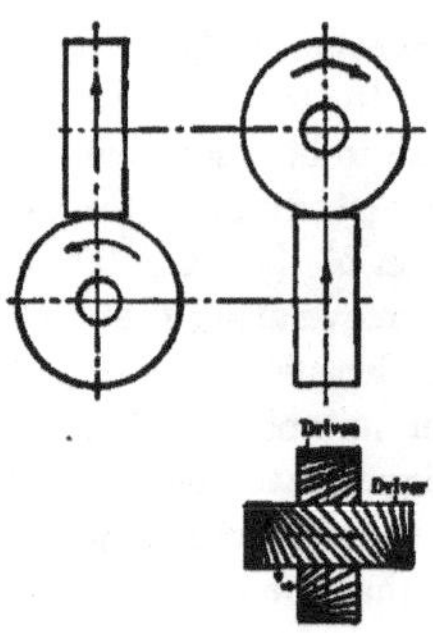

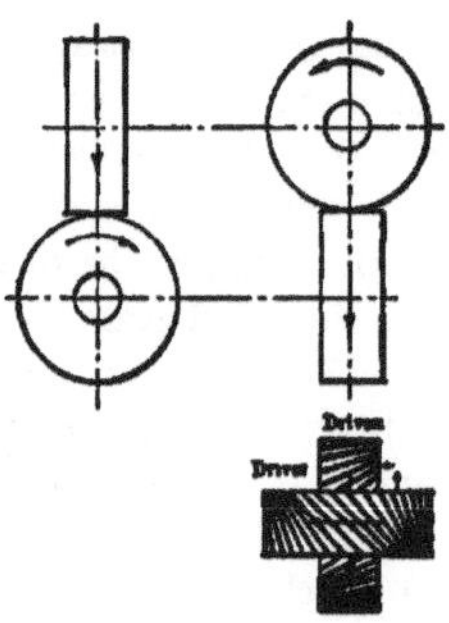

FIG. 182. FIG. 183.

LEFT-HAND SPIRAL GEARS.

cated, the gear marked the driver, which is now the driven gear, will rotate as shown, but the axial thrust of each gear would be as in Fig. 181. If the driven gear of Fig. 181 is made the driver, the lateral strains are as shown in Fig. 180. This is also true of the left-hand combinations shown in Figs. 182 and 183, originally published in the AMERICAN MACHINIST by William F. Zimmerman.

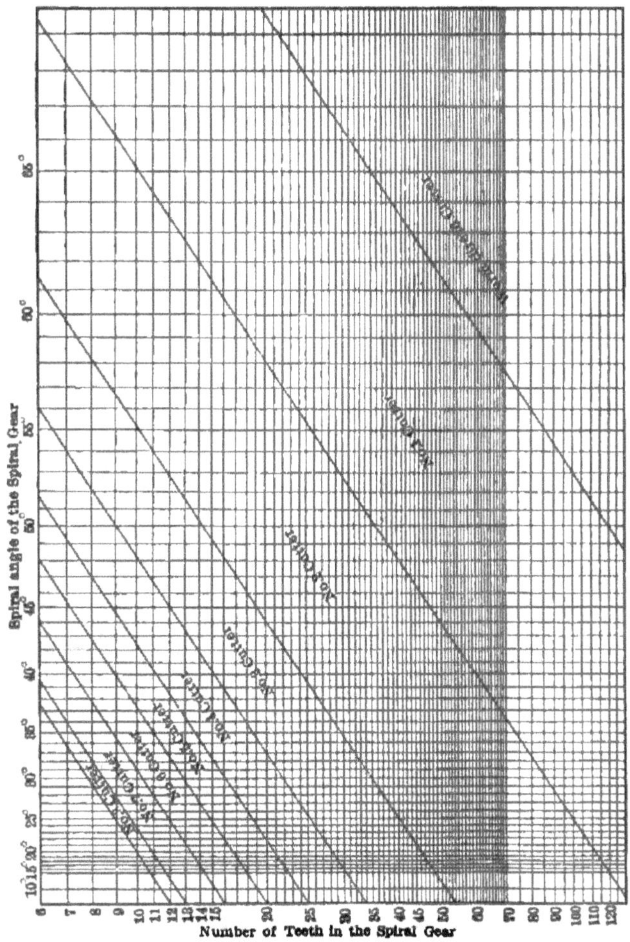

CHART 14. CUTTERS FOR SPIRAL GEARS.

SECTION IX

Skew Bevel Gears

Bevel gears which do not have their axes in the same plane, popularly known as skew bevel gears, present one of the most complicated constructions in gearing. Ordinarily, this arrangement consists of a common bevel pinion meshing with a bevel gear having teeth of a spiral type (see Fig. 184), the contact taking place on a plane parallel to that of, but somewhat removed from, the one in which lies the axis of the skew bevel gear.

FIG. 184. SKEW BEVEL GEARS.

The contact action of this combination is somewhat better than that in spiral gearing, as the sliding action of the latter is replaced by a combined rolling and sliding action. The pitch surfaces of the bevel pinion are frusta of a figure generated by the revolution of a straight line about an axis to which it is not parallel, a "hyperboloid of revolution."

A typical plan view of a skew bevel gear with its pinion in position is shown as Fig. 185, in which the dimension A represents the offset of the pinion. The apex point of the pinion lies in the perpendicular axis plane of the gear, to which point also converge the profile planes of the gear teeth actually in mesh with the pinion. It is obvious then that the profile planes of each succeeding tooth, when in mesh, must converge to the same point. This results in a circle of apexes for the gear having a radius equal to the offset of the pinion—*i.e.*, the succeeding converging tooth profiles, if prolonged, would all be tangent to a circle having a diameter equal to twice the offset of the pinion.

The pinion differs in no way from a regular bevel gear and, therefore, governs the proportions of the skew bevel gear. If the pinion was not offset, the combination simply a set of bevel gears, the pitch diameter of the gear

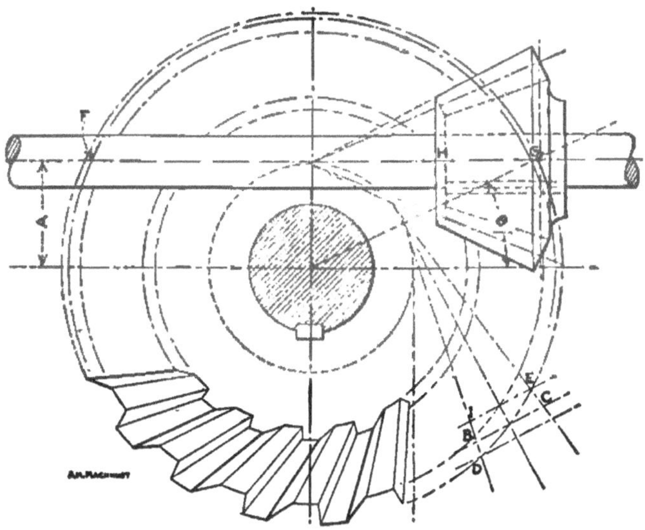

FIG. 185. DIAGRAM OF SKEW BEVEL GEARS.

would be *F-G*. The number of teeth in the skew bevel gear is therefore the same as would be required for an ordinary bevel gear of such pitch diameter. The actual pitch diameter of the skew bevel gear is considerably greater than this "equivalent pitch diameter," depending upon the amount which the pinion is offset.

The normal pitch of the gear (*B-C*) must conform to the circular pitch of the pinion, but the circular pitch (*D-E*) of the gear depends upon its actual pitch diameter, the number of teeth being fixed by the pinion and equal to the number of teeth required for a common bevel gear of a pitch diameter equal to *F-G*.

The sliding action of the teeth upon one another also depends upon the amount of offset to the pinion shaft. In the combination illustrated in Fig. 185 it is evident that sliding of the extreme end of the pinion teeth will occur from I to D on the gear. This sliding action is accentuated with any increase in the dimension A, the limiting condition being when A equals the radius of the gear, in which case there would be sliding contact only and no turning moment. On the other hand, when A equals zero the sliding action is eliminated and there is only true rolling contact.

The principal relationships of the various dimensions, angles, etc., of skew bevel gears follow:

NOTATIONS FOR SKEW BEVEL GEARS

p = diametral pitch.
n = number of teeth in pinion.
N = number of teeth in gear.
A = offset of pinion shaft.
θ = angle of offset.
d' = pitch diameter of pinion.
D'_e = equivalent pitch diameter of gear.
D' = pitch diameter of gear.
d = outside diameter of pinion.
D = outside diameter of gear.
p' = circular pitch of pinion.
$p_1{}^n$ = normal pitch of gear.
p'_1 = circular pitch of gear.
E_1 = center angle of pinion.
F_1 = face angle of pinion.
C_1 = cutting angle of pinion.
E_2 = center angle of gear.
F_2 = face angle of gear.
C_2 = cutting angle of gear.
E'_2 = contact angle of gear and pinion = E_1.
J = angle increment.
K = angle decrement.
V_1 = diameter increment of pinion.
s = addendum.

FORMULAS FOR SKEW BEVEL GEARS

$$d' = \frac{n}{p} \tag{1}$$

$$D'_e = \frac{N}{p} \tag{2}$$

$$Tan.\ \theta = \frac{2A}{D'_e} \tag{3}$$

$$D' = \frac{2A}{sin.\ \theta} \tag{4}$$

$$p'_1 = \frac{3.1416D'}{N} \tag{5}$$

$$Tan.\ E_1 = \frac{d'}{D'_e} \qquad (6)$$

$$Tan.\ J = \frac{2\ sin.\ E_1}{n} \qquad (7)$$

$$Tan.\ K = \frac{2.314\ sin.\ E_1}{n} \qquad (8)$$

$$F_1 = E_1 + J \qquad (9)$$

$$C_1 = E_1 - K \qquad (10)$$

$$s = \frac{d'}{n} \qquad (11)$$

$$V_1 = s\ cos.\ E_1 \qquad (12)$$

$$D = D' + 2V_1, \text{ or for greater accuracy } = D' + \frac{2V_1D'_e}{D'} \qquad (13)$$

$$E_2 = (90 - E_1) \qquad (14)$$

$$C_2 = (90 - E_1) - K \qquad (15)$$

$$F_2 = \frac{[(90 - E_1) + J]D'}{D'_e} \qquad (16)$$

DISCUSSION OF FORMULAS

The formulas for ascertaining the various dimensions and angles of the pinion are similar to those for any bevel gear, once the center angle is obtained.

The "equivalent pitch diameter" of the gear is the same as the pitch diameter of the regular bevel gear that would give the required speed ratio, and is obtained by dividing the number of teeth in the gear by the diametral pitch.

The angle of offset is the angle between the axis plane of the gear and a plane passing through the axis of the gear and the common contact point of the pitch diameters (outer) of the pinion and gear. Its tangent is obtained by dividing the offset of the pinion shaft by half the equivalent pitch diameter of the gear, or twice the offset of the pinion shaft divided by the equivalent pitch diameter of the gear.

The pitch diameter of the gear is then obtained by dividing twice the pinion shaft offset by the sine of the angle of offset.

The circular pitch of the gear is equal to the quotient of the pitch circumference by the number of teeth.

The tangent of the center angle of the pinion is found by dividing the pitch diameter of the pinion by the equivalent pitch diameter of the gear.

The outside diameter of the gear is usually found by adding to its pitch diameter twice the diameter increment of the pinion. This method is not quite accurate, as the diameter increment of the gear and pinion is only the same when there is no offset to the pinion shaft. The greater this offset is, the proportionally smaller does the diameter increment of the gear become.

A more accurate way of ascertaining the outside diameter of the gear is by the use of the second formula. This more accurate method is not absolutely correct, however, for it is based on the assumption that the decrease in diameter increment is proportional to the ratio of the equivalent pitch diameter to the pitch diameter of the gear, which is not absolutely true. The possible error is so small for any standard gear, however, as to be quite immaterial.

The center angle of each individual gear tooth is equal to the complement of the center angle of the pinion, so that the center angle of a skew bevel gear may be taken as the complement of the center angle of the pinion with which it is to run.

The cutting angle of the skew bevel gear is likewise the same for each individual tooth and is equal to the difference betwen the center angle of the gear and the angle decrement.

The face angle of the skew bevel gear as obtained by the formula given is not absolutely accurate, but the error is sufficiently trivial to be safely overlooked in practice, unless the face of the pinion is unusually wide and the pitch equally small. In such a case, the cut-and-try method of fitting the gear to the pinion is advisable, as the calculations involved for an accurate mathematical solution are extremely complex.

The face angle of a skew bevel gear would not be the same as that of a bevel gear matched to mate with the skew gear pinion unless the offset of the pinion shaft were zero. Such a condition, which would be that existing between a set of common bevels of proper proportions, would fix the minimum face angle for a skew bevel gear. The maximum face angle would occur with the pinion shaft's offset equal to half the pitch diameter of the gear and would be one of 90 degrees. Between these limits the face angle of a skew bevel gear may be anything, depending upon the difference in the pitch and equivalent pitch diameters of the gear. Formula (6) is derived on the assumption that the increase in face angle of the skew bevel gear, from that of a set of bevel gears of similar pitch diameters to the condition where there would be no rolling action, is governed by the ratio of the pitch diameter of the gear to its equivalent pitch diameter. This relationship is only approximately accurate, for the actual increase in face angle is not quite constant between its minimum and maximum values. For all practical shop requirements, however, formula (16) may be considered as correct. Any possible error that might arise would but slightly affect the total depth of tooth at the small end of the gear where it would be least noticeable or harmful.

EXAMPLE IN THE DESIGN OF SKEW BEVEL GEARS

Required a pair of skew bevel gears; 10 diametral pitch, 85 teeth in gear, 13 teeth in pinion; pinion shaft offset 1½ inches.

$$\text{Pitch diameter of pinion, } d' = \frac{13}{10} = 1.3''. \quad (1)$$

$$\text{Equivalent pitch diameter of gear, } D'_e = \frac{85}{10} = 8.5''. \quad (2)$$

$$\text{Angle of offset, } \theta,\ \tan.\ \theta = \frac{2 \times 1.5}{8.5} = 0.3529 \quad (3)$$

$$\theta = 19^\circ\ 26'.$$

$$\text{Pitch diameter of gear, } D' = \frac{2 \times 1.5}{0.33271} = 9.01'', \text{ say } 9.0''. \quad (4)$$

$$\text{Circular pitch of gear, } p'_1 = \frac{3.1416 \times 9}{85} = 0.33''. \quad (5)$$

$$\text{Center angle of pinion, } E_1,\ \tan.\ E_1 = \frac{1.3}{8.5} \quad (6)$$

$$= 0.1529$$

$$E_1 = 8^\circ\ 42'.$$

$$\text{Angle increment, } J,\ \tan.\ J = \frac{2 \times 0.15126}{13} \quad (7)$$

$$= 0.02327$$

$$J = 1^\circ\ 20'.$$

$$\text{Angle decrement, } K,\ \tan.\ K = \frac{2.314 \times 0.15126}{13} \quad (8)$$

$$= 0.02692$$

$$K = 1^\circ\ 33'.$$

$$\text{Face angle of pinion, } F_1 = (8^\circ\ 42') + (1^\circ\ 20') = 10^\circ\ 2'. \quad (9)$$

$$\text{Cutting angle of pinion, } C_1 = (8^\circ\ 42') - (1^\circ\ 33') = 7^\circ\ 9'. \quad (10)$$

$$\text{Addendum, } s = \frac{1.3}{13} = 0.1''. \quad (11)$$

$$\text{Diameter increment of pinion, } V_1 = 0.1 \times 0.1513 = 0.01513''. \quad (12)$$

$$\left.\begin{array}{r} \text{Outside diameter of gear, } D = 9 + 2 \times 0.01513 \\ \text{or, } D = 9 + \dfrac{2 \times 0.01513 \times 8.5}{9} \end{array}\right\} = 9.03''. \quad (13)$$

$$\text{Center angle of gear, } E_2 = (90 - 8^\circ\ 42') = 81^\circ\ 18'. \quad (14)$$

$$\text{Cutting angle of gear, } C_2 = (90 - 8^\circ\ 42') - 1^\circ\ 33' = 79^\circ\ 45'. \quad (15)$$

$$\text{Face angle of gear, } F_2 = \frac{[(90 - 8^\circ\ 42') + 1^\circ\ 20']9}{8.5} = 87^\circ\ 29'. \quad (16)$$

MACHINING SKEW BEVEL GEARS

Any of the machines used for cutting the ordinary type of bevel gear can be used for machining skew bevel gears, if simple adjustments or modifications are made. The carrying spindle of the machine must be offset from the plane of the cutting tool by a distance equal to the offset of the pinion shaft. The subsequent operations of cutting the teeth are similar to those employed in cutting plain bevel gears, the rotary adjustment of the gear being governed by the circular pitch of the gear, not its normal pitch which corresponds to the circular pitch of the pinion. The adjustments are somewhat more complicated than when cutting the simpler gears and must be performed with great care, as there is no common apex toward which to work. This adds to the difficulties of accurate workmanship and is the main reason why skew bevel gears are so seldom employed.

Another method of designing and cutting skew bevel gears that is sometimes employed is to make both the teeth of the gear and the pinion of spiral type. When this is done, the degree of obliquity of the teeth in the gear and in the pinion is made the same in order to facilitate manufacture and design. A layout for such gear is shown in Fig. 186.

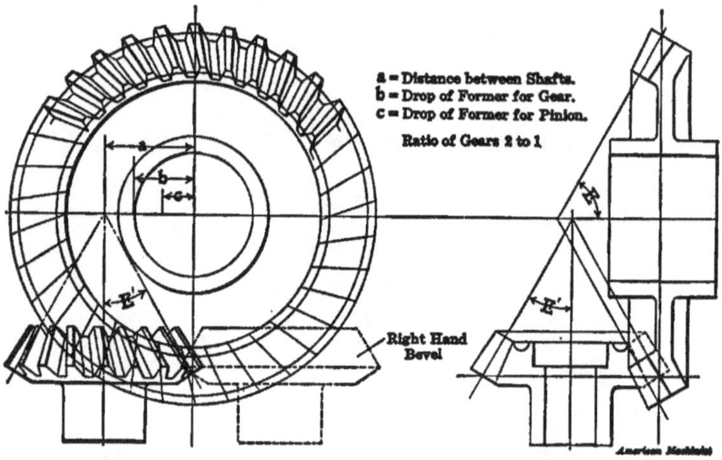

FIG. 186. LAYOUT FOR SKEW BEVEL GEARS.

These gears are turned up according to the dimensions for bevel gears of the same number of teeth, pitch, and ratio and no alteration in the diameters is usually made or is any alteration in the angles necessary, due to the fact that though the apex points of the two gears do not coincide, the converging conical surfaces are parallel to those of bevel gears with a common apex point.

Both gear and pinion are machined with the plane of the cutting tool offset from the carrying spindle of the machine. This offset is different for the two gears if their speed ratio is other than 1 to 1. For gears of similar dimensions, the total offset of the shafts would be divided in two and the correct offset between the spindle of the machine and the cutting tool plane would be one-half the total offset for both gears. For any other speed ratio, the total offset is divided proportionally to the ratio, the smaller offset being employed for cutting the pinion and the larger for machining the gear. For instance, when cutting skew bevel gears having a shaft offset of 2 inches and a speed ratio of 2 to 1, the machine drop or offset for the pinion would be $\frac{2}{2+1}$ = 0.666 inch and for the gear, 0.666 × 2 = 1.333 inches.

Skew bevel gears cut according to this apportioning method have proved very satisfactory, and the only criticism that can be advanced is on account of the decreased strength of the teeth as the gears are usually cut. The teeth being inclined to the circumference of the gear—that is, not being radial—the circular pitch must necessarily be greater than that of common bevel gears of similar proportions, for the circular pitch of the common bevels corresponds to the normal pitch of the skew bevel gears. This would necessitate an increase in diameters, the amount of increase depending upon the angularity of the teeth. If this is attended to, the full strength of the teeth will be developed.

The obliquity of the teeth of skew bevel gears of all varieties is the cause of one other annoyance, due to the unavoidable sliding action between the teeth. It has been found that if the common 14½-degree involute tooth is used the teeth do not clear properly. This has been overcome by makiug the angle 20 degrees. In extreme cases an even greater angle might have to be employed, but for any ordinary installation of skew bevel gears, the adoption of the 20-degree involute tooth will allow the teeth to clear satisfactorily.

SECTION X

Intermittent Gears

Intermittent gears are designed to allow the driven gear or follower one or more periods of rest during each revolution of the driver. This may be accomplished in a rough manner by cutting out a number of teeth in the follower as illustrated in Fig. 192, but the cut and try method must be employed to obtain a definite ratio. This type of intermittent gear is seldom used there being nothing but the spring *b* to keep the follower from moving during a period of rest, and the first tooth of the driver enters contact in a very uncertain manner, it sometimes being necessary to shorten the first tooth in the driver to prevent it from striking the top of the first tooth in the follower.

The proper design of intermittent gears is not as difficult as it first appears. The pitch and outside diameters are found as for an ordinary spur gear, the pitch desired must correspond to an even number of teeth. The blank space on the driving gear is milled to the pitch line, and the stops in the follower are cut by a cutter of a diameter corresponding to the pitch diameter of the driver. If no such cutter is at hand, use the nearest to that size to rough out the stops and finish them with a fly cutter which can be set to any desired radius.

It is well not to have the gears too near the same size; the driver should be the smallest in order to secure all the contact possible in the stops.

The simplest form of intermittent gear is shown by Fig. 193, the follower being moved but a short distance for each revolution of the driver. It will be noticed that a small amount of fitting will always be required at the point *a* to allow the point of the stop to clear.

A more complicated drive is shown in Fig. 194, the follower being moved one-sixth of a revolution for each revolution of the driver. Each of these gears is turned up as for a spur gear of 30 teeth 5 diametral pitch. The cutting operation would be as follows: Index for 30 teeth; cut the first three teeth, then index for two teeth without cutting, and so on around the blank. The six stops are then milled, with a cutter 6 inches in diameter, to a depth of 0.2 inch, or the addendum of the gear, which completes the follower. Four teeth are then cut in the driver, and the remainder of the blank milled to the pitch line. A little filing and the gears are complete.

A still simpler method of cutting these gears, and one that avoids the necessity of first laying them out, is as follows: Drop a cutter equaling the pitch diameter of the driver into the blank of the follower, to the depth of the addendum at the points stops are desired. Then cut the first tooth at a point midway between two of the stops, and continue cutting toward one of the stops until the point of the stop touching the outside circumference

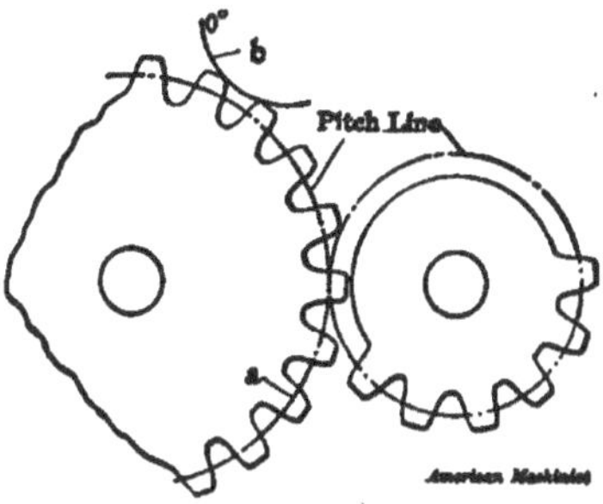

FIG. 192. POOR DESIGN OF INTERMITTENT GEARS.

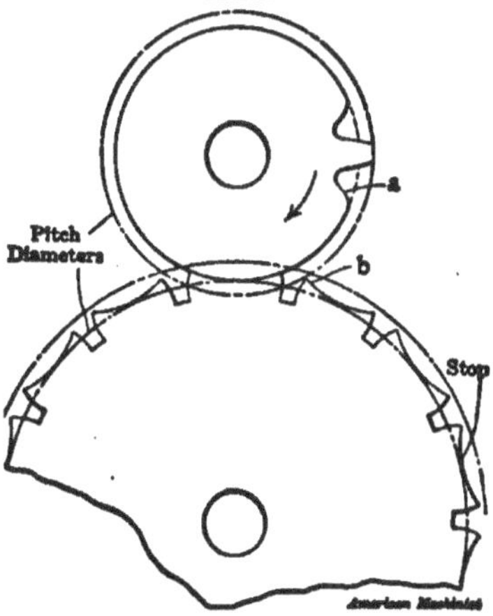

FIG. 193. INTERMITTENT GEARS WITH TWELVE STOPS.

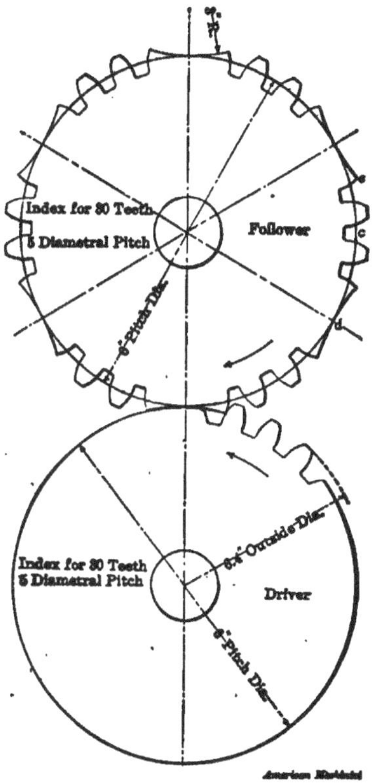

FIG. 194. INTERMITTENT GEARS WITH SIX STOPS.

of the blank is cut away, or, in other words, until there is no blank space on the gear between the last space cut and the point of the stop. The same number of teeth are then cut in the opposite direction until the same condition is met. If the stops are evenly spaced the cutting of the remaining teeth is a simple matter. If the stops are not evenly spaced the first tooth for each group must be located between each stop. The same number of teeth are then cut in the driver as there are spaces in the follower for each group and the remainder of the blank milled to the pitch line. For a pair of gears such as shown in Figs. 193 and 194, the cutting of the teeth by this process will be a simple matter.

The cutting of internal intermittent gears is a counterpart of the above. Bevel gears, while being more difficult to cut, are governed by the same rules (see Fig. 195).

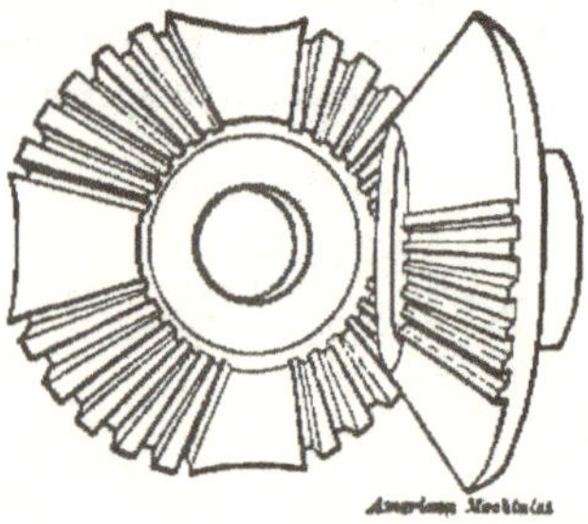

FIG. 195. INTERMITTENT BEVEL GEARS.

MODIFICATIONS OF THE GENEVA STOP

The accompanying engravings illustrate three highly ingenious and extremely interesting modifications of the device used in watches to prevent overwinding which have been applied by Mr. Hugo Bilgram to various automatic machines constructed at his works. The constructions have a family resemblance in principle, though they are entirely unlike from a structural standpoint.

Three main features characterize the constructions: First, the intermittent motion of the Geneva stop; second, the entire absence of shock at engagement or disengagement; and third, the positive character of the movement—the parts being locked in position both when they are in motion and when they are idle.

Fig. 196 represents the smallest departure from the watch mechanism. The interrupted disk *a* is the driver and revolves continually. The driven piece is seen at *b*, and the requirements are that the driven piece shall remain at rest during three-fourths of a revolution of the driver, and shall then make one-quarter of a turn during the remaining quarter turn of the driver. The driver may revolve in either direction, but supposing it to turn in the direction of the arrow, a roller *c* attached to the driver is about to enter one of four radial slots in the face of the driven piece. During the succeeding quarter turn of the driver the parts will move together, the motion of the follower ceasing when groove *d* has reached the position occupied in the figure by groove *e*, and this movement of the follower will obviously occupy 90 degrees of angle. It will be seen that the parts are so laid out that roller *c* enters and leaves the grooves tangentially, insuring absence of shock at both the commencement and the conclusion of the engagement.

At *fghi* on the follower is a series of rollers raised above the faces surrounding the grooves, and the circular part of the driving disk carries a circular groove *jkl* at such a radial distance as to engage these rollers in succession during the idle period of the follower and hence lock it in position. It is obvious that in the direction of motion supposed, this circular groove is just leaving roller *i*, and so disengaging it preparatory to movement by roller *c*. On the completion of the follower's movement, roller *f* will occupy the position of roller *g* in the illustration, while the end *l* of groove *jkl* will have turned to a position ready to embrace it and so lock the follower in position. With motion in the opposite direction, groove *jkl* in the position shown would be in the act of engaging roller *i* on the completion of the movement. Rollers *g* and *i* being at the same distance from the center *m* of the driver, both are engaged by the groove during a revolution, the locking taking place with one and the unlocking with the other, both rollers being in the groove during most of the time.

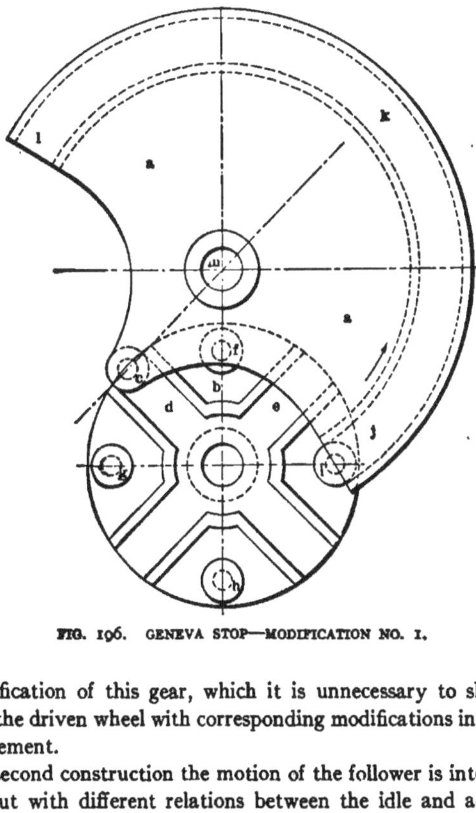

FIG. 196. GENEVA STOP—MODIFICATION NO. 1.

A modification of this gear, which it is unnecessary to show, has five grooves in the driven wheel with corresponding modifications in the character of the movement.

In the second construction the motion of the follower is intermittent like the last, but with different relations between the idle and acting periods. The driver runs continuously, the relationship being:

During $\frac{5}{6}$ turn of the driver the follower is at rest.

During $1\frac{1}{6}$ turn of the driver the follower makes one complete turn.

In other words, the follower makes one turn to every two turns of the driver, but this revolution of the follower occupies little more than a turn of the driver.

Pinned to the face of the driven gear is the plate *a*, Fig. 197, the arm *b* of which is fitted to embrace a hub *f* on the driver shaft, whereby, until released, the follower is locked in its idle position. Revolving with this hub is an arm *c* carrying a roller *d*, which is fitted to engage the slot *g*. As it does so, the notch *e* in hub *f* comes opposite finger *h*, thus disengaging the locking mechanism. Roller *d* enters tangentially without shock and accelerates the motion of the follower until the roller reaches the line of centers, when the gears engage and the motion goes on. The completion of the revolution of the driver finds the roller *d* again on the line of centers, but engaging slot *i*, and the continuance of the motion brings the parts again to the positions shown. It should be noted that in this mechanism not only is the starting and stopping of the driver without shock, but at the instant of engagement of the gear teeth the roller *d* has brought the velocity of the follower up to that due to the gears, so that the transition of the motion from the pin to the gears and back again from the gears to the pin is also without shock.

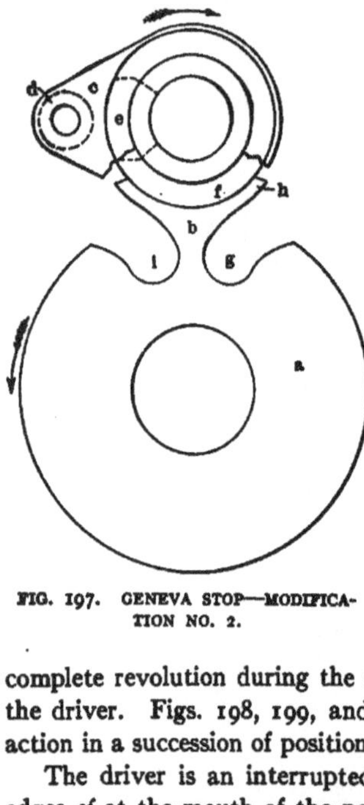

FIG. 197. GENEVA STOP—MODIFICATION NO. 2.

The most elaborate of these mechanisms is that shown in Figs. 198, 199, and 200. In this the driver—turning about *a*—is required to turn through about 73 per cent. of a revolution, while the follower stands still, the follower then making a complete revolution during the remaining 27 per cent. of the revolution of the driver. Figs. 198, 199, and 200 are side views intended to show the action in a succession of positions.

The driver is an interrupted disk *bcd*, Fig. 198, having cam-shaped edges *ef* at the mouth of the notch. Slightly in the rear of this disk is a toothed sector *g*. The incomplete driven gear *h* meshes with *g* during the acting periods, and the purpose of the remainder of the mechanism is to start *b* in motion and throw the teeth in mesh as well as to lock the follower in position during the period that it stands still. Fig. 198 shows the parts in position at the beginning of the movement of the follower, which is still locked in the position which it occupies during its idle period. A bar *i* on the front face of gear *h* rides on, and up to this point has been locked in the idle position by the disk *bcd*. Finger *j*—one of a pair *jk* —is attached to the rear of gear *h*, where it may turn freely between the sector *g* and the driving pulley *l*. This pulley *l* carries two rollers *mn* arranged to engage the fingers *jk* respectively. In the position of Fig. 198 the driving disk, moving

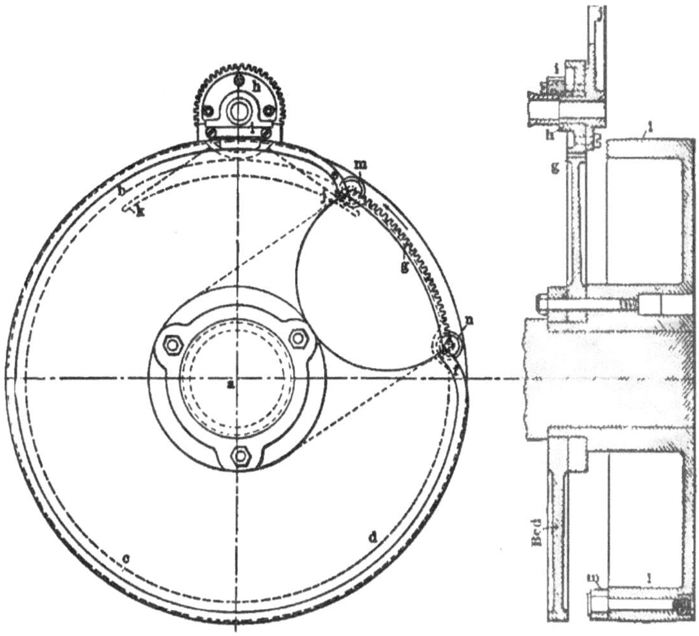

FIG. 198. GENEVA STOP—MODIFICATION NO. 3.

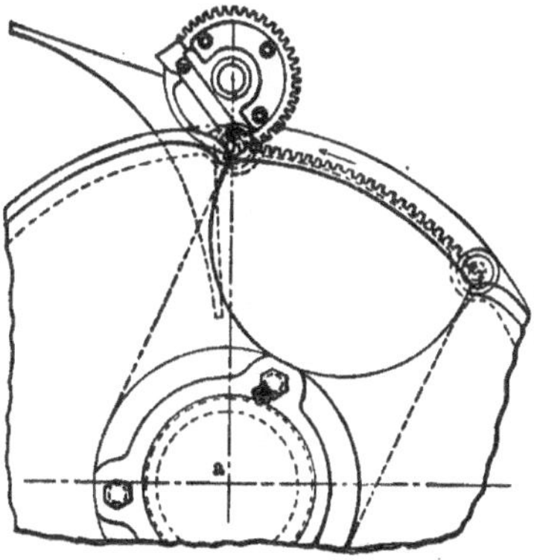

FIG. 199.

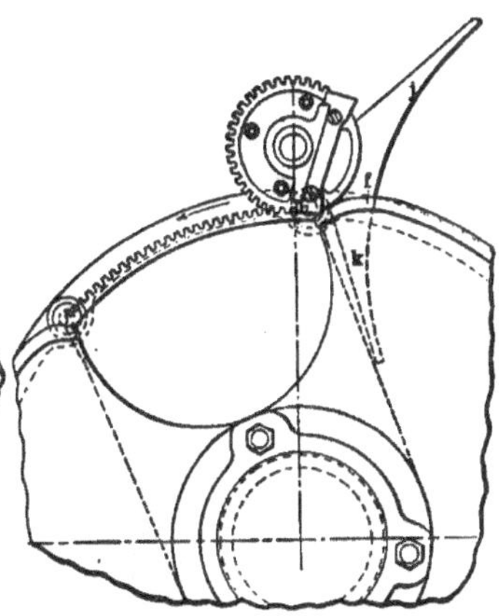

FIG. 200.

GENEVA STOP—MODIFICATION NO. 3.

in the direction of the arrow, has brought roller *m* into position, where it is about to engage finger *j*, the direction of the acting side of *j* being tangential to the motion of *m*, so that the movement begins without shock. To permit *j*, *b*, and *i* to turn, the edge *e* of the disk is dressed off, but to such a degree that contact is maintained between the right-hand end of *i* and the edge of the disk as the follower turns, so that the motion is positive without slack. As the movement progresses the speed on the follower increases until the position of Fig. 199 is reached, when—the roller being on the pitch circle of sector *g*—the speed of the follower is the same as that due to the gears and the teeth drop into mesh without shock. From this on the gears drive, and the arms *k* turn completely over, the position when the gears go out of mesh being

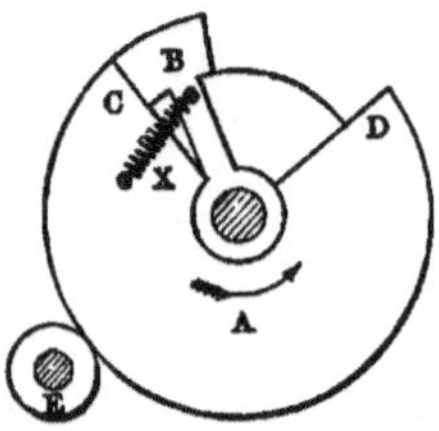

FIG. 201.

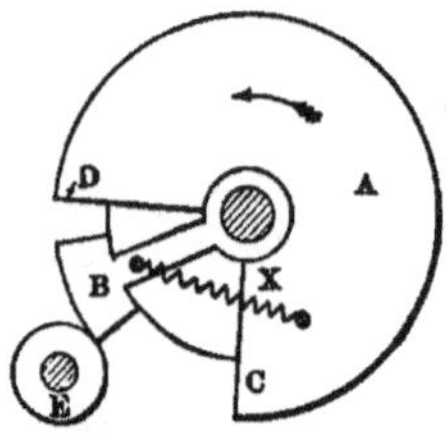

FIG. 202.

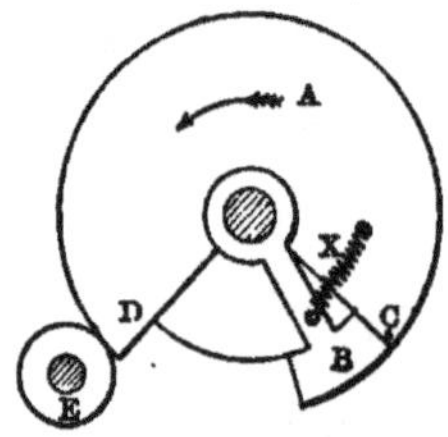

FIG. 203.

AN INTERMITTENT SPUR GEAR.

shown in Fig. 200. From this on the action is the reverse of that shown by Figs. 198 and 200, the driving piece being now the cam-shaped edge *f* of the disk, the finger *k* preserving the positiveness of the motion and preventing the driven pieces overrunning by momentum as they are brought to rest, the final stopping being accomplished as the roller slips out of action in a tangential direction, and again without shock. From the position of Fig. 200 to that of Fig. 198 the bar *i* simply rides on the edge of the driver disk and the follower remains at rest.

"It should be remembered that these mechanisms are not models designed to embody a pretty movement invented beforehand, but they are parts of machines, some of which are made in considerable numbers, and have been devised, as occasion arose, to accomplish certain required results. As such, they represent the art of invention carried to a high degree of perfection."

The line cuts (Figs. 201, 202, and 203) show a pair of intermittent spur gears in three positions. The peculiarity about this gear is that although a dwell occurs, the teeth of the gear and pinion are in mesh at all times.

The mutilated gear *A* is the driver and is secured to its shaft. A portion of its rim—dependent upon the length of dwell—is cut away. A segment *B* is mounted on the same shaft. This segment is free to swing in the cut-away portion of *A*, and is held in place against the side of *A* by a collar on the shaft. The teeth of *B* match with the teeth of *A* in both of its extreme positions. *B* is held against the face *C* by a spring *X*. This spring is elastic enough to allow *B* to move as far as *D*.

As the gear *A* moves in the direction of the arrow, it turns the pinion *E*. When *B* engages with *E*—the resistance of *E* being greater than the resistance of the spring *X*—the segment *B* remains stationary, while *A* moves till *D* comes in contact with *B*. During the time that *B* is at rest the pinion *E* is of course also at rest. As soon as *D* comes in contact with *B*, both *B* and *E* begin again to move. When *B* reaches a position where its teeth are no longer in engagement with the teeth of *E*, the spring *X* returns it to the face *C*. These gears were used as a feed gear for paper, the paper being cut during the dwell.

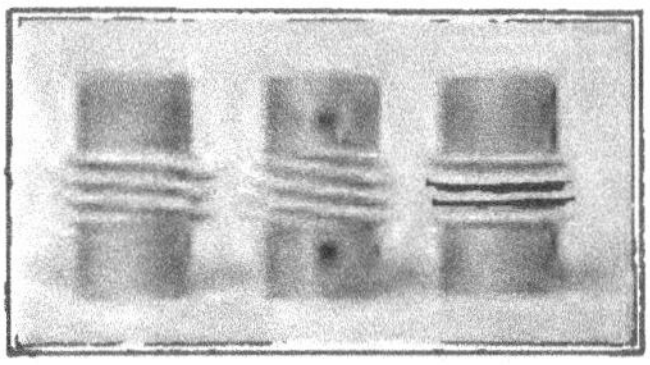

FIG. 204. AN INTERMITTENT WORM.

The half-tone, Fig. 204, shows three views of an intermittent worm used in a looping machine. The pitch is 1-6 inch. The dwell is two-thirds of a turn, and the advance the remaining third. It was cut on an ordinary 16-inch lathe, using a mutilated change-gear.

AN INTERESTING PAIR OF SPIRAL INTERMITTENT GEARS

Figs. 205 and 206 show a pair of intermittent gears having the peculiar characteristics that, if the large gear be rotated continuously in one direction, the pinion will rotate alternately three-quarters of a turn in one direction and one-quarter of a turn in the opposite direction, with a rest or dwell between each movement. Similar gears have been made as part of a certain machine, the nature of which we are not at liberty to mention. The angle of spiral of both gears is 45 degrees, and under ordinary conditions either gear could be the driver. The large gear, however, has around portions of its periphery two tongues—which are practically a continuation of certain of the teeth. These tongues fit into grooves in the pinion, and during their passage through the grooves lock the pinion at rest. Owing to this feature, the large gear must in this case be the driver.

The pinion has twelve teeth, divided into four groups of three teeth each, with one of the before-mentioned slots between each group. Two opposite groups of teeth are cut left hand, the two alternate groups are cut both left and right, leaving the teeth like a series of pegs.

Imagine the handle at a position opposite the pinion, then the left-hand teeth in the large gear will be at the left. There are nine teeth cut in this segment which, when the handle is turned—in the direction of the hands of a clock—engage first with the three teeth of a double-cut group on the pinion, then with the three left-hand full teeth of the next group, then with the three teeth of the other double-cut group. The pinion has then turned three-quarters of a revolution. The tongue then engages with the slot in the pinion and the rotation of the pinion is arrested. The large gear turns until the three right-hand teeth on its periphery come into mesh with the double-cut group first referred to, and reverse the direction of rotation of the pinion for a space of three teeth, or one-quarter turn, when the tongue again locks the pinion and the handle reaches the starting position. The large gear has thus made one turn and the pinion has advanced through three groups of three teeth each, equal to three-quarters of a turn, and has reversed through one group of three teeth, or one-quarter turn. Thus the total advance is but two groups of three teeth, or one-half turn, and to bring the gear and pinion into the

FIG. 205. THE RIGHT-HAND TEETH IN MESH WITH THE "PEG" SEGMENT.

FIG. 206. THE LEFT-HAND TEETH IN MESH.

same relative position as they were at the start the wheel must make another complete turn.

The blank for the large gear was turned to the extreme diameter across the top of tongue, mounted in the milling machine, and the left-hand teeth, which extend clear across the face, were gashed slightly below the level of the top of the tongue. The blank was then indexed halfway around from the central left-hand space, the table swung for right hand, and a deep right-hand gash made for locating. Then the blank was indexed halfway around from the central right-hand space, the table swung back for left hand, and the left-hand teeth section milled down to the proper gear diameter. Then the full-length left-hand teeth were cut, and also a gash made outside of the end teeth which join the tongue so as to get the curve of the tooth, but not going far enough to touch the tongue. The blank was then indexed back to the right-hand locating space; the table swung, blank milled down to the proper gear diameter, and the right-hand teeth were cut the same as the left-hand. The table was then set square, and an end mill was put into the spindle, and the stock milled away—leaving the tongue. It will be noticed that it was not necessary to remount the blank at all. The outsides of the teeth joining the tongue were chipped out as smooth and true to curve as possible.

An important feature was to get the angle of the end of the tongue just right, and not file it too far back, which would have allowed the large gear

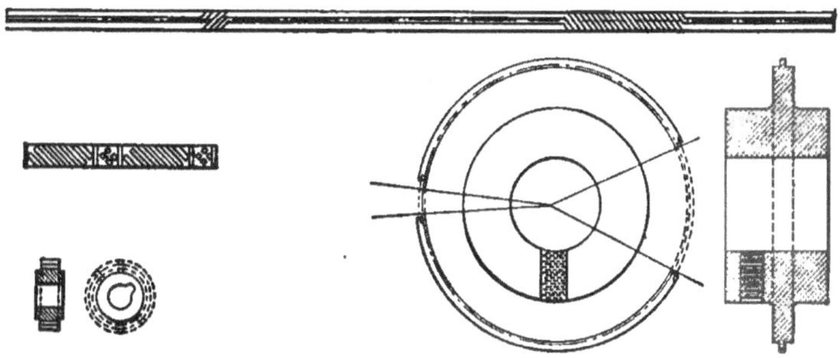

FIG. 207. LAYOUT OF A PAIR OF INTERMITTENT SPIRAL GEARS.

to carry the small one somewhat beyond the point where the tongue should enter the slot, thus causing the end of the tongue to strike the side of the small gear and arrest the movement or perhaps cause breakage. The cutting of the small gear was simply a matter of setting the cutter directly over the center of the blank and swinging the table for both hands; the slots were cut at the last machine operation, and then the little sharp projections which were left at the ends of the crosscut sections and near the slots were chipped off, as they were surplus and would interfere with the movement of the gears.

Fig. 207 shows layout of a set of the same style of gearing as Figs. 205 and 206, but of a different ratio.

The gears were cut by the Boston Gear Works, and were prepared preliminary to cutting into more expensive blanks. The work was accomplished by the usual methods, and the finished gearing accomplished the desired results and is now in successful use.

SECTION XI

ELLIPTICAL GEARS

Elliptic gears are in general use on shapers, planers, slotters and similar machine tools to transmit a quick-return motion to the ram. The cost of production is more than for gears having circular pitch lines, but they are undoubtedly the cheapest quick-return motion among known mechanical movements. The method to be outlined is not new, but is as accurate as any and the cost of the tools is not high. This method is in general use in many shops to the exclusion of other methods not considered as good.

In order more clearly to describe the process it will be well to take an actual case and carry it through from start to finish. Assume that a pair of gears is ordered to transmit a 3 to 1 quick-return motion and the centers are 8 inches apart. About 3 diametral pitch is specified.

METHOD OF LAYING OUT*

Fig. 208 represents diagrammatically the method of laying out the gears. Lines AA' and BB' are first drawn perpendicular to each other. The major axis of the pitch line of the gear is the same as the given center distance, 8 inches, and is laid off as shown. The focus points XX' are drawn in so that AX is in the same ratio to $A'X$ as the given quick-return ratio. The points A and B' are located by setting the dividers at one-half the major axis BA' and cutting the minor axis with arcs having centers at X and X'. Arcs having radii equal to OB' and OA' are drawn covering one quadrant, as shown. This quadrant is divided into six equal divisions, as shown by radii from the center O. These radii are marked a, a^1, a^2, etc. Next intersecting lines are drawn from the intersection points of a, a^1, a^2, etc., with the circles, as shown, and the intersection points of these perpendiculars are points on a perfect ellipse.

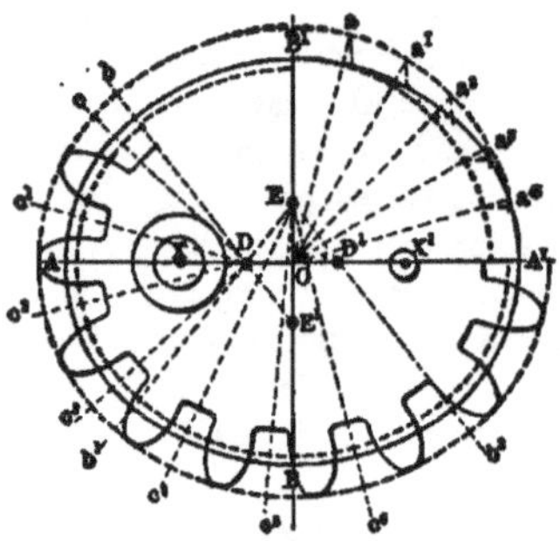

FIG. 208. LAYOUT FOR AN ELLIPTIC GEAR.

"With a center on AA^1 find an arc that will very nearly cut the points from a^4, a^3, a^2, and A^1. Also find a center on BB^1 about which an arc can be drawn cutting points B^1, a, a^1, and a^2. These centers are used when cutting

* W. E. Thompson.

the teeth and should be laid out as accurately as possible. After finding two centers and measuring the respective distances from O, the other two centers may be put in and an ellipse drawn, as shown. This line is then stepped off with dividers set at the corrected tooth thickness for the desired pitch and the number of divisions noted.

"It is preferable to have an odd number of teeth for convenience in cutting, so if the pitch line does not divide into an even number of divisions, half of which is an odd number, when the dividers are set to the chordal-tooth thickness of the desired pitch, the pitch line may be reduced or enlarged, as is found necessary. When the pitch is given, as it was in this case, the pitch line may be divided and the divider division measured. The corresponding

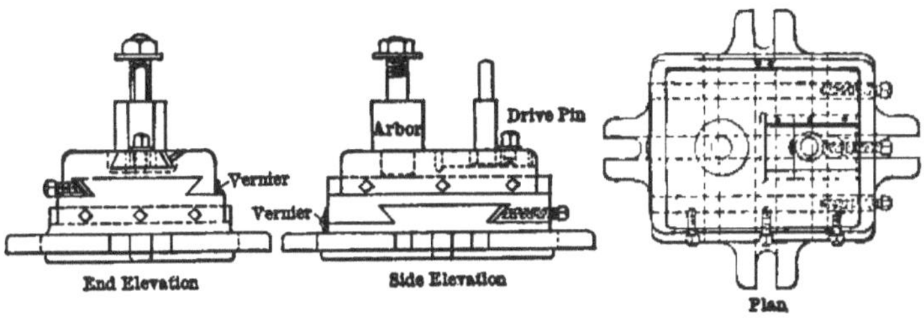

FIG. 209. FIXTURE FOR CUTTING ELLIPTICAL GEARS.

pitch is used in selecting a cutter. In this case the line was divided into 30 divisions which corresponded very close to 2¾ pitch, so these cutters were used.

"Radial lines common to two centers of the elliptic arcs are drawn, as shown at b, b^1, b^2, and in the other quadrant not drawn. These lines are the dividing points between two different arcs and are drawn before laying out or cutting the teeth. Radial lines from the four centers are drawn, through the centers of the space divisions, as shown at c, c^1, c^2, etc. These lines are for the purpose of starting the first cut and checking the rest to prevent large errors.

"The base circles of the two different curves are drawn, as shown, and a few teeth laid out by an accurate odontograph on each curve, as shown, or cutters may be selected by measuring the pitch radii DA and EB and figuring the proper number of teeth to cut. Cutters are selected by one or the other of these methods and the gear, previously having the shaft and driving pin holes X and X^1 bored and the addendum outline roughed, is ready for cutting.

METHOD OF CUTTING

"The fixture shown at Figs. 209 to 211 is bolted onto a circular milling attachment. The fixture consists of a base plate carrying two slides moving at right angles to each other. These slides are equipped with verniers registering zero when the center of the arbor is in line with the center of rotation of the milling attachment. The drive pin is riveted into a separate slide that is adjustable and is in line with the arbor and line of motion of the large slide. This fixture is placed in position on the machine and the rough-gear blank clamped in place.

The point D^1, Fig. 208, is then set over the center of rotation of the attachment by means of the verniers and the center of the cutter brought into the line AA^1. In this case the cutter is an end mill used to finish the outside of

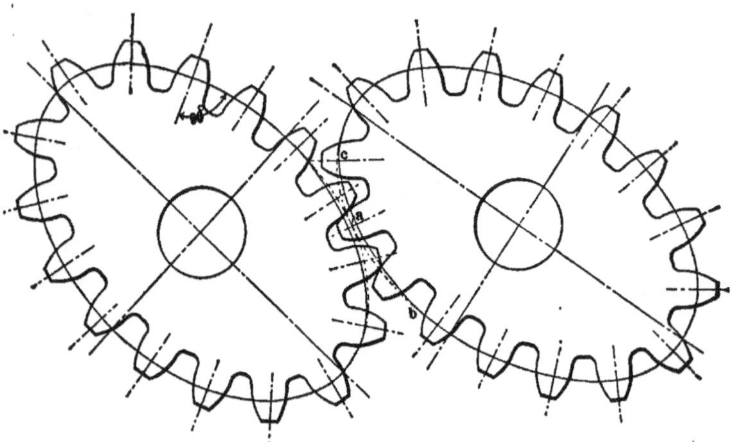

FIG. 210. ELLIPTICAL GEARS WITH BORE IN CENTER, SHOWING SEPARATION OF TRUE ELLIPTICAL PITCH LINES.

the gear. By moving around each quadrant setting the corresponding centers over the center of rotation for each quadrant the outside is milled true with the pitch line. A cutter for the ends is then put in the machine and in line with AA^1. The first space may then be cut to depth with two or more cuts, depending on the tests for alignment. A piece of glass having a center line and a number of short lines scratched equal distances apart and from the center line is used for testing the cuts. The center line is placed over the line AA^1 on the first cut, and the short lines brought intersecting the pitch line. By using a glass a variation of 0.001 inch may be readily seen and remedied. After the first space is cut the gear-tooth calipers may be set and used to check the cutting. By checking with both the radial center lines and the tooth calipers errors are reduced to a minimum. After the first gear is cut it may be used as a templet for all others, or the indexing may be noted and repeated."

Several years ago the author had an experience cutting elliptical gear that will be of interest. The driven gear was required to have two variations of speed per revolution; therefore the bore was put in the center of the gears instead of at the foci, which is usual. The gears cut are shown in Fig. 214. When the ellipse is very flat the four-arc method, described above, cannot be employed, as the pitch line cannot be described even approximately correct by four arcs. Therefore the gears were cut as illustrated by Fig. 211, the blank being set for each tooth cut. The teeth were first located on a templet which was secured to the face of one of the gears; they were both cut at one

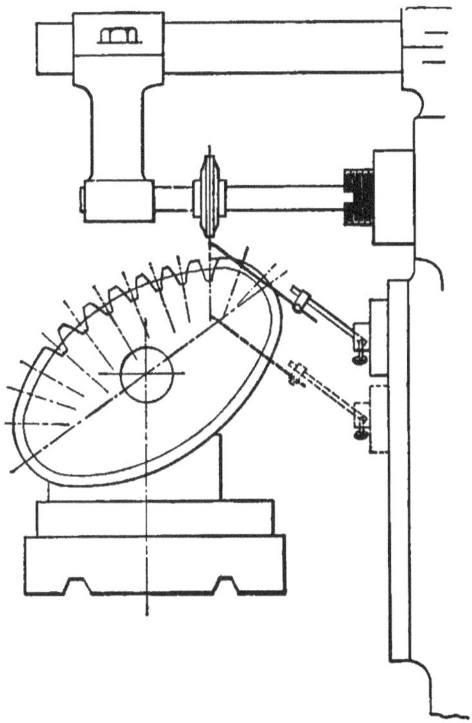

FIG. 211. CUTTING THE TEETH.

time. The blank was manipulated until the center line of each tooth space was brought in line with the center line of the cutter.

This is accomplished with a surface gauge placed against the front ways of the milling machine as illustrated in Fig. 211. The depth of the teeth was obtained by first bringing the cutter to the outside of the templet and raising the table a distance equaling the depth of the tooth, or by locating the pitch line of the templet with another surface gauge from the cutter spindle.

A cutter made to finish the outside diameter as the teeth are cut is a decided improvement over first milling or slotting the outer surface, although milling off the points of the teeth after they are cut is the next best plan. When the gears were mounted as shown in Fig. 210, it was found that the pitch lines separated at four points of the ellipse, as between *b* and *c*, therefore there was excessive backlash between the teeth at these points.

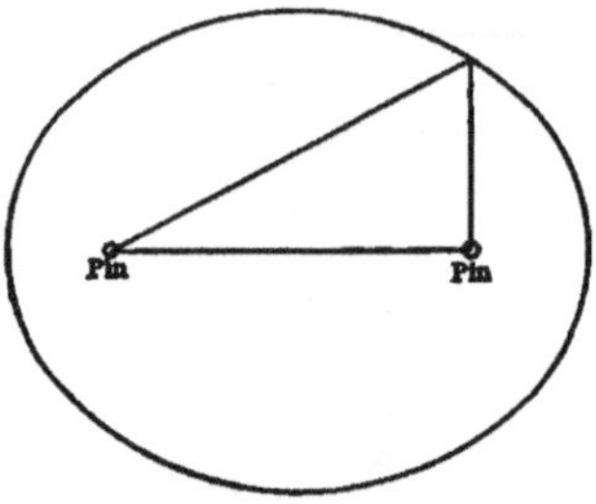

FIG. 212. THE GARDENERS' ELLIPSE.

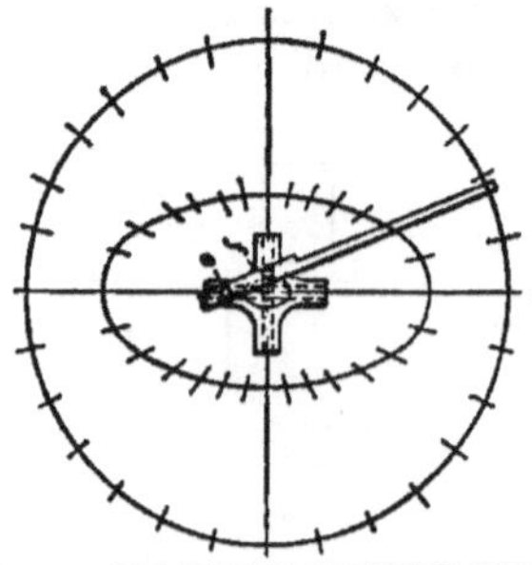

FIG. 213. THE METHOD EMPLOYED IN LAYING OUT GEARS SHOWN IN FIGS. 214 AND 215.

The gears had been laid out with a gardeners' ellipse, using a piece of silk thread looped around pins set in the foci; the loop being adjusted until the describing point passed through the intersections of both the major and minor

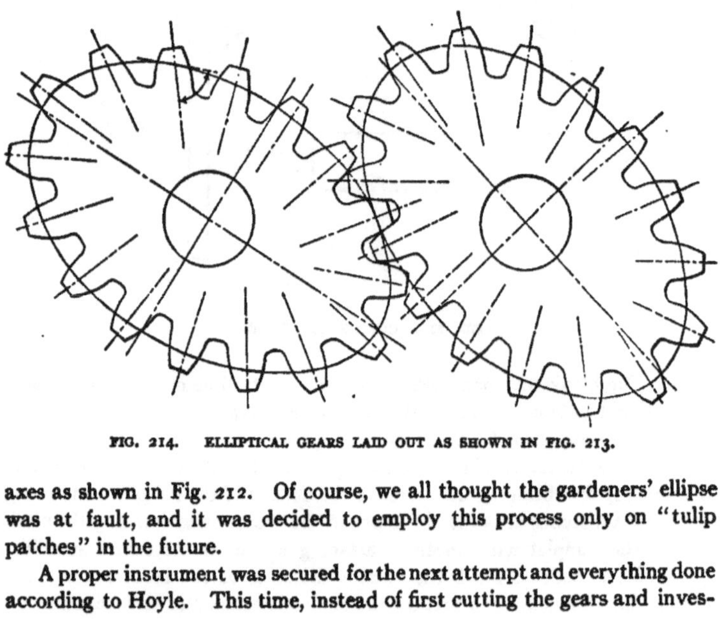

FIG. 214. ELLIPTICAL GEARS LAID OUT AS SHOWN IN FIG. 213.

axes as shown in Fig. 212. Of course, we all thought the gardeners' ellipse was at fault, and it was decided to employ this process only on "tulip patches" in the future.

A proper instrument was secured for the next attempt and everything done according to Hoyle. This time, instead of first cutting the gears and investigating afterward, two templets were laid out to represent the pitch lines of the gears. These templets developed the same error found in the first gears, so the pitch lines were corrected as per dotted lines in Fig. 210 and the gears were satisfactory.

Later on several gears of the same size were required and a search was made for a simpler method of laying them out. Noticing the attachment

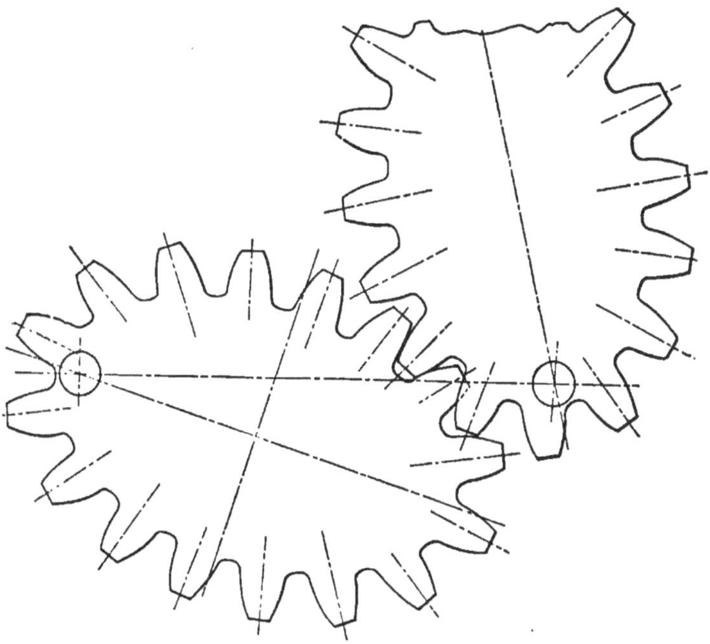

FIG. 215. INTERFERENCE OF GEARS SHOWN IN FIG. 214, WHEN BORE IS AT FOCI.

described by Mr. George B. Grant in his "Treatise on Gears," section 150, an attempt was made to apply this principle. A circle whose radius equalled the radii of both the major and minor axes was first drawn and divided into the number of teeth to be cut in the gear. The ellipse was then described from the same center. The center lines of the teeth were then projected from the points located on the outer circle to the ellipse, the edge of the blade being on a line with the points *e* and *f* as shown in Fig. 213. The center line of the tooth thus projected did not always cross the pitch line at right angles, except at the major and minor axes, and doubts were expressed as to the success of gears cut on these lines, but it was thought worth a trial at least. The pitch lines were corrected in the same manner as before and the gears were cut.

They not only ran better than the first gears, but it was found necessary to remove part of the correction between the points *b* and *c*. This is accounted for by the fact that the circular thickness of the teeth increased between these points owing to the obliquity of the teeth; the thickness of the tooth being measured on the normal section (see Fig. 214).

Gears with the bore located at the foci will not operate when cut in this manner; the center lines of the teeth must be at right angles to the pitch line at all points. The interference of the teeth when thus mounted is shown in Fig. 215.

When the bore is in the center of an elliptical gear it is better to have an even number of teeth, otherwise the gears must be cut separately. For an odd number of teeth the tooth centered on the major axis must engage a space located at the minor axis of the engaging gear. It is better to use an even number and place the edge of the teeth on the major and minor axes, for gears of this type.

For a complete treatment of the subject of elliptical gears Grant's "Treatise on Gear Wheels" is to be recommended.

SECTION XII

Epicyclic Gear Trains

CALCULATIONS RESPECTING EPICYCLIC GEAR TRAINS

DERIVATION OF FORMULAS FOR SEVERAL USUAL TYPES, AND EXTENSION OF THE METHOD OF ANALYSIS TO A SOMEWHAT COMPLEX EPICYCLIC TRAIN*

This form of gearing, which is really that in which one gear revolves around the center of one with which it is in contact, has received considerable attention, and one notices its use in several directions. We will therefore look into some of the calculations respecting it, leading from the simpler to the more complex.

SIMPLE PAIR OF GEARS IN FIXED BEARINGS

Example I.

If in Fig. 216 R and N are two gears in mesh, r and n being their respective numbers of teeth, their bearings being fixed, then:

$$\frac{\textit{Velocity of drive N gear N}}{\textit{Velocity of drive R gear R}} = \frac{r}{n};$$

or,

$$\textit{N's velocity} = \textit{R's velocity} \times \frac{r}{n}.$$

If, however, R revolves in a positive direction, n must revolve in the opposite, that is, in a negative direction.

$$\therefore \textit{N's velocity} = \textit{R's velocity} \times \frac{r}{n}. \qquad (1)$$

In all these calculations it is essential that great care be taken in order to obtain the correct sign of the resulting velocity.

GEARS IN FIXED BEARINGS, WITH AN IDLER

Example II.

An intermediate gear I is placed in contact with both N and R, Fig. 217. The effect will be that of giving N motion in the same direction as R.

$$\therefore \textit{N's velocity} = \textit{R's velocity} \times \frac{r}{n}. \qquad (2)$$

* Francis J. Bostock.

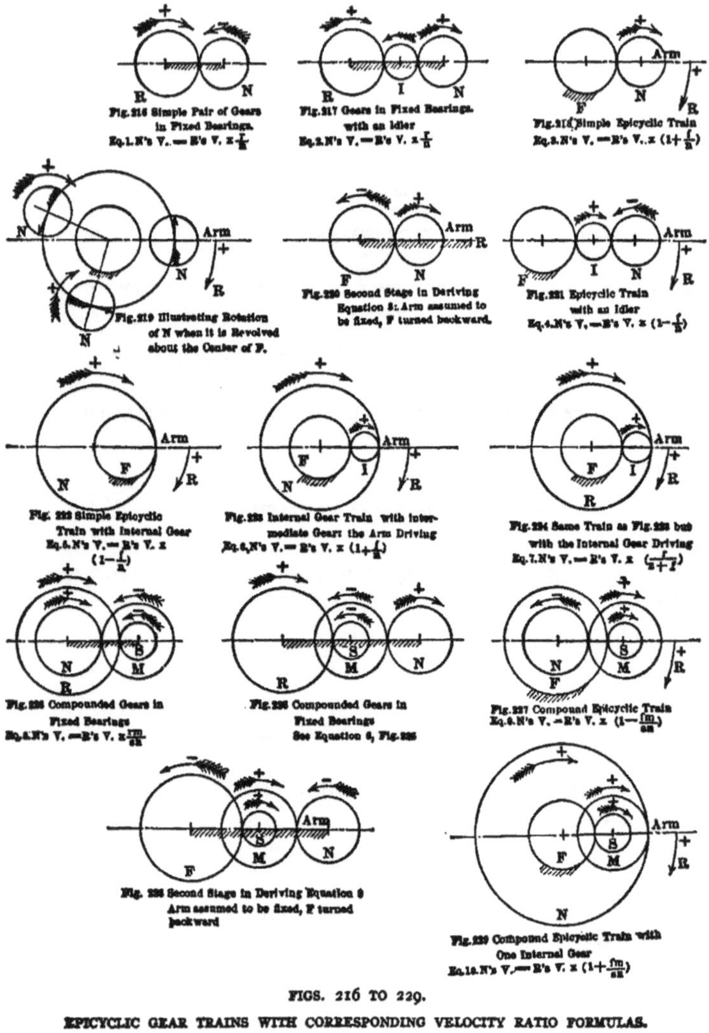

Fig. 216 Simple Pair of Gears in Fixed Bearings
Eq. 1. N's V. = R's V. x r/n
Fig. 217 Gears in Fixed Bearings with an Idler
Eq. 2. N's V. = R's V. x r/n
Fig. 218 Simple Epicyclic Train
Eq. 3. N's V. = R's V. x (1 + f/n)
Fig. 219 Illustrating Rotation of N when it is Revolved about the Center of F.
Fig. 220 Second Stage in Deriving Equation 3: Arm assumed to be fixed, F turned backward.
Fig. 221 Epicyclic Train with an Idler
Eq. 4. N's V. = R's V. x (1 − f/n)
Fig. 222 Simple Epicyclic Train with Internal Gear
Eq. 5. N's V. = R's V. x (1 − f/n)
Fig. 223 Internal Gear Train with intermediate Gear: the Arm Driving
Eq. 6. N's V. = R's V. x (1 + f/n)
Fig. 224 Same Train as Fig. 223 but with the Internal Gear Driving
Eq. 7. N's V. = R's V. x (r/(n + f))
Fig. 225 Compounded Gears in Fixed Bearings
Eq. 8. N's V. = R's V. x rm/sn
Fig. 226 Compounded Gears in Fixed Bearings
See Equation 8, Fig. 225
Fig. 227 Compound Epicyclic Train
Eq. 9. N's V. = R's V. x (1 − fm/sn)
Fig. 228 Second Stage in Deriving Equation 9
Arm assumed to be fixed, F turned backward
Fig. 229 Compound Epicyclic Train with One Internal Gear
Eq. 10. N's V. = R's V. x (1 + fm/sn)

FIGS. 216 TO 229.

EPICYCLIC GEAR TRAINS WITH CORRESPONDING VELOCITY RATIO FORMULAS.

SIMPLE EPICYCLIC TRAIN

Example III.

Two gears, *F* and *N*, are in mesh, the centers of which are on the arm *R*, which is capable of revolving around the center of *F*. It is required to find the velocity ratio between *R* and *N* when *R* revolves around the fixed gear *F*;

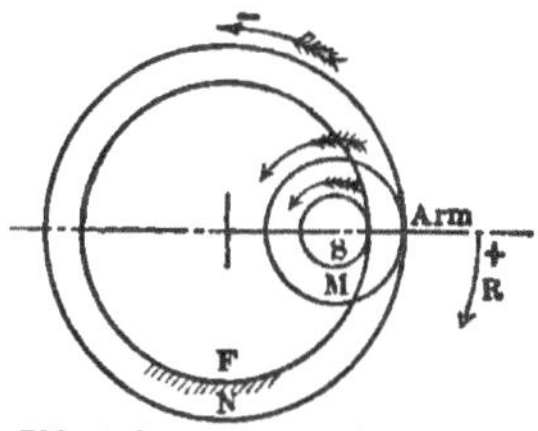

FIG. 230. Compound Epicyclic Train with Two Internal Gears See Eq. 8, same as Fig. 227

NOTATION

R — Denotes Driving Gear, or, in some cases, Arm.
r — Number of Teeth in Driving Gear.
N — Denotes Driven Gear or Arm.
n — Number of Teeth in Driven Gear.
I, S and M denote Intermediate Gears.
F — Denotes Fixed Gear.
f — Number of teeth in it.
V — Angular Velocity.
Denotes part which is fixed.

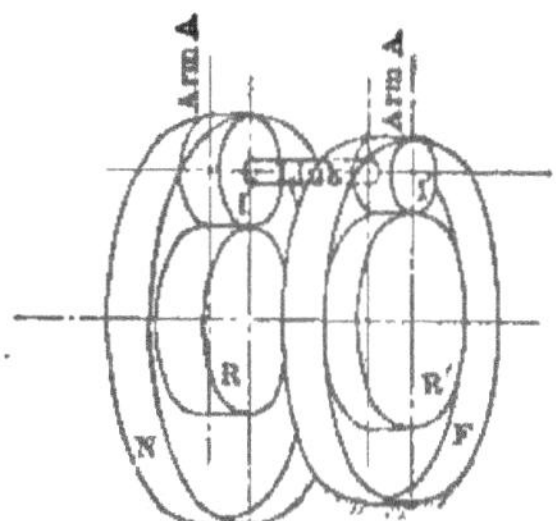

FIG. 231 An Epicyclic Train Consisting of Two Central Gears, One Arm carrying Two Planetary Gears, and Two Internal Gears, One of which is Fixed.

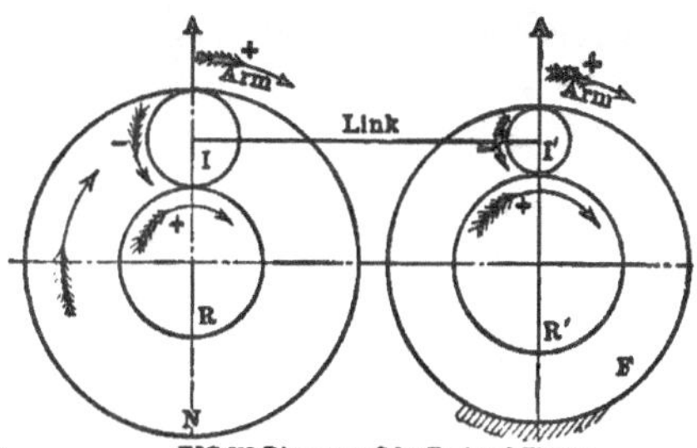

FIG. 232 Diagram of the Train of Fig 231 Eq. 11. N's V. = R's V. x $\left(\frac{r'n-rf}{n(r+f)}\right)$

FIGS. 230 TO 232.

EPICYCLIC GEAR TRAINS WITH CORRESPONDING VELOCITY RATIO FORMULAS.

Fig. 218 shows the arrangement. The gear *N* is subject to two motions due to the following two conditions:

a. The fact of its being fixed to the arm *R*.
b. The fact that it is in contact with the gear *F*.

We will therefore in the first place suppose that they are not in gear, and that *N* cannot rotate on the arm *R*. Then if *R* makes one revolution around *F* it is obvious that *N* must also make one revolution around *F*, as in Fig. 219.

$$\therefore \textit{N's velocity, due to condition a,} = \textit{R's velocity,}$$

the direction being the same as *R's*.

Secondly, if instead of *R* making one revolution around *F* in a + direction, we cause *F* to make one in the opposite, that is, negative direction, we shall have exactly the same effect. Therefore place F and N in mesh, and fix the arm R, as in Fig. 220.

Then if F makes -1 revolution, N will make $+\frac{f}{n}$ revolutions. (According to Equation 1.)

But -1 of $F = +1$ of R.

$$\therefore 1 \text{ revolution of } R = \frac{f}{n} \text{ revolutions of } N.$$

or,

$$N\text{'s velocity due to conditions } b = \frac{f}{n} R\text{'s velocity.}$$

By addition we obtain the total impulses given in N, that is:

$$\begin{aligned} N\text{'s velocity} &= R\text{'s velocity} + \frac{f}{n} R\text{'s velocity} \\ &= R\text{'s velocity}\left(1 + \frac{f}{n}\right). \end{aligned} \tag{3}$$

EPICYCLIC TRAIN WITH IDLER

Example IV.

If an intermediate gear I be inserted between F and N, as in Fig. 221, we have a similar case to the above; but the intermediate gear has the effect of changing the direction of revolution of N (Equation 2), due to its contact with F through I.

$$\therefore N\text{'s velocity} = R\text{'s velocity} \times \left(1 - \frac{f}{n}\right) \tag{4}$$

It will be seen that if $f = n$, N will not have any motion of rotation at all; and it will have a positive one if $f < n$ and negative if $f > n$. Thus by the adjustment of f and n one can obtain great reduction in speed by means of few moving parts.

SIMPLE EPICYCLIC TRAIN WITH INTERNAL GEARS

Example V.

Instead of the driven gear N being external, it might have been internal, as shown in Fig. 222. The effect will be the same as inserting an intermediate gear in Example III, giving the same result as Case IV, namely:

$$N\text{'s velocity} = R\text{'s velocity} \times \left(1 - \frac{f}{n}\right). \tag{5}$$

In this case $n > f$.

$\therefore$ The final direction is always $+$.

INTERNAL GEAR EPICYCLIC WITH INTERMEDIATE GEAR

Example VI.

Fig. 223 shows a still further modification of this condition, I being an intermediate gear. The result is:

$$N\text{'s velocity} = R\text{'s velocity} \times \left(1 + \frac{f}{n}\right). \tag{6}$$

THE SAME TRAIN WITH THE INTERNAL GEAR DRIVING

Example VII.

With the above type, one often arranges the outer internal gear to be the driver, imparting motion to the arm carrying the intermediate gear (see Fig. 224).

We have seen by equation 6 that:

$$\frac{N\text{'s velocity (driven)}}{R\text{'s velocity (driver)}} = \frac{1}{\left(1 + \frac{f}{n}\right)}$$

$$\therefore N\text{'s velocity} = R\text{'s velocity} \div \left(1 - \frac{f}{r}\right)$$

$$= R\text{'s velocity} \times \left(\frac{r}{r + f}\right). \qquad (7)$$

The last two examples constitute what is known as the "Sun and Planet" gear, which is largely used in many mechanisms. All the above examples show "simple" gearing, but they can be compounded with great advantage.

COMPOUND GEARS IN FIXED BEARINGS

Example VIII.

Gears compounded together are shown in Figs. 225 and 226, 226 being a diagram of 225. One repeats the well-known rule that:

$$\frac{\text{Velocity of driven gear}}{\text{Velocity of driver gear}} = \frac{\text{Product of number of teeth of driver gears}}{\text{Product of number of teeth of driven gears}}$$

$$\text{or, } N\text{'s velocity} = R\text{'s velocity} \times \frac{r \times m}{s \times n}. \qquad (8)$$

The direction is the same as N's, namely, $+$.

COMPOUND EPICYCLIC TRAIN, WITHOUT INTERNAL GEAR

Example IX.

We will now arrange to fix one of the gears F, and by means of the arm R revolve the others around it, thereby causing N to revolve as shown in Figs. 227 and 228. As before, we will assume the gears M and S to be out of mesh, so that when the arm R, carrying with it the gear N, makes one revolution around F, N must also make one revolution relatively to F. Also when they are in mesh, the arm R being fixed and F makes one revolution in a negative direction (see Fig. 228), N will make $-\frac{fm}{sn}$ revolutions. (Equation 8.)

Now the total motion imparted to N must be the sum of these two, namely:

$$1 \text{ revolution of } R = 1 - \frac{fm}{sn} \text{ revolutions of } N,$$

or,

$$N\text{'s velocity} = R\text{'s velocity} \times \left(1 - \frac{f \times m}{s \times n}\right). \qquad (9)$$

COMPOUND EPICYCLIC TRAIN WITH ONE INTERNAL GEAR

Example X.

Fig. 229 shows a slight modification of the last case, N being an internal instead of an external gear. Obviously the only difference will be in the direction of N's motion, that is:

$$N\text{'s velocity} = R\text{'s velocity} \times \left(1 + \frac{fm}{sn}\right). \qquad (10)$$

COMPOUND EPICYCLIC TRAIN WITH TWO INTERNAL GEARS

Example XI.

A further modification, however, is one in which both F and N are internal gears (Fig. 230), the effect of such being a change of sign in the equation.

$$\therefore N\text{'s velocity} = R\text{'s velocity} \times \left(1 - \frac{fm}{sn}\right). \qquad (9)$$

The type shown in Figs. 227 and 230 is, perhaps, one of the best methods of obtaining a good reduction of speed in an easy and cheap manner.

There are several combinations of the examples shown, but as they are all somewhat similar we will take another typical case as a guide for future calculations.

AN EPICYCLIC TRAIN CONSISTING OF TWO CENTRAL GEARS, ONE ARM CARRYING TWO PLANETARY GEARS, AND TWO INTERNAL GEARS, ONE OF WHICH IS FIXED

Example XII.

The writer has successfully used the arrangement shown in Figs. 231 and 232, in which R and R' are two spur gears mounted on one shaft; I and I' are two "planet" pinions, while F and N are two internal gears, the former being fixed. R and R' are made to revolve, which has the effect of giving N a very slow speed.

A SCHEME FOR FINDING THE VELOCITY RATIO

As this is somewhat complicated, we will work it out in stages:

1. Obtain the revolutions of the arm A when R' makes one revolution, F, of course, being fixed.

2. Obtain N's revolutions when the arm A is fixed and R makes one revolution.

3. Assume R fixed, and that the arm makes one revolution; obtain, then, N's revolutions.

4. Then if N makes so many revolutions to one of the arm, as given by stage 3, we can by proportion obtain how many will be caused by the amount given by Stage 1.

5. Add the results of 2 and 4 together, and obtain the motion given to N by one revolution of R, which is the desired result.

THE SCHEME WORKED OUT

Working the above out we obtain:

1. When F is fixed and R' makes one revolution, the arm A must make $+\frac{r'}{r'+f}$. (According to Equation 7.)

2. R makes one revolution, arm A being fixed; then N must make $-\frac{r}{n}$ revolutions. (According to Equation 2. Negative sign used because of the internal gear.)

3. When R is fixed and arm A makes one revolution, N will make $+\left(r+\frac{r}{n}\right)$ revolutions. (According to Equation 6.)

4. With one revolution of arm, N makes $1+\frac{r}{n}$ revolutions, from Stage 3; $\therefore$ with $\frac{r'}{r'+f}$ revolutions of the arm, as derived in Stage 1, N will make

$$\left(1+\frac{r}{n}\right)\times\left(\frac{r'}{r'+f}\right).$$

5. The aggregate is the sum of the effects derived in Stages 4 and 2, namely, to one of R, N makes:

$$\left(1+\frac{r}{n}\right)\times\left(\frac{r'}{r'+f}\right)+\left(-\frac{r}{n}\right)=\frac{(n+r)r'}{n(r'\text{Œ}f)}-\frac{r}{n}$$
$$=\frac{rr'+r'n-rr'-rf}{n(r'+f)}=\frac{r'n-rf}{n(r'+f)}.$$

The final direction of revolution of N will depend upon the relation which $r'n$ bears to rf; if the former be greater, then the direction will be positive (+), and *vice versa*. The formula for this combination is then:

$$N\text{'s velocity} = R\text{'s velocity}\times\left(\frac{r'n-rf}{n(r'+f)}\right). \qquad (11)$$

SOME NUMERICAL EXAMPLES IN EPICYCLIC GEARING

In order to illustrate the above examples we will take one or two cases.

If in Example and Fig. 218, $f = 30$, $n = 25$, then to one revolution of R, N will make $\left(1+\frac{f}{n}\right) = 1 + 30/25 = 2\frac{1}{5}$ revolutions.

It will be obvious that with $f = n$, N would revolve at twice the speed of R.

In the type shown in Fig. 7, $f = 60$, $n = 65$; then

$$\frac{\textit{Velocity of N}}{\textit{Velocity of R}}=\frac{1-\frac{1}{n}}{1}=\frac{1-60/65}{1}=5/65=1/13.$$

The arrangement of Fig. 227 is much used. Let $n = 60, f = 61, s = 40, m = 41$.

Then the velocity ratio between N and R is $1 - \frac{fm}{sn}:1$

$$= 1 - \frac{61 \times 41}{40 \times 60} = 1 - \frac{2,501}{2,400} = \text{say } 1:24, \text{ in a minus direction.}$$

Illustrating Example XII, Fig. 231, let $r = 90, r' = 91, f = 120, n = 121$.

$$\frac{\textit{Velocity of } N}{\textit{Velocity of } R} = \frac{r'n - rf}{n(r' + f)} = \frac{91 \times 121 - 90 \times 120}{121(91 + 120)}$$

$$\frac{11,011 - 10,800}{121 \times 211} = \frac{211}{121 \times 211} = \frac{1}{121}.$$

DIRECTION OF ROTATION OF GEARS

The following, in reference to epicyclic gear calculations, is by Oscar J Beale, AMERICAN MACHINIST, July 9, 1908:

"A very valuable article relative to epicyclic gears is by Prof. A. T. Woods in the AMERICAN MACHINIST for February 14, 1889. This article is well-nigh perfect. It is so clear and comprehensive that it was of great help to me. I have read a number of later articles, and I have always gone back to this in order to clear up the subject. I think that many of your readers would like to see it reprinted.

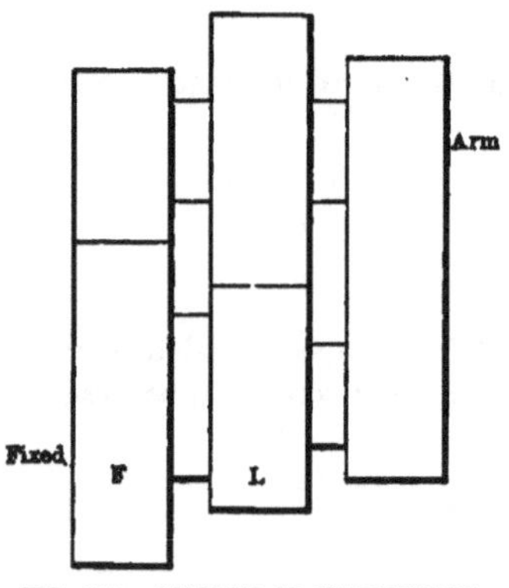

FIG. 233. DIAGRAM OF AN EPICYCLIC TRAIN

"About the best way to determine the direction of rotation of epicyclic gears is by careful inspection of the position of the members; then, if you make a correct statement of the effect of each member, Professor Woods' methods will bring the answer right. One must be sure to give the result the proper sign; that is, one must be able to add and to subtract algebraically.

"I have sometimes used a sort of mental key in cases like this sketch, Fig. 233. If the pitch circle of L is smaller than the pitch circle of F, the rotation of L will be opposite that of the arm. If L is greater than F, the rotation of L will be the same as that of the arm.

"This 'mental key' may help some; but after all, it is usually better to reason mathematically as in Professor Wood's article."

We reprint herewith Professor Woods' article to which Mr. Beale refers because of its value in determining the direction of rotation of epicyclic trains.

EPICYCLIC TRAINS*

"An epicyclic train consists of a number of gear wheels, or pulleys, and belts, some of which are carried upon a revolving arm. For example, in Fig. 236 the wheel *F* is fastened to the shaft *B*, about which arm *A* turns. This arm carries the axes of *C* and *L*, *C* being an idle wheel gearing with *F* and *L*. The motion of the wheel *L* is thus composed of three motions: (1) That which it has by reason of its revolution about *B* as a center, (2) that due to the revolution of the arm *A* about *F*, and (3) that due to its connection with *F* by means of the wheel *C*. We will consider the effect of these motions sepa-

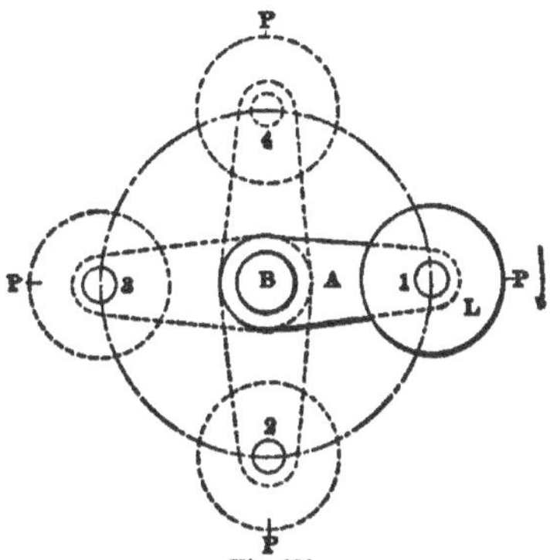

Fig. 234

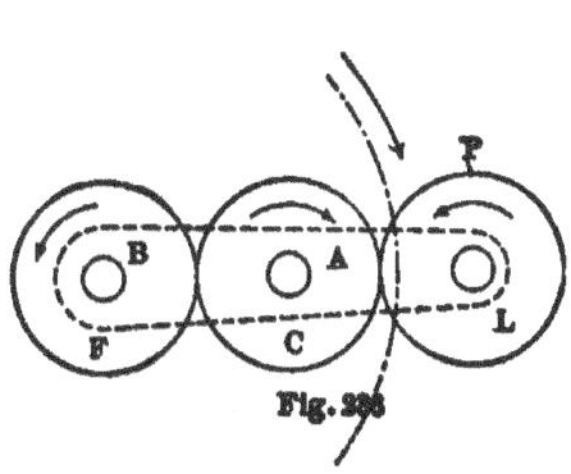

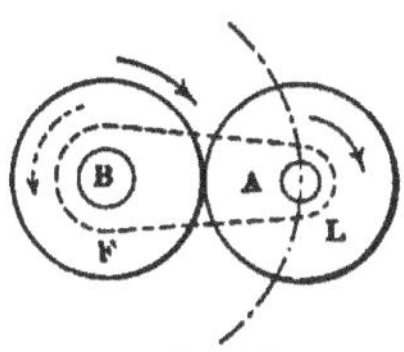

Fig. 235

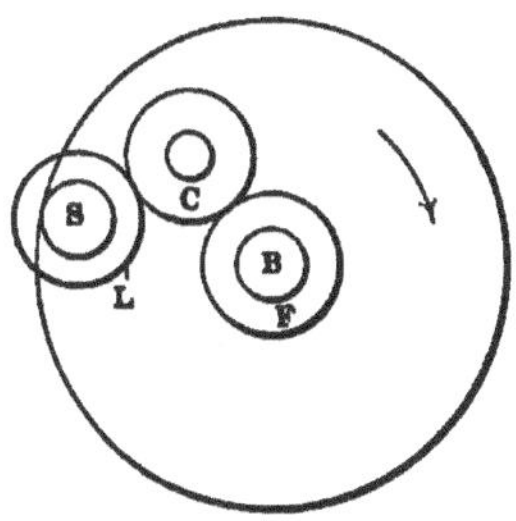

Fig. 237

DIAGRAMS OF EPICYCLIC GEARS

rately, and will begin with the simplest possible arrangement. In Fig. 234 let *A* be an arm which revolves about a center *B*, and carries a wheel *L*, which we will suppose to be fastened to it. If the arm be turned through one revolution, the wheel *L* will in effect revolve once about its own center. This will be clear by an examination of the successive positions shown in dotted lines, the revolution of the arm being in the direction of the arrow. For example, follow the motion of a point such as *P*; at 1 it is to the right of the center, at 2 below it, at 3 to the left, and at 4 above the center, finally returning to its first position on the right. We thus see that *L* has practically made one revolution

* Prof. A. T. Woods.

about its own center, just as it would have if it had been fixed at *B* concentric with the arm. If *L* has not revolved by reason of the revolution of *A*, the point *P* would have remained horizontally to the right of the center during the revolution of *A*. This is, of course, the same motion as that of a crank-pin and crank, and will be still more clear if we remember that, if the pin did not in effect revolve about its own center, it would not turn in the brasses, and they could be dispensed with.

Now, considering the second motion of *L*, that due to the revolution of the arm about *F*, refer to Fig. 235, and let the wheel *F* be fixed, or a dead wheel, concentric with the arm *A*. Let *F* and *L* have the same number of teeth. Then while the arm revolves once in the direction indicated, *L* will revolve in the same direction, the result being the same as if the arm had remained fixed, and *F* had been revolved once in the opposite direction. The final motion of *L*, while the arm revolves once, is therefore one revolution, as in Fig. 234, supposing *F* to be removed, and one revolution, as in Fig. 235, supposing *F* to be in place and fixed; or, while the arm revolves once, *L* revolves twice in the same direction.

Now, instead of *F* being fixed, let it revolve once in the direction of the dotted arrow. The effect of this will be to give one additional revolution to *L*, resulting in three revolutions of *L* to one of the arm. In order, then, to get at the resultant motion of the last wheel in epicyclic trains, we must consider the three independent motions separately: *First*, suppose the first wheel *F*, which is concentric with the arm, to be removed; *second*, suppose the first wheel to be in place and fixed; and *third*, suppose the arm to be fixed and the first wheel to revolve as intended. The final motion of *L* is the sum of these three motions.

For the sake of brevity we will designate revolution in the direction of the hands of a watch ahead, or + ; and that in the opposite direction backward, or −. Thus expressed, the revolutions of *L* in Fig. 235, as we have just discussed it, will be:

(a) + 1 due to the revolution of the arm,
+ 1 due to the revolution about *F''*,
+ 1 due to the revolution of *F*.
+ 3 revolution of *L* to one of the arm.

As a further illustration assume that in Fig. 235, *F* has 40 teeth and *L* 30, then if *F* is a fixed wheel, *L* will revolve:

(b) + 1 due to the revolution of the arm,
+ $40/30$ due to the revolution about *F*,
0 due to the revolution of *F*,

+ $7/3$ revolutions of *L* to one of the arm.

If *F* makes one revolution backward while the arm makes one ahead, we will have for *L*:

(c) $+ 1$ due to the revolution of the arm,
$+ 40/30$ due to the revolution about F,
$+ 40/30$ due to the revolution of F,

$+ 11/3$ revolutions of L to one of the arm.

If F makes one revolution ahead, or in the same direction as the arm, the result is to balance the effect of the revolution about F, and we have for L:

(d) $+ 1$ due to the revolution of the arm,
$+ 40/30$ due to the revolution about F,
$- 40/30$ due to the revolution of F,

$+ 1$ revolution of L to one of the arm, or the same as in Fig. 234. Similarly, if we take the conditions the same as (c) and let F have 30 teeth and L 40, we will have $+ 21/2$ revolutions of L to one of the arm.

We will now consider the effect of introducing an idle wheel, as shown in Fig. 236. In the first place, let L equal F, and let F be a fixed wheel. The revolutions of L will be:

(e) $+ 1$ due to the revolution of the arm,
$- 1$ due to the revolution about F,
0 due to the revolution of F,

0 revolution of L, or, in other words: A point P, which is, say, vertically over the center of L, will remain so throughout the revolution of the arm. If we assume the same condition as (a), the resulting revolution of L will be:

(f) $+ 1$ due to the revolution of the arm,
$- 1$ due to the revolution about F,
$- 1$ due to the revolution of F,

$- 1$ revolution of L to one of the arm.

If we let F have 40 teeth and L 30, and let F be a dead wheel, we will have for L:

(g) $+ 1$ due to the revolution of the arm,
$- 40/30$ due to the revolution about F,
0 due to the revolution of F,

$- 1/3$ revolution of L to one of the arm.

Or, reversing the position of the wheels, making $F = 30$ and $L = 40$, the revolutions of L will be:

(h) $+ 1$ due to the revolution of the arm,
$- 40/30$ due to the revolution about F,
0 due to the revolution of F,

$+ 1/4$ revolution of L to one of the arm.

The arrangement shown at (e) is used in one form of rope-making machinery. The "arm" *A*, Fig. 236, is then the revolving frame which carries the bobbins on which the strands or wire have been wound. *B* is the center of this frame, and on it the wheel *F* is fixed. A small yoke or frame, which carries a bobbin, is fixed on the axis of *L*, there being as many of these wheels and bobbins as there are to be strands in the rope. Then if *F* and *L* have the same number of teeth as at (e), the axes of the bobbins always point in one direction, and the rope is laid up without twisting the separate strands. If *L* has a few less teeth than *F*, the strands will be given a slight twist, making the rope harder.

Arrangements such as Figs. 235 and 236 are applicable to boring bars having sliding head. In such cases *B* would be the dead center on which the bar turns and on which the wheel *F* is fastened, being, therefore, a dead wheel. The wheel *L* is fastened to the end of the feed-screw in the bar, as shown at *S*, in Fig. 237, which represents the end view of the bar. While the arrangement is an epicyclic train, such as we have discussed, the explanation of it is extremely simple, because the motion to be determined is that of the screw *S* *with regard to the bar*, not with regard to the lathe, or any stationary object.

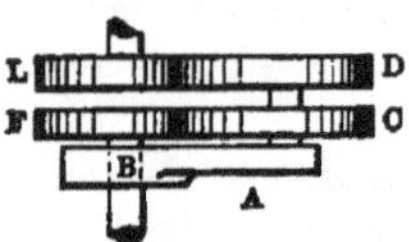

Fig. 238

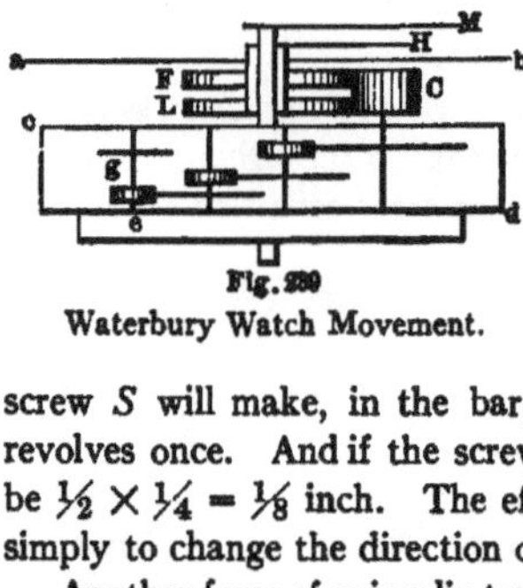

Fig. 239
Waterbury Watch Movement.

As *F* is fixed, the effect on *L* of one revolution of the bar is the same as if the bar remained stationary and *F* revolved once. Thus, if *F* has 20 teeth and *L* 40, the screw *S* will make, in the bar, $^{20}/_{40} = \frac{1}{2}$ of a revolution, while the bar revolves once. And if the screw has four threads to the inch, the feed will be $\frac{1}{2} \times \frac{1}{4} = \frac{1}{8}$ inch. The effect of the idle wheel (shown in Fig. 237) is simply to change the direction of the feed.

Another form of epicyclic train is that shown in Fig. 238 in which the last wheel is concentric with the arm and first wheel. This does not change the resultant motion of *L* in any way, but only makes a more convenient form for transmitting motion from *L* to other parts of the machine. If *F* and *L* have 40 and 30 teeth, while the wheels *C* and *D* are equal, we will have the same motion as at (g) and (h), supposing *F* to be a dead wheel. A recent and novel application of this form of train is to be found in the Waterbury watch, the principle of which is shown in Fig. 239. In this figure *ab* is the face of the watch and *c d* the frame which carries the principal train of wheels, that from *C* to the balance wheel *g*. This frame turns about the center shown below it and the bearing in the face. It is driven by the spring *e* and carries the minute hand *M*, and hence revolves once each hour. Between the frame and the face is a pinion C, having 8 teeth, which is connected to the balance wheel g by the train of wheels shown, and itself gears with two wheels F and L, having, respectively, 44 and 48 teeth. F is fastened to the face, and so is a dead wheel, and L is fastened to the hour hand H by the tube, as shown. It remains to show that the hour hand will revolve once in 12 hours, as required by means of this connection. We have here an epicyclic train FCL in which the first wheel is fixed. The revolutions of L during one revolution of the arm, as we have called it, or the frame cd are, therefore:

$+1$ due to the revolution of the arm,
$-\frac{44}{48}$ due to the revolution about F,
0 due to the revolution of F,

$+\frac{4}{48} = +\frac{1}{12}$ revolution during one revolution of the arm or minute hand, which is, of course, as it should be. The remainder of the train of wheels in this watch do not differ in principle from that ordinarily employed, a peculiarity being, however, that the entire "works," held in the frame cd, revolve within the case every hour.

Another peculiar adaptation of epicyclic trains is for the production of very slow velocities, using a small number of wheels. For example, in Fig. 238, let the numbers of teeth on the several wheels be $F = 19$, $C = 20$, $D = 21$, and $L = 20$, and let F be a dead wheel. Working out this train as we have the others, it will be found that, while the arm A makes one revolution, the last wheel I will make but $\frac{1}{400}$ of a revolution. In the same way, if we take the number of teeth in order as above, as 27, 40, 37 and 25, the last wheel will make but one revolution, while the arm makes 1,000.

We have chosen examples in which the first wheel is the dead wheel, as these are the simplest and most common. By adjusting the speed of the first wheel, however, it becomes possible to transmit velocities by means of epicyclic trains, which would be practically impossible by ordinary means. As an illustration, suppose it is required to have one shaft make 641 revolutions to one of another. As 641 is a prime number, this ratio could not be transmitted exactly by ordinary gearing on account of the large number of teeth required for a single wheel; but by means of an epicyclic train it can be readily accomplished. Of course, the necessity for such ratios as this rarely occurs in machinery.

A method of solving problems involving epicyclic trains, which will be more convenient for many than that which we have followed, is by means of a general formula. Let v = the value of the train of gears, or the product of the number of teeth on the drivers divided by the products of the numbers of teeth on the followers, which would be, in Fig. 238,

$$\frac{F \times D}{C \times L}.$$

In case of pulleys, v = the product of the diameters of the drivers divided by the product of the diameters of the followers. Let f, l and a represent the number of revolutions of the first wheel, last wheel, and arm, respectively, in the same time.

Then,

$$v = \frac{l - a}{f - a}$$

If one direction is represented as +, the other will be represented as −. If the last wheel is to revolve in the same direction as the first, supposing the arm to be fixed, v is +, and if the opposite direction, it is −. For example, take the data as at (b); then.

$$v = {}^{40}\!/_{30} = \frac{l - a}{0 - a}, \text{ whence } l = \tfrac{7}{3}\, a.$$

Again, let it be required that L shall make one revolution to 1,000 of A (Fig. 238),

$$v = \frac{1 - 1{,}000}{0 - 1{,}000} = + \frac{999}{1{,}000} = \frac{27 \times 37}{40 \times 25} = \frac{F \times D}{C \times L}.$$

TABLE OF PROPORTIONS OF DIFFERENTIAL BACK-GEARS

The following table, originally published in AMERICAN MACHINIST by Ernest J. Lees, gives data for ready reference. For a back-gear on drill presses and other light machinery, the differential back-gear as originally

SIZE NO.	1	2	3	4	5	6	7	8	9	10	11	12
Diameter of Pulley	8	8	10	10	12	12	15	15	15	15	18	18
Face of Pulley	3	3	3½	3½	4½	4½	5½	5½	6½	6½	7½	7½
Width of Belt	2½	2½	3	3	4	4	5	5	6	6	7	7
Approximate H. P. at 300 r. p. m.	2½	2½	4¼	4¼	7½	7½	11	11	13	13	18	18
Shaft Diameter D	1½	1½	1½	1½	1⅝	1⅝	1¾	1¾	2	2	3	3
Pitch Diameter pinion A	2¼	4½	2 4/7	5 1/7	2 4/7	5 1/7	2 4/7	6	3	9	5	13
Number of teeth in A	18	36	18	36	18	36	18	42	18	54	15	39
Pitch Diameter Internal Gear B	9	9	10 2/7	10 2/7	10 2/7	10 2/7	12 6/7	12 6/7	18	18	25	25
Number of teeth in B	72	72	72	72	72	72	90	90	108	108	75	75
Pitch Diameter Idler C	3⅜	2¼	3 6/7	2 4/7	3 6/7	2 4/7	5 1/7	3 3/7	7½	4½	10	6
Number of teeth in C	27	18	27	18	27	18	36	24	45	27	30	18
Number of Idlers	3	3	3	3	3	3	3	3	3	3	3	3
Diameter Pitch of Gears	8	8	7	7	7	7	7	7	6	6	3	3
Face of Gears	1¼	1¼	1½	1½	1⅝	1⅝	1¾	1¾	2	2	3	3
Ratio	5 to 1	3 to 1	5 to 1	3 to 1	5 to 1	3 to 1	6 to 1	3.142 to 1	7 to 1	3 to 1	6 to 1	2.923 to 1

TABLE 24. PROPORTIONS OF DIFFERENTIAL BACK-GEARS.

designed. For a heavy drive and continuous service there is a better method of arrangement. This consists of using three idlers in place of one, these being equally spaced in order to retain the balance of the whole when locked up and driving direct. It will be readily seen that this arrangement calls for the following conditions in the gearing: That the number of teeth in pinion, idler,

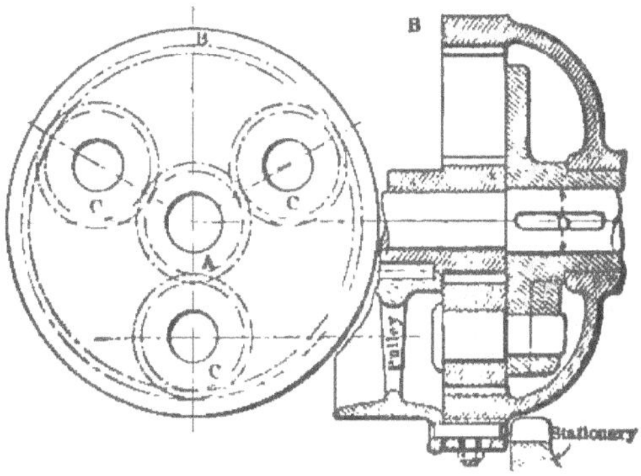

FIG. 240. DIFFERENTIAL BACK-GEARING.

and internal gear must each be divisible by three, at the same time having correct diameters and pitch.

A table is given herewith showing drives from 2½ to 18 horse-power in 12 sizes. This is arranged so that there are only 6 different diameters of internal gears, and by using different pinion and idlers one size can be used for two different ratios.

Fig. 240 shows the principle on which these back-gears are operated. The locking device should not be copied, however, as it would be rather inconvenient to use, and as there are various methods of operating clutches the writer has not gone into detail on this point, but gives the general dimensions, leaving the rest to be worked out by the designer to suit the requirements and conditions of the machine on which it is to be used.

SECTION XIII

Friction Gears*

A friction drive, as the term is here employed, consists of a fibrous or somewhat yielding driving wheel working in rolling contact with a metallic driven wheel. Such a drive may consist of a pair of plain cylindered wheels mounted upon parallel shafts, or a pair of beveled wheels, or of any other arrangement which will serve in the transmission of motion by rolling contact. The use of such drives has steadily increased in recent years, with the result that the so-called paper wheels have been improved in quality, and a considerable number of new materials have been proposed for use in the construction of fibrous wheels.

THE WHEELS TESTED

Choosing materials which have been used for such purposes, driving wheels of each of the following materials have been tested: straw fiber, straw fiber with belt dressing, leather fiber, leather, leather faced iron, sulphite fiber, tarred fiber.

The straw fiber wheels are worked out of the blocks which are built up usually of square sheets of straw board laid one upon another with a suitable cementing material between them and compacted under heavy hydraulic pressure. In the finished wheel the sheets appear as disks, the edges of which form the face of the wheel. The material works well under a tool, but it is harder and heavier than most woods and takes a good superficial polish. The wheel tested was taken from stock.

The wheel of straw fiber with belt dressing was similar to that of straw fiber, except that the individual sheets of straw board from which it was made had been treated, prior to their being converted into a block, with a "belt dressing" the composition of which is unknown to the writer.

The leather fiber wheel was made up of cemented layers of board, as were those already described; but in this case the board, instead of being of straw fiber, was composed of ground sole leather cuttings, imported flax and a small percentage of wood pulp. The material is very dense and heavy.

The leather wheel was composed of layers or disks of sole leather.

The leather faced iron wheel consisted of an iron wheel having a leather strip cemented to its face. After less than 300 revolutions the bond holding the leather face failed and the leather separated itself from the metal of the

* Abstract of paper presented to the American Society of Mechanical Engineers, December, 1907, by W. M. Goss, Professor, University of Illinois.

wheel. This wheel proved entirely incapable of transmitting power and no tests of it are recorded.

The wheel of sulphite fiber was made up of sheets of board composed of wood pulp. The sulphite board is said to have been made on a steam-drying continuous process machine in the same way as is the straw board.

The tarred fiber wheel was made up of board composed principally of tarred rope stock, imported French flax, and a small percentage of ground sole leather cuttings.

Each of the fibrous driving wheels was tested in combination with driven wheels of the following materials: Iron, aluminum, type metal. All wheels tested, both driving and driven, were 16 inches in diameter. The face of all driving wheels was 1¾ inches while that of all driven wheels was ½ inch.

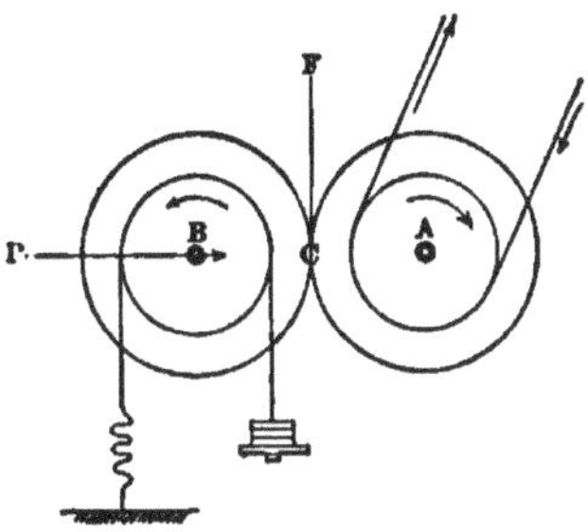

FIG. 241. DIAGRAM OF TESTING MACHINE FOR FRICTION WHEELS.

The purpose of the experiments was to secure information which would permit rules to be formulated defining the power which may be transmitted by the various combinations of fibrous and metallic wheels already described. To accomplish this it was necessary to determine for each combination of driving and driven wheel the coefficient of friction under various conditions of operation; also the maximum pressures of contact which can be withstood by each of the fibrous wheels.

The testing machine used is shown diagrammatically by Fig. 241. The principles involved will be made clear by assigning the functions of the actual machine to the several parts of this figure. The shaft A runs in fixed bearings and carries the fibrous friction wheel. This wheel is the driver. Its shaft A carries, besides the friction wheel, two belt pulleys, one on either side, which, from any convenient source of power, serve to give motion to the driver. The shaft B carries the driven wheel, which in every case was of metal. The bearings of this shaft are capable of receiving motion in a horizontal direction and by means of suitable mechanism connected therewith, the metal driven wheel may be made to press against the fibrous driver with any force desired. The pressure transmitted from B to A is hereinafter referred to as the "pressure of contact" and is frequently represented by the symbol P. The tangential forces which are transmitted from the driver to the driven wheel are received, absorbed and measured by a friction brake upon the shaft B. In action, therefore, the driven wheel always works against a resistance, which resistance may be modified to any desired degree by varying the load upon the brake. The theory of the machine assumes that the energy absorbed by the brake equals that transmitted from the driver to the driven wheel at the contact point C. Accepting this assumption, the forces developed at the periphery of the brake wheel may readily be reduced to equivalent forces acting at the circumference of the driven wheel. The force, which is directly transmitted from the driver to the driven wheel, is hereinafter designated by the symbol *F*. It will be apparent from this description that the functions of the apparatus employed are such as will permit a study of the relationship existing between the contact pressure *P* and the resulting transmitted force

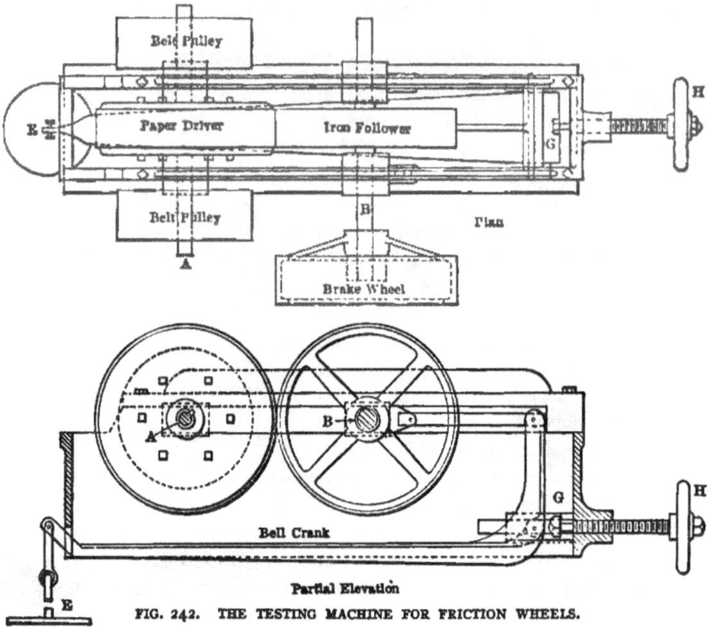

FIG. 242. THE TESTING MACHINE FOR FRICTION WHEELS.

F, which relation is most conveniently expressed as the coefficient of friction. It is,

$$f = \frac{F}{P}.$$

It is obvious, in comparing the work of two friction wheels, that the one which develops the highest coefficient of friction, other things being equal, can be depended upon to transmit the greatest amount of power.

The actual machine as used in the experiments is shown by Fig. 242. Its construction satisfies all conditions which have been defined except that shaft *B*, Fig. 241, does not run in bearings which are absolutely frictionless, as is required by a rigid adherence to the theoretical analysis already given. These bearings, however, are of the "standard roller bearing" type and of ample size, and it is believed that the friction actually developed by them is so small compared with the energy transmitted between the wheels that it may be neglected.

The bearings of the fixed shaft *A* are secured to the frame of the machine. The bearings of the axle *B* are free to move horizontally in guides to which they are well fitted. Those bearings are connected by links to the short arm of a bell crank lever, the arm of which projects beyond the frame of the machine at the right-hand end and carries the scale pan and weights *E*. The effect of the weights is to bring the driven wheel in contact with the driver under a predetermined pressure, the proportions of the bell crank lever being such as to make this pressure in pounds equal,

$$P = 10\,W + 73,$$

where *W* is the weight on the scale pan *E*.

The fulcrum of the bell crank lever is supported by a block *G* which may be adjusted horizontally by the hand wheel *H* at the rear of the machine, so that whatever may be the diameter of the driven wheel, the long arm of the bell crank may be brought to a horizontal position. The constants employed in calculating the coefficient of friction from observed data are as follows:

Diameter of friction wheels (inches)	16
Effective diameter of brake (inches)	18.35
Ratio of diameter of friction wheel to that of brake wheel	1.145
Effective load on brake	F'
Coefficient of friction	$1.145\,\frac{F'}{P}$

The slippage between the friction wheels was determined from the readings taken from the counters connected to each one of the shafts.

THE TESTS

In proceeding with a test, load was applied to the scale pan *E*, Fig. 242, to give the desired pressure of contact, after which the hand wheel *H* at the back of the machine was employed to bring the bell crank to its normal position. This accomplished, with the driving wheel in motion, the driven wheel would roll with it under the desired pressure of contact. A light load was next placed upon the brake to introduce some resistance to the motion of the driven shaft, and conditions thus obtained were continued constant for a considerable period. Readings were taken simultaneously from the counters and time noted. After a considerable interval the counters were again read, time again noted, and the test assumed to have ended. From the readings of the counters and from the known diameters of the wheels in contact, the percentage of slip attending the action of the friction wheels was calculated. Three facts were thus made of record, namely: (*a*) The pressure of contact, (*b*) the coefficient of friction developed, and (*c*) the percentage of slip resulting from the development of said coefficient of friction.

This record having been completed, the load upon the brake was increased and observations repeated, giving for the same pressure of contact a new

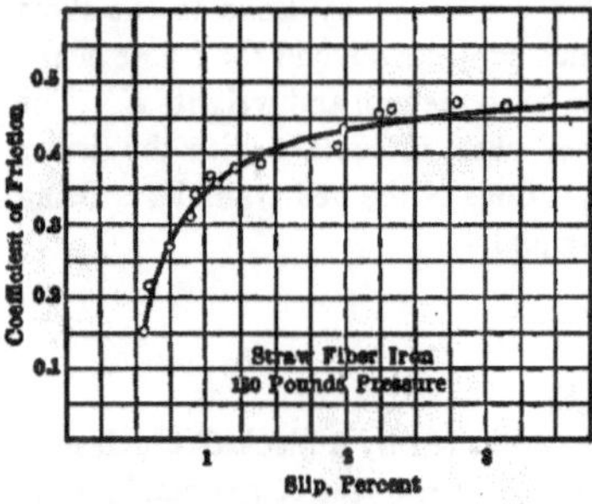

FIG. 243.

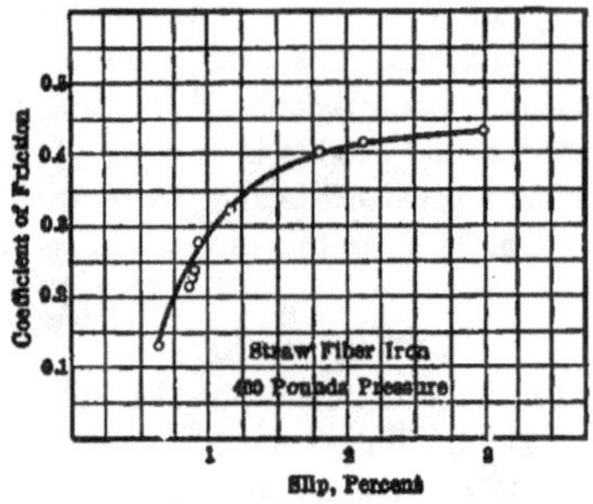

FIG. 244.

CURVES FOR STRAW FIBER AND IRON, TYPICAL FOR ALL CURVES PLOTTED FROM THE FRICTION TESTS

coefficient of friction and a higher percentage of slip. This process was continued until the slippage became excessive and in consequence thereof the rotation of the driver ceased. By this process a series of tests was developed disclosing the relation between slip and coefficient of friction for the pressure in question. Such a series having been completed, the load upon the weight holder *E* was changed, giving a new pressure of contact, and the whole process repeated. As the work proceeded, curves showing the relation of coefficient of friction and slip for pressures per inch width of face in contact of 150 pounds and 400 pounds, respectively, were secured. The curves shown by Figs. 243 and 244 for the straw fiber driving wheel in contact with the iron driven wheel are typical in their general form of those obtained from all combinations of wheels, but the curves of no two combinations were alike in their numerical values.

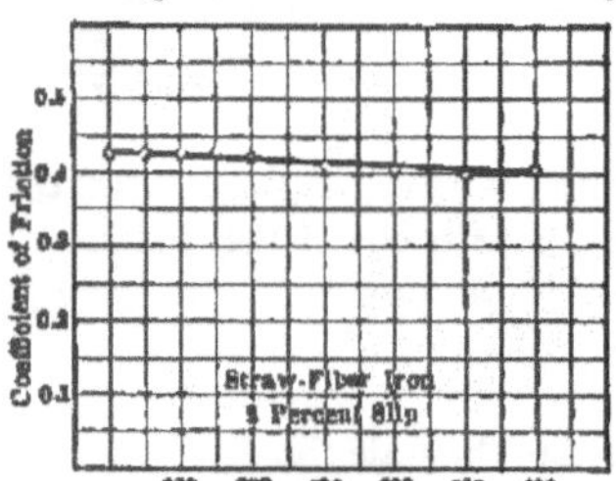

FIG. 245. CURVE FOR STRAW FIBER AND IRON WITH CONSTANT SLIP.

Having completed this series of tests at constant pressure, a series was next run for which the coefficient of slip was maintained constant at 2 per cent., and the pressure of contact varied from values which were low to those which are judged to be near the maximum for service conditions, with the results, which in all cases were similar in character with those given for the straw fiber and iron wheels, as set forth by Fig. 245. The numerical values of points for other combinations were not the same as those shown by Fig. 245, but in the case of most of the combinations the coefficient of friction at constant slip gradually diminishes as the pressure of contact is increased.

As the series of tests involving each combination of wheels proceeded, the increase in pressure of contact was discontinued when the markings made upon the driving wheel by the metallic follower became so distinct as to suggest that a safe limit had been reached; but when all other data had been secured, tests were run for the purpose of determining the ultimate resistance of the fibrous wheel to crushing. The details of these will be described later.

COEFFICIENT OF FRICTION DEVELOPED BY THE SEVERAL COMBINATIONS OF WHEELS—STRAW FIBER AND IRON

The results of experiments involving a straw fiber driver and an iron driven wheel are shown graphically in Figs. 243, 244, and 245. Figs. 243 and 244 illustrate the relation between slip and coefficient of friction when the two wheels are working together under pressures per inch width of 150 and 400 pounds, respectively.

The figures show that although the values of the coefficient of friction are slightly lower than corresponding ones for 150 pounds pressure, the curves are sufficiently similar to establish the fact that the law governing change in coefficient friction with slip is independent of the pressure of contact. When the slippage is 2 per cent. the coefficient of friction is 0.425 for a contact pressure of 400 pounds. That the coefficients of friction for all pressures between the limits of 150 pounds and 400 pounds are practically constant is well shown by the diagram Fig. 245. The pressure of 400 pounds is the maximum at which tests of this combination of wheels were run, though straw fiber was successfully worked up to a pressure of 750 pounds.

STRAW FIBER AND ALUMINUM

By curves plotted from values for a straw fiber driver and aluminum driven wheel, it can be shown that when the working pressure is 150 pounds per inch width and the slippage is 2 per cent. the coefficient of friction is 0.455; also, that for all pressures ranging from 100 to 400 pounds, the coefficient of friction is practically constant when the rate slip is constant. The maximum pressure at which tests involving this combination of wheels were run was 400 pounds per inch width.

STRAW FIBER AND TYPE METAL

By curves plotted from values for a straw fiber driver and a type metal driven wheel it can be shown that when the two wheels are operated under a pressure of contact of 150 pounds per inch width and when the slip is 2 per cent. the coefficient of friction is 0.310; also, that for all pressures of contact ranging from 100 to 400 pounds, the coefficient of friction is practically constant when the slip is constant.

STRAW FIBER WITH BELT DRESSING AND IRON

Curves plotted from values for a straw fiber driver treated with belt dressing, and an iron driven wheel show that when the two wheels are worked together under a pressure of 150 pounds per inch width and when the slip is 2 per cent. the coefficient of friction is 0.12; also, that for all pressures up to 400 pounds per inch width, the coefficient of friction remains constant. The greatest pressure at which tests of this combination of wheels were run was 500 pounds per inch width.

LEATHER FIBER AND IRON

Curves plotted from the results of tests involving a leather fiber driver and an iron driven wheel show that when the two wheels are worked together under pressure of 150 pounds per inch in width and when slip is 2 per cent. the coefficient of friction is 0.515. When the contact pressure is 300 pounds per inch width, the coefficient of friction is 0.510. The greatest pressure at which tests of this combination of wheels were run was 350 pounds per inch width, although leather fiber was successfully worked up to a pressure of 1,200 pounds per inch width.

LEATHER FIBER AND ALUMINUM

Curves plotted from the results of experiments involving a leather fiber driver and an aluminum driven wheel show that under a contact pressure of 150 pounds per inch width and a slip of 2 per cent. the coefficient of friction is 0.495.

This value remains practically constant under all pressures. The maximum pressure used in tests of this combination of wheels was 400 pounds.

LEATHER FIBER AND TYPE METAL

Curves plotted from the results of experience involving a leather fiber driver and a type metal driven wheel show that when the wheels are operated under a contact pressure of 150 pounds per inch width and when the slip is 2 per cent. the coefficient of friction remains constant for all pressures up to 400 pounds per inch width.

TARRED FIBER AND IRON

Curves plotted from the results of the experiments involving a tarred fiber driver and an iron driven wheel show that the change in the value of the coefficient of friction with change of slip is practically independent of the pressure of contact. When the slip is 2 per cent., the coefficient of friction is 0.220 for a pressure of contact of 150 pounds and 0.250 for a pressure of contact of 400 pounds per inch width.

Tests of this combination were made also under different speeds when the wheels were working together under a pressure of contact of 250 pounds per inch width and when the slip was 2 per cent., with the result that the coefficient of friction was found to remain nearly constant for speeds of 450 and 3,350 feet per minute, respectively. The greatest pressure at which tests of this combination of wheels were run was 400 pounds per inch width, although tarred fiber was successfully worked up to a pressure of 1,200 pounds per inch width.

TARRED FIBER AND ALUMINUM

Curves plotted from the results of experiments involving a tarred fiber driver and an aluminum driven wheel show that when the slip is 2 per cent. and the pressure of contact 150 pounds per inch width, the coefficient of friction is 0.305; also, that for a pressure of 400 pounds per inch width, the coefficient of friction is 0.295. The greatest pressure at which tests of this combination were run was 400 pounds per inch width.

TARRED FIBER AND TYPE METAL

Curves plotted from the results of experiments involving a tarred fiber driver and a type metal driven wheel show that when the slip is 2 per cent. the coefficient of friction developed under 150 pounds pressure per inch width is 0.275; and under 400 pounds pressure per inch width, the coefficient of friction is 0.270. The maximum pressure at which tests of this combination of wheels were run was 400 pounds per inch width.

LEATHER AND IRON

Curves plotted from the results of experiments involving a leather driver and an iron driven wheel show that when the slip is 2 per cent. the coefficient of friction under a pressure of contact of 150 pounds per inch in width is 0.225 and under a pressure of 400 pounds, 0.215. The maximum pressure at which tests of this combination of wheels were run was 400 pounds per inch width, although the leather driver was successfully operated up to a pressure of 750 pounds per inch width.

LEATHER AND ALUMINUM

Curves plotted from the results of experiments involving a leather driver and an aluminum driven wheel show that when the pressure is 150 pounds per inch in width and the slip is 2 per cent. the coefficient of friction is 0.260; and when the pressure is 300 pounds per inch in width, the coefficient of friction is 0.295. The maximum pressure at which tests of this combination of wheels were made was 350 pounds per inch width.

LEATHER AND TYPE METAL

Curves plotted from the results of the experiments involving a leather driver and a type metal driven wheel show that when the slip is 2 per cent. and the contact pressure 150 pounds per inch width, the coefficient of friction developed is 0.410. The greatest pressure at which tests of this combination of wheels were run was 350 pounds per inch width.

SULPHITE FIBER AND IRON

Curves plotted from the results of the experiments involving a sulphite fiber driver and an iron driven wheel show that when the slip is 2 per cent. and the pressure 150 pounds per inch width, the coefficient of friction is 0.550. The maximum pressure at which tests of this combination of wheels were run was 350 pounds per inch width, although the sulphite fiber wheel was successfully operated up to a pressure of 700 pounds per inch width.

SULPHITE FIBER AND ALUMINUM

Curves plotted from the results of the experiments involving a sulphite fiber driver and an aluminum wheel show that when the slip is 2 per cent. and the pressure 150 pounds per inch width, the coefficient of friction developed is 0.410. The greatest pressure used in tests of this combination of wheels was 350 pounds per inch width.

SULPHITE FIBER AND TYPE METAL

Curves plotted from the results of the experiments involving a sulphite fiber driver and a type metal driven wheel show that when the slip is 2 per cent. and the contact pressure 150 pounds per inch width, the coefficient of friction is 0.515. The maximum pressure used in tests of this combination of wheels was 350 pounds per inch width.

RESISTANCE TO CRUSHING

Upon the completion of tests designed to disclose the frictional qualities of the several combinations, each fibrous wheel was subjected to test for the purpose of determining the maximum pressure per inch width of the face which could be sustained by it. This was accomplished by placing the wheel to be tested in the machine under a pressure of contact of 200 pounds per inch width. The load on the brake was then adjusted to give a 2 per cent. slip, and this brake load was maintained without change throughout the remainder of the tests. Thus adjusted, the machine was operated until the driver had completed 15,000 revolutions. This accomplished, and for the purpose of determining the reduction, if any, in the diameter of the fibrous wheel, the brake load was removed and the operation of the machine continued without load for a period of 6,000 revolutions, the readings of the counters being taken at the beginning and at the end of the period. Under conditions of no load, the actual slip was assumed to be zero and the apparent slip observed was used for determining the reduction in diameter of the fibrous wheel which had been brought about by the previous running under pressure. This accomplished, the pressure of contact was increased, usually by 100-pound increments, and the whole operation repeated. This process was continued until failure of the fibrous wheel resulted. It will be seen that the ultimate resistance to crushing, as found by the process described, is that pressure which could not be endured during 15,000 revolutions.

A summary of results is as follows:

A CONCLUSION AS TO METAL WHEELS

An examination of Table 25, which presents a comparison of values representing the coefficient of friction of the several combinations of wheels tested, reveals the fact that the relative value of the metal driven wheels is not the same when operated in combination with different fibrous driving wheels. It appears that those driving wheels which are the more dense work more efficiently with the iron follower than with either the aluminum or type metal followers; but in the case of the softer and less dense driving wheels, and especially in the case of those in which an oily substance is incorporated, driven wheels of aluminum and type metal are superior to those of iron. Finely powdered metal which is given off from the surface of the softer metal wheels seems to account for this effect, and the character of the driving wheels is perhaps the only factor necessary to determine whether its presence will be beneficial or detrimental. Finally, with reference to the use of soft metal driven wheels, it should be noted that no combination of such wheels with a fibrous driver appears to have given high frictional results. Except when used under very light pressures, the wear of the type metal was too rapid to make a wheel of its material serviceable in practice.

CONCLUSIONS AS TO FIBROUS WHEELS

The relative value of the different fibrous wheels when employed as drivers in a friction drive may be judged by comparing their frictional qualities as set forth in Table 25 and their strength as set forth in Table 26. The results show at once that the addition of belt dressing to the composition of a straw fiber wheel is fatal to its frictional qualities. The highest frictional qualities are possessed by the sulphite fiber wheel, which, on the other hand, is the weakest of all wheels tested. The leather fiber and tarred fiber are exceptionally strong; and the former possesses frictional qualities of a superior order. The plain straw fiber, which in a commercial sense is the most available of all materials dealt with, when worked upon an iron follower possesses frictional qualities which are far superior to leather, and strength which is second only to the leather fiber and the tarred fiber.

	Coefficient of friction when contact pressure is 150 pounds per inch		
	Iron	Aluminum	Type metal
Sulphite Fiber	0.550	0.530	0.515
Leather Fiber	0.515	0.495	0.350
Straw Fiber	0.425	0.455	0.310
Tarred Fiber	0.250	0.305	0.275
Leather	0.225	0.360	0.410
Straw Fiber with belt dressing	0.120		

TABLE 25—COEFFICIENT OF FRICTION.

	Load in pounds	Decrease in diameter	
Straw Fiber	200	0.000	Wheel failed before running 15,000 revolutions under 750 pounds pressure.
	650	0.053	
	750	0.125	
Leather Fiber	200	0.000	Wheel failed before running 15,000 revolutions under 1200 pounds pressure.
	300	0.005	
	400	0.013	
	500	0.021	
	600	0.027	
	700	0.040	
	800	0.051	
	900	0.068	
	1000	0.099	
	1100	0.125	
	1200	0.200	
Tarred Fiber	200	0.000	Wheel failed before running 15,000 revolutions under 1200 pounds pressure.
	300	0.026	
	400	0.038	
	500	0.052	
	600	0.071	
	700	0.098	
	800	0.138	
	900	0.182	
	1000	0.250	
	1100	0.295	
	1200		
Leather	350	0.047	Wheel failed before running 15,000 revolutions under 750 pounds pressure.
	450	0.090	
	550	0.015	
	650	0.240	
	750		
Sulphite Fiber	200	0.010	Wheel failed before running 15,000 revolutions under 700 pounds pressure.
	300	0.032	
	400	0.056	
	500	0.088	
	600	0.146	
	700	0.258	

TABLE 26—STRENGTH OF VARIOUS FIBER WHEELS.

THE POWER CAPACITY OF FRICTION GEARS

A review of the data discloses the fact that several of the friction wheels tested developed a coefficient of friction which in some cases exceeded 0.5. That is, such wheels rolling in contact have transmitted from driver to driven wheels a tangential force equal to 50 per cent. of the force maintaining their contact. These wheels also were successfully worked under pressures of contact approaching 500 pounds per inch in width. Employing these facts as a basis from which to calculate power, it can readily be shown that a friction wheel a foot in diameter, if run at 1,000 revolutions per minute, can be made to deliver in excess of 25 horse-power for each inch in width. It is certainly true that any of the wheels tested may be employed to transmit for a limited time an amount of power which, when gauged by ordinary measures, seems to be enormously high; but obviously, performance under limiting conditions should not be made the basis from which to determine the commercial capacity of such devices. In view of this fact, it is important that there be drawn from the data such general conclusions with reference to pressures of contact and frictional qualities as will constitute a safe guide to practice.

WORKING PRESSURE OF CONTACT

The results of these experiments do not furnish an absolute measure of the most satisfactory pressure of contact for service conditions. Other things being equal, the power transmitted will be proportional to this pressure, and hence it is desirable that the value be made as high as practicable. On the other hand, it has been noted as one of the observations of the test that as higher pressures are used, there appears to be a gradual yielding of the structure of the fibrous wheels; and it is reasonable to conclude that the life of a given wheel will in a large measure depend upon the pressure under which it is required to work. After a careful study of the facts involved, it has been determined to base an estimate of the power which may be transmitted upon a pressure of contact which is 20 per cent. of the ultimate resistance of the material as established by the crushing tests already described. This basis gives the following results:

SAFE WORKING PRESSURES OF CONTACT

	PRESSURE
Straw fiber	150
Leather fiber	240
Tarred fiber	240
Sulphite fiber	140
Leather	150

COEFFICIENT OF FRICTION

The coefficient of friction for all wheels tested approaches its maximum value when the slip between driver and driven wheel amounts to 2 per cent.

and, within narrow limits, its value is practically independent of the pressure of contact. A summary of maximum results is shown by Table 30. In view of these facts, it is proposed to base a measure of the power which may be transmitted by such friction wheels as those tested upon the frictional qualities developed at a pressure of 150 pounds per inch of width, when operating under a load causing 2 per cent. slip. For safe operation, however, deductions must be made from the observed values. Thus, the results of the experiments disclose the power transmitted from wheel to wheel, while in the ordinary application of friction drives some power will be absorbed by the journals of the driven axle so that the amount of power which can be taken from the driven shaft will be somewhat less than that transmitted to the wheel on said shaft. Again, under the conditions of the laboratory, every precaution was taken to keep the surfaces in contact free of all foreign matter. It was, for example, observed that the accumulation of laboratory dust upon the surfaces of the wheels had a temporary effect upon the frictional qualities of the wheels, and friction wheels in service are not likely to be as carefully protected as were those in the laboratory. In view of these facts, it has been thought proper to use as the basis from which to determine the amount of power which may be transmitted by such wheels as those tested, a coefficient of friction which shall be 60 per cent. of that developed under the conditions of the laboratory. This basis gives the following results:

COEFFICIENT OF FRICTION WORKING VALUES

	COEFFICIENT OF FRICTION
Straw fiber and iron	0.255
Straw fiber and aluminum	0.873
Straw fiber and type metal	0.186
Leather fiber and iron	0.309
Leather fiber and aluminum	0.297
Leather fiber and type metal	0.183
Tarred fiber and iron	0.150
Tarred fiber and aluminum	0.183
Tarred fiber and type metal	0.165
Sulphite fiber and iron	0.330
Sulphite fiber and aluminum	0.318
Sulphite fiber and type metal	0.309
Leather and iron	0.135
Leather and aluminum	0.216
Leather and type metal	0.246

HORSE-POWER

Having now determined a safe working pressure of contact and a representative value for the coefficient of friction, it is possible to formulate equations expressing the horse-power which may be transmitted by each combination of wheels tested. Thus, calling *d* the diameter of the friction wheel in inches, *W* the width of its face in inches, and *N* the number of revolutions per minute, the equations become, for combinations of,

	HORSE-POWER
Straw fiber and iron	0.00030 *dWN*
Straw fiber and aluminum	0.00033 *dWN*
Straw fiber and type metal	0.00022 *dWN*
Leather fiber and iron	0.00059 *dWN*
Leather fiber and aluminum	0.00057 *dWN*
Leather fiber and type metal	0.00035 *dWN*
Tarred fiber and iron	0.00029 *dWN*
Tarred fiber and aluminum	0.00035 *dWN*
Tarred fiber and type metal	0.00031 *dWN*
Sulphite fiber and iron	0.00037 *dWN*
Sulphite fiber and aluminum	0.00035 *dWN*
Sulphite fiber and type metal	0.00034 *dWN*
Leather and iron	0.00016 *dWN*
Leather and aluminum	0.00026 *dWN*
Leather and type metal	0.00029 *dWN*

The accompanying chart gives a convenient means of determining the value of any one of the variable factors in the formula horse-power = 0.0003 *dWN* for the straw fiber friction wheel working in combination with an iron follower, the remaining factors being known or assumed. To transform values thus found to corresponding ones for the other possible combinations of wheels, it is necessary only to multiply by the proper factor chosen from the table of multipliers given with the chart.

APPLICATION OF RESULTS TO FORM OTHER THAN THOSE EXPERIMENTED UPON FACE FRICTION GEARING

A fibrous driving wheel, acting upon the face of a metal disk, constitutes a form of friction gear which is serviceable for a variety of purposes. If the driver is so mounted that it may be moved across the face of the disk, the velocity ratio may be varied and the direction of the disk's motion may be reversed. The contact is not one of pure rolling. If the driver is cylindrical in form, the action along its line of contact with the disk is attended by slip, amount of which changes for every different point along the line. The recognition of this fact is essential to a discussion of the power-transmitting capacity of the device.

Experiments involving the spur form of friction wheels already described have shown that slip greatly affects the coefficient of friction; that the coefficient approaches its maximum value when the slip reaches 2 per cent., and that when the slip exceeds 3 per cent., the coefficient diminishes. It is known that reductions in the value of the coefficient with increments of slip beyond 3 per cent. are at first gradual, although the characteristics of the testing machine have not permitted a definition of this relation for slip greater than 4 per cent. The experiments, however, fully justify the statement that for maximum results the slippage should not be less than 2 per cent. nor more than 4 per cent. It is the maximum limit with which we are concerned in considering the amount of power which may be transmitted by face friction gearing.

From the discussion of the previous paragraph, it should be evident that, for best results, the width of face of the friction driver and the distance between the driver and center of disk should always be such that the variations in the velocity of the particles of the disk having contact with the driver will not exceed 4 per cent. A convenient rule which, if followed, will secure this condition is to make the minimum distance between the driver and the center of the driven disk twelve times the width of the face of the driver. For example, a driver having a ¼-inch width of face should be run at a distance of 3 inches or more from the center of the disk. Similarly, drivers having faces ½, 1, or 2 inches in width should be run at a distance from the center of the disk of not less than 6, 12, or 24 inches, respectively. When these conditions are met, all formulas for calculating the power which may be transmitted apply directly to the conditions of face driving.

It may not infrequently happen that friction wheels must be run nearer the center of the disk than the distance specified; there is, of course, no objection to such practice, but it should not be forgotten that as the center of the disk is approached, the coefficient of friction, and consequently the capacity to transmit power, diminishes.

CONDITIONS TO BE OBSERVED IN THE INSTALLATION OF FRICTION DRIVES

Whatever may be the form of the transmission, the fibrous wheel must always be the driver. Neglect of this rule is likely to result in failure which will appear in the unequal wear of the softer wheel, occasioned by slippage.

The rolling surfaces of the wheel should be kept clean. Ordinarily they should not be permitted to collect grease or oil, nor be exposed to excessive moisture. Where this cannot be prevented, a factor of safety should be provided by making the wheels larger than normal for the power to be transmitted.

Since the power transmitted is directly proportional to the pressure of contact, it is a matter of prime importance that the mechanical means employed in maintaining the contact be as nearly as possible inflexible. For example, arrangements of friction wheels which involve the maintenance of contact through the direct action of a spring have been found unsatisfactory, since any defect in the form of either wheel introduces vibrations which tend to impair the value of the arrangement. It is recommended that springs be avoided and that contact be secured through mechanism which is rigid and which when once adjusted shall be incapable of bringing about any release of the pressure to which it is set.

EXPLANATION OF CHART

Chart 15 is plotted for the most common materials used for friction gearing, straw fiber and cast iron, and gives means of determining the variable factors for the fiber wheel in the formula *horse-power* = $0.0003\ dWN$, in which d is the diameter of the wheel in inches, W its width of face in inches, and N the number of revolutions per minute.

To use the chart for other friction materials multiply the values obtained from the chart by the proper factor selected from the table below:

Straw fiber and aluminum	1.10
Straw fiber and type metal	0.73
Leather fiber and cast iron	1.97
Leather fiber and aluminum	1.90
Leather fiber and type metal	1.17
Tarred fiber and cast iron	0.97
Tarred fiber and aluminum	1.17
Tarred fiber and type metal	1.03
Sulphite fiber and cast iron	1.23
Sulphite fiber and aluminum	1.17
Sulphite fiber and type metal	1.13
Leather and cast iron	0.53
Leather and aluminum	0.87
Leather and type metal	0.97

(*a*) *To find the total horse-power* which can be transmitted by a wheel, having given the diameter of the wheels in inches, the width of its face in inches and the revolutions per minute, locate the intersection of the vertical line representing the given speed with the diagonal line representing the given diameter. Follow the horizontal line passing through this point, to the right or left as the case may be, until it intersects the vertical line representing the given width of the face. The diagonal line through this point will give total horse-power required from the scale so marked.

(*b*) *To find the speed in revolutions per minute* for a wheel, having given its diameter in inches, its width of face in inches, and the total horse-power to be transmitted, locate the intersection of the vertical line representing the width of face with the diagonal line representing the total horse-power to be transmitted. Follow the horizontal line passing through this point, to the right or left as the case may be, until it intersects the diagonal line representing the diameter in inches. The vertical line passing through this point indicates on the scale at the bottom of the chart the speed required.

(*c*) *To find the width of face in inches* for a wheel, having given the total horse-power to be transmitted, its diameter in inches and its speed in revolutions per minute, locate the intersection of the vertical line representing the given speed with the diagonal line representing the given diameter. Follow

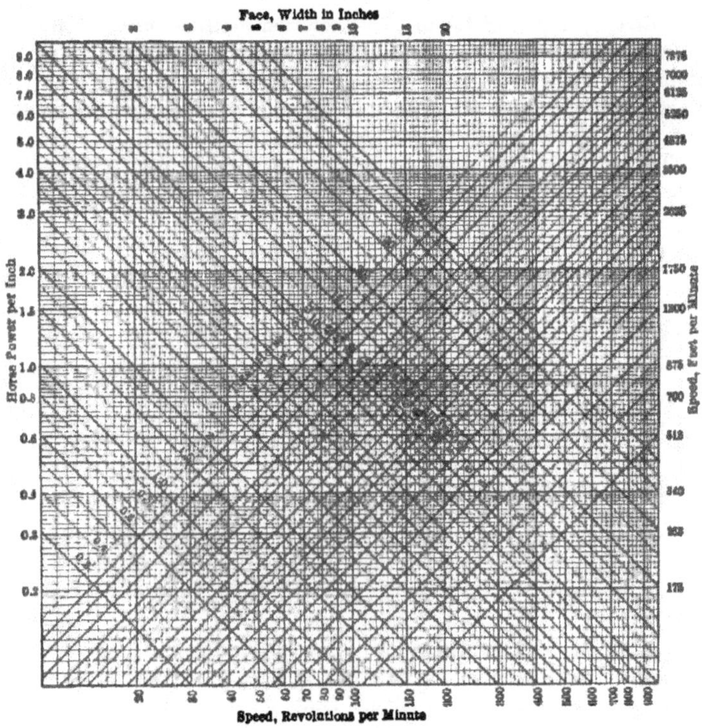

CHART 15. PROPORTIONS OF FIBROUS FRICTION GEARING.

the horizontal line passing through this point, to the right or left as the case may be, until it intersects the diagonal line representing the given total horse-power. The vertical line passing through this point will indicate the width of face required on the scale at the top of the chart.

(*d*) *To find the diameter in inches* for a wheel, having given the horse-power to be transmitted, its width of face in inches, and its speed in revolutions per minute, locate the intersection of the vertical line representing the width of face with the diagonal line indicating the total horse-power. Follow the horizontal line passing through this point, to the right or left as the case may be, until it intersects the vertical line representing the speed. The diagonal line passing through this point represents the diameter which is required.

(*e*) *To find the surface speed* of a wheel, having given its diameter in inches and its speed in revolutions per minute, locate the intersection of the vertical line representing the speed in revolutions per minute with the diagonal line representing the given diameter. The horizontal line passing through this point represents the surface speed in feet per minute which is required, and which is read on the vertical scale at the right of the chart.

(*f*) *To find the horse-power per inch of face* for a wheel, having given the total horse-power transmitted and the width of the face in inches, locate the intersection of the vertical line representing the width of face with the diagonal line representing the total horse-power. The horizontal line passing through this point represents the horse-power per inch of face required and may be read on the vertical scale at the left of the chart.

FRICTION DRIVE ON A FORTY-FOUR FOOT PIT LATHE*

The machine here described was designed to meet the demands of an establishment manufacturing the heaviest type of electrical machinery. The ever-increasing dimensions of this class of machinery make it particularly desirable that the existing heavy machine tools should be capable of extension of capacity with a view to probable future requirements, and that a pit lathe is peculiarly adapted to such extension will, doubtless, be readily admitted.

The face-plate of this machine measures 30 feet in diameter, and the present dimensions of the pit will admit of swinging 44 feet on centers, with a maximum width of 12 feet. The large face-plate is built up of twelve segments. The rim is of box section, the ends of the rim in each section being finished to make the joint, and the segments being held together at the rim by body-bound bolts. The arms are slotted for bolts, and the space between segments is also shaped to receive the usual square-headed bolts, as the inner end of each segment is fastened to the smaller face-plate by several body-bound bolts.

A feature of interest in connection with this machine is the method of drive adopted, which is a friction roller, 18 inches diameter, made of compressed paper, while the rim of the large face-plate, 15 inches wide, affords the necessary contact surface for driving.

Power is supplied by a 75 horse-power motor, quadruple-geared, the use of the multiple voltage system giving the machine a range covering all diameters from 6 feet to the present capacity, though the gear train is designed to admit of two changes of back gear in addition.

* Extract from a paper presented at the New York meeting of the American Society of Mechanical Engineers by John M. Barney.

Fig. 246 shows the assembled pit lathe driven by the friction roller while taking a heavy facing cut, on which occasion four tools were employed. The picture also shows the driving motor with its train of gears and the mechanism employed for adjusting the pressure on the friction roller.

FIG. 246. FRICTION-DRIVEN LATHE.

AN INVESTIGATION OF FIBER FRICTIONS

The Kelsey Motor Company, during 1921, conducted an exhaustive investigation of the properties and behavior of friction drives, the results of which were presented in a paper read before a meeting of the Friction Drive Engineering Society by Philip Kriegel.* Abstracts from this address are:

"In order that a comprehensive idea might be secured of the functioning of the friction fibers under different conditions of speed, slip and twist, the friction wheels were brought in contact with the disk at four different positions with regard to the disk. These were so placed that the middle of the working face of each fiber was at approximately 7¾ inches, 6 inches, 4 inches, and 2 inches from the center of the dish. With the motor running at about 1,625 revolutions per minute, as was the average throughout the test, the speeds in the four positions would be, at an assumed zero slip, about

1480 r. p. m. or 6,600 feet per minute
1140 r. p. m. or 5,100 feet per minute
760 r. p. m. or 3,400 feet per minute
380 r. p. m. or 1,700 feet per minute

Beginning with a contact pressure of 250 feet a load was sent through the transmission, the most that could be absorbed without excessive slipping.

*Experimental Engineer, Kelsey Motor Co.

A complete record was taken of the horse-power input, the speed of both disk and friction wheel and the torque on the dynamometer scale. Pressures increasing in increments of 25 pounds from 250 to 500 pounds, and increments of 50 pounds up to 1,050 pounds were applied. The runs in each position were repeated.

A general idea of the type of fibers investigated, their composition, characteristics and dimensions can be gained from Table 27. The disk, 17 inches in diameter, used for Fibers No. 1 to No. 6, was of "amelite" (an aluminum-zinc-copper alloy). For Fibers No. 7 to No. 10 it was found necessary to face the disk from the periphery inward 4 inches with boiler plate steel.

TABLE 27

FIBER NO.	DIMENSIONS, INCHES		TYPE OR NAME	COMPOSITION AND REMARKS
	WIDTH OF WORKING FACE	OUTER DIAM. OF FILLER		
1	1.55	16.95	"Tarred" fiber	Paper—Kraft—100%—tarred, compressed in rings
2	1.55	16.80	Cotton duck fiber	Radial pieces of cotton duck shaped into a ring under high compression
3	1.55	16.80	Metal and brake lining fiber	Strips of brake lining (containing asbestos) in alternate rings with copper mesh
4	1.35	16.95	"Micarta" fiber	A chemical—(bakelite?) laid in rings between layers of cotton gauze
5	1.50	16.95	Black split fiber	Paper—100% rag—chemically treated, made in two split rings
6	1.60	17.00	Paper fiber	Paper—rag—35%; sulphite 25%, rope 40%. Very soft. Improperly finished
7	1.30	16.95	Molded fiber	Believed to be molded asbestos which had probably been treated with cement
8	1.30	17.00	"Egyptian" fiber	A rag fiber, chemically treated, vulcanized, and hardened under pressure
9	1.45	16.95	"Oil celoron" fiber	Paper containing rag—5%; sulphate—95%. Treated with oil and compressed
10	1.50	16.95	"Chrome friction" fiber	Rawhide, presumably treated with chrome solution

In Table 28 are shown the average horse-power transmitted by the fibers in each of the four positions indicated, the speeds of the friction wheel, and the corresponding road mileage at a gear reduction of 4.5 to 1 (now used in the Kelsey automobile). Every effort was made to transmit the same or as much horse-power as possible at the minimum contact pressure applied and to keep the horse-power input fairly constant with increasing pressures. The variations in the amount of horsepower transmitted must give rise to certain indications which will be discussed later.

TABLE 28—SHOWING PER CENT. TRANSMISSION EFFICIENCY AND PER CENT. SLIPPING AT FOUR POSITIONS ON THE DISK

PRESSURE OF CONTACT, POUNDS	PER CENT.	FIBER NUMBER																			
		ONE				TWO				THREE				FOUR				FIVE			
		DISTANCE OF MIDDLE OF FIBER FROM CENTER OF DISK, INCH																			
		7.72	5.97	3.97	1.97	7.72	5.97	3.97	1.97	7.72	5.97	3.97	1.97	7.82	6.06	4.06	2.06	7.75	6.00	4.00	2.00
250	Efficiency	90.3	89.2	....	....	88.3	78.5	....	....	88.9	....	....	....	88.5	....	....	....	61.4	....	....	....
	Slip	3.8	6.5	....	....	10.0	20.5	....	....	10.2	....	....	....	11.5	....	....	....	53.2	....	....	....
300	Efficiency	93.7	92.0	87.7	61.3	90.8	89.1	67.6	64.2	90.0	78.2	....	....	94.0	93.8	....	....	66.4	....	....	....
	Slip	3.5	5.5	14.0	38.5	7.2	13.1	39.2	20.1	8.2	29.1	....	....	8.0	7.2	....	....	47.2	....	....	....
350	Efficiency	93.0	92.0	90.6	65.2	89.0	90.1	74.8	72.5	92.0	91.4	45.5	....	94.8	94.2	67.0	....	78.3	70.4	....	....
	Slip	3.2	4.7	12.2	25.0	7.0	11.4	27.8	16.2	7.5	6.0	57.1	....	7.0	6.0	16.0	....	24.7	34.3	....	....
375	Efficiency	95.0	90.5	90.8	68.9	88.6	89.8	77.7	76.2	90.6	91.7	49.8	73.3	96.5	96.2	77.8	74.4	88.0	74.2	41.6	71.0
	Slip	3.5	4.7	11.0	21.0	6.8	10.5	22.0	15.0	6.5	4.8	53.4	24.6	6.0	5.2	12.0	16.1	21.0	33.4	63.6	27.7
400	Efficiency	95.4	90.4	90.0	69.9	88.6	89.4	79.4	77.7	90.1	91.6	53.4	73.8	95.0	95.2	79.4	77.9	95.4	79.1	45.5	72.6
	Slip	3.2	4.6	9.7	19.7	6.6	9.1	20.0	16.2	6.5	4.8	49.6	21.3	6.0	5.1	10.5	15.5	17.0	26.0	58.0	23.7
425	Efficiency	92.7	90.3	88.5	67.1	88.4	88.4	82.5	72.5	90.0	90.2	59.5	75.8	94.4	94.6	81.2	79.0	95.4	83.2	50.0	75.4
	Slip	3.0	4.7	9.4	19.8	6.7	9.1	19.5	15.0	6.0	4.8	47.6	19.3	6.0	5.4	9.5	13.7	11.5	20.9	57.1	19.0
450	Efficiency	92.0	90.0	88.0	66.0	88.8	88.2	82.0	68.1	90.0	89.7	63.3	76.0	93.7	93.3	87.4	77.4	95.0	84.5	58.0	77.8
	Slip	3.2	4.5	9.5	17.7	6.2	7.8	18.0	15.0	5.7	4.8	42.2	18.0	6.0	5.5	9.0	13.9	11.0	19.1	52.0	16.6
475	Efficiency	90.2	89.4	88.0	65.2	88.6	87.0	81.9	67.0	89.2	88.4	66.0	78.0	93.7	92.7	90.5	72.0	95.0	87.1	73.3	77.6
	Slip	3.0	4.5	9.2	17.6	6.0	7.8	17.5	15.0	5.5	4.7	39.9	15.8	6.0	4.9	7.7	13.3	10.5	16.4	36.0	13.3
500	Efficiency	89.6	89.1	87.0	63.6	88.2	84.3	81.8	65.8	88.9	88.4	70.0	79.0	93.5	92.2	90.2	71.0	93.4	91.2	90.9	75.0
	Slip	3.2	4.5	9.0	17.0	6.1	8.5	16.5	13.6	5.5	4.7	37.4	16.4	6.0	5.2	7.0	13.7	9.5	12.7	18.5	12.4
550	Efficiency	89.0	87.2	86.4	60.6	87.7	82.3	80.9	61.7	88.0	88.7	70.1	73.1	93.0	91.6	90.9	70.4	93.2	91.1	91.2	74.0
	Slip	3.1	4.5	9.0	14.9	6.6	7.9	15.0	12.4	5.5	4.4	34.5	16.0	5.5	4.8	7.0	13.2	9.5	11.4	16.8	11.8

600	Efficiency	87.8	85.8	85.7	59.3	86.3	80.1	80.5	60.6	88.0	88.6	74.7	72.4	92.5	91.0	89.2	68.3	93.5	91.5	88.0	74.0
	Slip	3.4	4.5	9.0	12.8	5.4	7.3	13.5	12.4	5.0	4.4	32.3	14.7	5.5	5.4	6.5	12.7	9.0	11.3	17.0	11.8
650	Efficiency	87.5	85.6	83.1	58.3	86.2	77.6	78.8	55.0	88.4	88.5	75.2	72.0	91.1	86.2	87.9	66.4	93.0	93.0	87.9	71.6
	Slip	3.4	4.5	9.0	12.8	5.5	7.3	13.0	12.4	5.0	4.1	28.8	14.7	5.0	4.8	6.5	12.6	9.0	11.0	16.8	10.8
700	Efficiency	85.9	84.7	82.9	57.7	83.8	77.5	78.3	53.5	87.5	87.3	76.4	70.0	90.1	85.0	87.2	66.0	93.2	91.1	86.8	68.7
	Slip	3.3	4.5	8.8	12.9	5.2	6.1	13.0	13.0	5.0	4.1	27.6	14.7	5.0	4.4	6.5	12.6	9.0	10.5	16.5	11.5
750	Efficiency	85.8	83.4	82.2	56.0	83.2	76.5	77.2	49.2	87.5	87.1	82.5	70.0	88.2	84.3	84.6	65.2	92.8	89.6	85.9	66.8
	Slip	3.2	4.5	8.5	13.0	5.0	6.4	11.7	13.0	4.8	4.3	22.9	14.7	5.0	4.2	6.5	12.8	8.6	10.0	15.5	11.1
800	Efficiency	84.0	82.6	81.7	55.2	81.4	76.0	77.1	48.0	87.5	86.8	84.5	69.7	88.0	83.6	83.6	64.8	92.5	87.9	85.9	65.0
	Slip	3.6	4.5	8.0	13.0	5.2	7.0	11.5	12.4	5.0	4.4	20.0	14.7	5.2	4.3	6.5	12.7	8.0	10.1	15.0	12.6
850	Efficiency	83.2	81.6	81.0	53.5	79.9	75.3	76.2	47.5	87.1	85.9	85.8	69.2	87.4	82.5	83.4	63.9	93.2	86.8	85.6	64.1
	Slip	3.6	4.7	7.8	13.3	5.1	7.0	11.0	12.4	4.5	4.3	18.0	14.3	5.0	4.4	6.0	12.0	8.0	10.1	14.6	12.7
900	Efficiency	83.0	81.0	80.4	52.1	77.9	74.5	75.4	44.7	86.8	84.4	87.1	68.6	87.3	82.3	82.7	63.9	92.5	86.8	85.0	62.5
	Slip	3.7	4.7	8.0	14.4	5.2	5.8	11.0	11.1	4.6	4.0	14.4	14.7	5.0	4.5	6.0	12.0	8.0	10.1	14.5	12.4
950	Efficiency	83.0	80.2	78.4	51.0	77.1	74.0	74.1	44.7	86.0	84.4	86.2	68.3	86.8	80.6	82.0	63.5	92.1	85.5	85.0	59.4
	Slip	3.6	5.0	8.0	14.5	5.0	6.1	10.0	11.1	4.5	4.0	13.5	14.3	4.6	4.5	6.0	12.6	7.6	10.4	14.5	14.0
1000	Efficiency	81.8	78.8	77.2	48.3	75.3	73.5	72.0	44.3	85.6	83.8	85.3	68.2	86.8	80.3	80.8	63.2	92.0	85.5	81.7	58.3
	Slip	3.6	5.0	7.8	13.7	4.5	6.1	10.0	11.1	4.5	4.0	12.7	14.7	4.5	4.2	6.0	11.8	7.5	10.5	14.0	13.5
1050	Efficiency	81.8	78.3	76.8	47.6	74.0	73.0	71.9	42.3	85.4	82.9	83.5	67.2	87.0	80.3	78.0	62.1	90.6	84.2	79.4	57.7
	Slip	3.6	5.3	8.0	13.6	4.5	5.0	9.4	11.1	4.4	4.0	12.5	14.3	4.3	4.2	5.5	11.5	7.3	10.4	13.5	13.7

TABLE 28

Pressure of contact, pounds	Per cent.	Six 7.68	Six 5.93	Six 3.93	Six 1.93	Seven 7.84	Seven 6.09	Seven 4.09	Seven 2.09	Eight 7.84	Eight 6.09	Eight 4.09	Eight 2.09	Nine 7.78	Nine 6.03	Nine 4.03	Nine 2.03	Ten 7.72	Ten 5.97	Ten 3.97	Ten 1.97
		Fiber number																			
		Distance of middle of fiber from center of disk, inch																			
250	Efficiency	85.7	91.1	....	....	93.8	....	....	....	84.0	....	....	....	86.0	90.0	....	....	86.0	85.6	75.5	....
	Slip	8.8	....	....	....	5.5	....	....	....	29.5	....	....	....	9.2	5.9	....	....	7.0	6.0	17.1	....
300	Efficiency	86.8	92.7	82.0	74.3	97.5	94.8	95.0	70.4	85.5	84.2	63.5	....	90.5	92.0	62.5	....	87.0	87.0	77.0	70.7
	Slip	8.0	....	....	....	3.6	6.2	11.0	17.2	23.4	11.4	43.2	....	7.5	5.8	32.5	....	6.4	6.4	10.5	27.7
350	Efficiency	88.0	92.8	85.0	73.6	96.0	95.3	94.2	72.0	88.5	90.4	75.5	72.2	90.5	93.2	82.2	....	87.7	86.4	83.4	70.9
	Slip	7.2	....	....	....	2.3	5.2	9.0	16.6	15.5	10.3	28.2	20.9	6.7	5.8	22.5	....	6.3	6.4	9.0	24.8
375	Efficiency	86.7	92.1	84.0	73.2	95.5	95.4	94.0	75.5	90.5	90.4	87.5	75.0	86.4	92.1	85.4	....	87.8	84.4	84.0	72.3
	Slip	7.0	....	....	....	2.0	5.0	6.5	14.3	12.2	10.0	21.2	21.4	6.3	5.8	13.4	....	6.0	6.0	9.0	24.2
400	Efficiency	85.4	88.9	83.2	73.3	95.3	94.1	93.8	79.6	93.0	92.1	90.0	79.0	85.6	92.6	83.4	....	87.8	82.7	84.0	73.3
	Slip	7.0	....	....	....	2.2	3.1	4.5	11.8	10.5	10.0	18.3	13.1	6.0	5.5	11.5	....	6.0	6.1	9.0	21.9
425	Efficiency	85.1	88.1	82.2	73.5	95.2	93.8	93.7	80.1	95.0	92.4	91.8	80.0	85.7	93.0	82.3	....	87.8	82.7	84.0	73.9
	Slip	7.0	....	....	....	2.0	2.8	4.3	11.3	9.8	10.0	17.0	10.6	6.0	5.0	11.2	....	6.0	6.1	9.0	21.9
450	Efficiency	82.9	85.9	81.5	72.6	95.0	92.3	93.6	81.8	94.7	90.7	90.8	80.0	85.6	93.6	81.5	....	87.6	82.6	84.0	74.0
	Slip	7.0	....	....	....	2.0	2.8	4.0	10.9	9.4	10.0	17.0	10.6	5.6	5.0	11.0	....	6.0	6.1	8.4	21.9
475	Efficiency	82.7	84.5	81.2	71.8	94.6	91.7	93.5	82.0	94.0	88.6	90.7	78.5	85.6	93.8	81.0	....	86.5	81.9	82.5	72.1
	Slip	7.0	....	....	....	1.7	2.6	4.0	9.6	9.5	9.8	17.0	10.6	5.5	4.0	11.0	....	6.0	6.1	8.0	21.9
500	Efficiency	82.7	83.4	80.5	70.7	94.4	89.4	93.6	82.0	93.5	88.0	90.5	76.7	85.4	93.8	80.8	....	85.9	81.9	82.8	71.9
	Slip	5.6	....	....	....	1.5	2.9	4.0	9.6	9.0	9.3	17.0	10.1	5.5	3.9	11.0	....	6.0	6.1	7.5	21.9
550	Efficiency	81.6	81.2	79.8	70.0	93.9	87.4	92.4	78.4	93.3	87.0	89.0	75.0	85.0	93.8	80.0	....	84.0	80.9	82.0	71.9
	Slip	5.5	....	....	....	2.0	3.0	3.7	9.1	8.0	8.5	14.4	7.6	5.5	3.8	11.0	....	6.0	5.8	7.8	21.9

600	Efficiency	78.7	79.5	78.6	67.0	93.6	86.8	91.7	76.1	93.0	86.2	88.5	73.9	83.5	92.1	79.7	....	83.8	80.9	82.0	70.6
	Slip	5.0	....	....	....	2.0	2.6	3.4	9.4	7.6	8.2	14.5	8.0	5.0	3.5	11.0	....	6.0	5.7	7.5	19.2
650	Efficiency	77.0	78.1	74.0	65.3	93.5	84.2	92.0	73.4	92.5	85.2	88.2	72.8	83.5	88.9	79.5	....	83.5	80.0	80.0	70.3
	Slip	5.0	....	....	....	2.0	2.5	3.3	9.6	7.0	8.2	14.5	7.6	5.0	4.0	10.6	....	6.0	5.7	7.0	19.2
700	Efficiency	76.0	76.3	71.5	64.9	93.5	84.2	90.0	71.3	92.4	84.5	87.5	70.4	82.0	88.9	77.0	....	83.1	79.1	80.0	70.0
	Slip	5.0	....	....	....	2.0	2.9	3.4	9.6	7.0	8.2	14.0	7.6	5.0	4.6	10.8	....	5.0	5.5	6.5	18.8
750	Efficiency	75.0	75.0	67.6	63.5	93.5	84.0	89.6	69.6	92.0	84.5	87.0	68.9	81.8	88.0	74.5	....	83.0	78.7	78.3	70.0
	Slip	5.0	....	....	....	2.0	2.7	3.3	9.2	7.0	8.0	13.0	5.0	5.0	4.2	10.7	....	5.0	5.5	6.5	18.8
800	Efficiency	73.6	72.4	61.6	60.2	91.7	80.0	88.4	66.9	92.2	84.5	86.6	67.9	81.5	85.1	73.0	....	81.8	77.8	77.3	68.7
	Slip	5.0	....	....	....	2.0	2.7	3.4	9.1	7.0	7.9	13.0	5.0	4.5	4.3	10.8	....	5.0	5.6	6.0	18.6
850	Efficiency	72.5	70.9	60.0	59.7	90.7	77.0	88.0	66.2	91.5	84.3	85.8	67.9	81.2	84.0	70.5	....	81.5	77.0	75.8	66.6
	Slip	4.0	....	....	....	2.0	2.6	3.3	8.8	6.5	7.9	12.5	5.0	4.6	4.4	11.0	....	5.0	5.3	6.0	16.9
900	Efficiency	70.5	66.6	58.2	57.8	88.6	74.5	87.4	65.1	89.2	84.0	85.3	64.8	79.7	83.3	70.2	....	81.0	76.1	74.2	65.8
	Slip	4.0	....	....	....	2.0	2.4	3.0	8.6	7.0	7.9	11.5	7.2	4.4	3.5	10.5	....	4.4	5.4	5.5	16.9
950	Efficiency	67.5	64.4	56.5	56.4	88.4	74.0	87.0	64.6	88.6	82.3	84.5	63.7	79.4	82.3	70.4	....	81.0	74.6	74.2	64.7
	Slip	3.7	....	....	....	1.7	2.1	3.0	9.0	6.6	7.6	10.5	6.8	4.5	3.6	10.0	....	4.3	5.5	5.0	16.9
1000	Efficiency	64.0	62.3	55.0	53.9	86.6	71.7	86.3	65.0	88.5	82.3	83.6	58.3	75.6	80.6	70.0	....	80.6	74.2	73.0	61.6
	Slip	3.5	....	....	....	2.0	2.2	3.0	8.9	6.0	7.6	10.5	2.0	4.6	4.0	9.0	....	4.2	5.8	5.0	16.9
1050	Efficiency	63.5	62.1	52.0	53.1	86.4	70.0	81.5	64.5	88.5	80.1	83.6	54.7	74.2	78.7	69.0	....	79.0	73.6	72.2	60.3
	Slip	2.6	....	....	....	1.8	2.2	3.0	9.1	6.0	7.8	10.5	4.0	4.3	3.7	8.7	....	3.5	5.2	5.0	16.8

TABLE 29

Fiber No.	Middle of Fiber in Contact with Disk at Inch	Average Horse Power Transmitted	Average Speed of Friction Wheel R.P.M.	Average Miles per Hour (Gear Reduction 4.5 to 1)	Fiber No.	Middle of Fiber in Contact with Disk at Inch	Average Horse Power Transmitted	Average Speed of Friction Wheel R.P.M.	Average Miles per Hour (Gear Reduction 4.5 to 1)
1	7.72	12.3	1,440	30.5	6	7.68	12.7	1,400	29.7
	5.97	13.1	1,100	23.2		5.93	12.3	1,120	23.7
	3.97	14.0	700	14.7		3.93	12.5	720	15.2
	1.97	3.4	310	6.6		1.93	6.1	370	7.8
2	7.72	13.6	1,370	29.0	7	7.84	10.7	1,450	30.7
	5.97	13.1	1,065	22.6		6.09	10.4	1,180	25.0
	3.97	13.6	630	13.3		4.09	7.8	770	16.3
	1.97	3.7	340	7.3		2.09	4.1	425	8.9
3	7.72	13.0	1,380	29.4	8	7.84	12.2	1,340	28.3
	5.97	13.0	1,090	23.1		6.09	9.0	1,035	21.9
	3.97	9.8	740	15.6		4.09	8.8	650	13.7
	1.97	3.9	325	7.0		2.09	3.5	360	7.6
4	7.82	13.3	1,420	30.1	9	7.78	6.3	1,450	31.0
	6.06	12.7	1,125	23.8		6.03	5.3	1,115	23.5
	4.06	11.3	720	15.2		4.03	3.2	780	16.5
	2.06	5.5	340	7.3		2.03	....	....	....
5	7.75	13.2	1,340	28.2	10	7.72	11.0	1,375	29.1
	6.00	10.5	1,050	22.2		5.97	8.5	1,100	23.2
	4.00	10.2	690	14.7		3.97	7.0	750	15.8
	2.00	5.4	340	7.3		1.97	3.5	300	6.4

POWER TRANSMITTING EFFICIENCY

In finding the actual transmission efficiency between the driving disk and the driven friction wheel it was found necessary to make proper allowance for the mechanical losses in the driving end of the apparatus. These were accurately determined by noting the amount of horse-power required to run the disk idly at all the speeds desired. The percentage transmission efficiency therefore resolved itself into the simple formula:

$$E = \frac{(P_t - P_d)K}{P_o} \times 100$$

where

E = the percentage efficiency.
P_t = total horse-power.
P_d = horse-power required to run disk idly.
P_o = horse-power transmitted.
K = constant for motor efficiency.

In Table 29 are given the actual percentage efficiency values. From a study of these it will be seen that in the case of the tarred paper Fiber No. 1 the maximum transmission efficiency was reached in the four positions at a contact pressure of 350–400 pounds (at 230–260 pounds per inch of working face). With the middle of the fiber at about 7.75 inches from the center of the disk, the efficiency rose steadily from 90 per cent. at a pressure of 250 pounds to 95.5 per cent. at a pressure of 400 pounds every increase in pressure from there on tending to produce a decrease in the actual transmission efficiency, so that at 1,050 pounds it was but 82 per cent. This decrease in efficiency with each additional increase in pressure beyond that required to produce maximum results was true not only of this fiber and in this position with relation to the disk but also of the remaining fibers and positions. As Fiber No. 1 was brought in toward the center of the disk, the efficiency values decreased, giving peak values of 92 per cent., 90.6 per cent. and 69.9 per cent. respectively. The per cent. slip increased from 3 per cent. at the 7.75-inch or high speed position to about 20 per cent. at the 2-inch position.

In the case of Fiber No. 2, the pressures required to produce maximum efficiency results are somewhat less than those in Fiber No. 1. It must be noted that here the efficiency values themselves are all decidedly smaller. The slip is about twice as great.

Except for the particularly great pressure which was required to secure the maximum value of 87.1 per cent. in the 4-inch position, the combination of brake lining and copper mesh Fiber No. 3 showed rather average results. The falling off in efficiency at high speed from the maximum at a pressure of 350 pounds to that at 1,050 pounds was very gradual, being but 6 per cent. The slip at normal maximum running pressure in high speed position was 7 per cent.

Fiber No. 4 had a narrower working face than most of the other fibers. This may account for the high maximum efficiency value of 96.5 per cent. which occurred at 375 pounds and at 6 per cent. slip in high speed. As will be observed from Table 29, the results in the 6-inch position are almost as great as those in the 7.75-inch position. In low speed the pressure required to transmit the maximum horse-power appears to be 50–100 pounds higher than in high speed. This fiber gave smoother results than any of the others tested.

Both Fibers Nos. 5 and 6 appear to have been entirely too soft to withstand the rigorous testing to which they were submitted. In Fiber No. 6 the results were extremely erratic. In Fiber No. 5 slipping was considerably greater than 20 per cent. at pressures up to 375 pounds in the high-speed position and at pressures up to 500 pounds in the low-speed position. The corresponding efficiency results were therefore very low. For Fiber No. 6 the efficiency results proved to be far worse than those of any other material. In all four positions the maximum seems to have been reached at 350 pounds pressure. From there on, each increase in pressure produced a very great diminution in transmission. Furthermore, the maximum efficiency results themselves were extremely low. On account of the rapid disintegration of the material, it was impossible to secure other than the most unreliable speed readings of the friction wheel. The values for percentage slip are therefore omitted from Table 30 for all except the high-speed position.

Fiber No. 7, also narrow in working face, gave efficiency values of 97.5 per cent. at high speed and 95 per cent. at low speed (4 inches) both at a pressure of 300 pounds. In this fiber, as in Fiber No. 2, the pressure required to produce the maximum figure was very low. Slipping was almost negligible, increasing from 2 per cent. in the high-speed position to a comparatively small value of 9.6 per cent. in the 2-inch position.

Fiber No. 8 behaved almost as uniformly as did the one immediately preceding it. The slipping was far in excess of the latter in all positions; the pressures required to produce the maximum transmission efficiency were greater; the percentage values were smaller but the actual decrease in efficiency as the fiber was brought in contact with the disk at points closer to the center was as uniform as it was in Fiber No. 7. This fiber showed maximum efficiencies of 95 per cent., 92.4 per cent., 91.8 per cent. and 80 per cent., respectively, for the four positions at 425 pounds pressure. The decrease in efficiency with increase in pressure was not in any way very pronounced.

On account of the tendency for the oil to ooze out to the surface, Fiber No. 9 presented results which were very inconsistent. The maximum transmission efficiency of 90 per cent., a comparatively low figure in the high-speed position, was attained at 300–350 pounds pressure. The decrease in efficiency from this pressure to that at 1,050 pounds was also greater than that shown by any of the fibers previously discussed, with the possible exception of No. 6. The values shown in Table 29 cannot be taken as a criterion of the way this fiber would behave under similar repeated tests. In fact, the high pressures in runs at both high and low speed forced the oil to the surface and caused the lubrication of the working face at 2 inches from the center of the disk the friction wheel slipped as much as 70–95 per cent.

The rawhide Fiber No. 10, not unlike the other soft friction materials, gave very low maximum efficiency, results ranging from 87.8 per cent. in high speed to 70 per cent. in the lowest speed, at 350 pounds pressure. The drop in efficiency with increase in pressure also appears to be small. The slip, averaging 6 per cent. in high speed, does not seem to be greater than that shown by any of the treated paper or cotton fibers with the possible exception of the tarred Fiber No. 1.

A study of all of the above information does not bring out an absolute correlation between the amount of slipping and the degree of transmission efficiency. The increase in the latter does not appear to be entirely commensurate with the decrease in the former. Furthermore, it does not necessarily follow that, in the case of any two fibers, the one possessing the greater transmission efficiency is submitted to less slipping. For instance, at a pressure of 375 pounds in high speed Fiber No. 1 shows an efficiency of 95 per cent. and a slip of 3.5 per cent., while Fiber No. 4 shows an efficiency of 96.5 per cent. and a slip of 6 per cent. It is quite apparent, therefore, that factors other than the degree of slipping influence the transmission efficiency of any friction fiber.

PHYSICAL PROPERTIES

The length of time of operating each fiber and the pressures to which they were subjected were approximately the same throughout the test. In judging the durability of the materials, therefore, no undue stress can be laid on a particular condition. The fibers were closely observed for physical appearance before, during and after operation, for temperature and noise.

Fiber No. 1. Was subjected to no glazing or burnishing action under high slipping such as is usually encountered at contact pressures up to 300 pounds. As the pressures were increased from 450 pounds up to 1,050 pounds, especially in the low-speed positions, the fiber grew hotter. After the check run in the 2-inch position it gave slight evidence of burning.

Fiber No. 2. Showed a decided tendency to burning at both high slipping and with increase in pressure above that required for normal operation. It was noticed that after cooling, the fiber had become somewhat harder than it was originally and the radial segments had twisted out of their original position, the degree of twist increasing as the fiber was brought closer to the center of the disk.

Fiber No. 3. In spite of the great amount of slipping, especially when tested in the 4-inch position, this fiber showed but little tendency to become hot. The copper mesh in the fiber had a decided cutting effect upon the disk. This fiber, like all of the remaining hard fibers, produced considerable noise in operation.

Fiber No. 4. At a pressure of 300 pounds the slip was so great as to produce rapid burning and a resultant odor of formalin. However, when this fiber not permitted to slip excessively, it does not heat up like any of the paper material, even after the application of the higher pressures of contact. Because of the failure on the part of the manufacturer to provide a width for the total thickness of the filler in excess of that of the working face and also bevelled edges on either side as retaining walls, this fiber showed a tendency to break down over the supporting flanges.

Fiber No. 5. Burned quickly when subjected to slip even for a very short period of time. In the course of operation, especially in the low-speed positions, the tendency to smoke was pronounced. At the end of the test the material showed a deep groove running through the middle of the ring, caused presumably by charring.

Fiber No. 6. Proved to be entirely too soft to withstand the higher pressures of contact. It also suffered from a poor, uneven finish given the working face by the manufacturer. In order to prolong the life of the fiber, every precaution was taken to reduce slipping to a minimum. At pressures of 700 pounds in the high-speed position, the fiber began to smoulder. The speed run in the same position produced broken uneven surfaces. In the low check positions the effect of burning was so great as to cause almost complete disintegration, great tufts of the material being virtually torn out of the filler.

Fiber No. 7. Showed slight tendency to slip in the short runs at low pressure and therefore suffered but little of the effects attending such a condition. There was also almost a negligible amount of heating at the high contact pressures. However, the fiber was extremely hard, glazed somewhat, and produced a peculiar siren-like noise when operated in the high-speed positions.

Fiber No. 8. Appeared in many respects to behave like Fiber No. 7. The effect of heating was very slight. It showed a glazed surface and produced rather more noise in operation than any of the fibers except No. 3 and No. 7.

Fiber No. 9. Is not believed to have been constructed with due regard to the severe conditions under which variable speed disk drives operate. In the runs in both high- and low-speed positions, the oil in the filler constantly came out to the surface, lubricating the disk and working face, causing considerable slipping and the consequent generation of great heat.

Fiber No. 10. Was one of the softest fibers of the ten. Under the ordinary operating conditions it behaved in many respects like Fiber No. 5, becoming just as hot. It expanded considerably, curled over the retaining wall and caused an unevenness and increase in width of the working face.

RESULTS OF ENDURANCE TESTS

In order to secure a better idea of the durability of the materials, Fibers Nos. 1, 2, 4, 7, and 8, considered at this time to be the best of the ten fibers under investigation, were subjected to fairly strenuous runs in the position 2 inches from the center of the disk. In this position it was believed the greatest amount of slipping and twisting would cause the most rapid breaking down of the materials. The motor speed averaged 1,650–1,725 r. p. m., the speed of the friction wheel at the supposed zero slip was calculated to be about 390–405 r. p. m. The pressure of contact used was 425 pounds, which it was previously found had given the average maximum efficiency values in the 2-inch position. Results of these runs are submitted in Table 30 below:

TABLE 30

FIBER NO.	TOTAL NO. HOURS OF RUN	TOTAL NO. REVOLUTIONS MADE BY FRICTION WHEEL	TOTAL NO. OF MILES (INTERNAL GEAR RATIO 4.5 TO 1)	AVERAGE R.P.M. OF FRICTION WHEEL	AVERAGE TRANSMISSION EFFICIENCY, PER CENT.	AVERAGE SLIP, PER CENT.	AVERAGE REDUCTION IN O.D. OF WHEEL, INCH
1	28.0	466,422	164.4	277	79.4	28.7	0.035
2	32.1	513,586	181.6	267	70.7	33.3	.....
4	27.2	485,107	170.9	298	81.5	23.7	0.030
7	31.6	339,122	119.5	177	56.0	55.3	0.085
8	32.2	578,422	204.5	300	76.3	26.3	0.050

The run of the tarred Fiber No. 1 was not continuous. It extended over four periods each of approximately seven hours' duration. Fluctuations in transmission efficiency and slip, evident in some of the other materials submitted to this endurance test, were not quite so apparent in this fiber. Starting with the friction members cold, the efficiency value was comparatively low. However, as the fiber warmed up gradually the percentage efficiency increased so that for the last average hour of operation in the period the efficiency was at a maximum. The amount of slip increased with increase in efficiency, thereby producing a decrease in the relative speed of the friction wheel. This phenomenon also held true in the case of the remaining fibers so tested. Examination of the fiber at the end of twenty-eight hours showed almost no perceptible reduction in its outer diameter, an increase of about 4 per cent. in width of working face and gave evidence of some burning on the inner edge of the face. Taking into consideration the fact that it was operated at a position on the disk much closer to the center than is ever approached in ordinary practice, the fiber proved to be of extraordinarily good material and showed very good efficiency results.

A careful study of the table shows readily that the compressed cotton duck Fiber No. 2 fell somewhat below fiber No. 1 in its transmission efficiency. Although operated under conditions almost identical with those of the material just described, this fiber showed decidedly worse results in every respect; a greater average slip, a more pronounced tendency to slip with each continuous run and the consequent decrease in the speed, and lower efficiency throughout.

The fiber remained unchanged in width of working face. The changes in outer diameter were as follows: Before test—outer edge, 16.687 inches; inner edge, 16.687 inches; after test—outer edge, 16.672 inches; inner edge 16.703 inches It is to be noted that where the outer edge of the fiber was worn away about 0.015 inch, the inner edge increased in diameter by fully 0.016 inch. This can be attributed to the fact that the twisting action on the fiber threw the radial pieces out of the lateral position at a point about ½ inch from the inner edge, the twist increasing at points closer to the inner edge. No doubt such action served to relieve the compression under which the cotton duck pieces had been laid into the wheel and to produce the consequent increase in outer diameter.

The "micarta" Fiber No. 4 appeared to be the best of all the fibers submitted to the endurance test, showing a fair uniformity in transmission efficiency and degree of slip in the four seven-hour runs. The increase in efficiency from the first to the seventh hour averaged only 5 per cent. As in the case of Fiber No. 1, the slip increased as the day's run progressed. However, the average efficiency and average slip proved to be the highest and lowest, respectively, of the five fibers tested. Because of poor design, especially lack of the bevelled edges previously discussed, this material became frayed somewhat. The slight burning action decreased the outer diameter uniformly over the whole working face by 0.03 inch. The relatively favorable behavior under these rather arduous conditions lead to the belief that this fiber would prove to stand up very well in every respect in practice.

Fiber No. 7 shattered every favorable impression previously created in regard to good transmission efficiency and durability. In three runs, each of nine hours, the efficiency fell from a value of about 85 per cent. for the first hour to about 25 per cent. for the ninth hour; the slip increased gradually from about 25 per cent. to about 85 per cent.; the speed of the friction wheel decreased from approximately 300 r. p. m. to 50 r. p. m. On one occasion the test was interrupted after a continuous run of four hours for several minutes and then resumed. The result of this was to bring the efficiency and speed back to the maximum point. Atmospheric temperature changes such as were produced by opening a window near the apparatus caused decided reductions in the speed and transmission efficiency. A feature of this fiber was the increasing noise in the course of operation. It commenced as a dull, heavy sound, developed into a sort of crunching noise and finally produced a rumble not unlike that of a heavy stone-crushing mill. After completion of the test, accurate calipering of the working face revealed a decrease in diameter of 0.07 inch on the outer edge and 0.10 inch on the inner edge.

Fiber No. 8 showed results favorably comparable in every respect with those of Fiber No. 1. The variation in transmission efficiency was almost negligible, being for the most part very close to the average indicated in Table 30. The percentage slip, almost constant throughout, was less than that of the tarred fiber and consequently the decrease in speed in the course of the eight- or nine-hour runs was very slight. Like Fiber No. 1, this material was subjected to slight burning, especially on the inner edge of the working face. This was evidenced by the following diminutions in outer diameter: Before test—17.00 inches; after test, outer edge—16.96 inches; inner edge—16.94 inches.

COEFFICIENT OF FRICTION

For disk drives wherein the increase in the amount of horse-power transmitted varies with the increase in contact pressure, *i.e.*, up to the pressure producing the maximum transmission efficiency, the value of the coefficient of friction is practically independent of the pressure of contact. Furthermore (as is claimed on authority), "excluding positions at the extreme center, it is independent of the position of the fiber wheel on the disk." For this reason and in order to adhere to a strict comparison of the coefficient values it was deemed advisable to base the calculations on the results secured from tests in the 4-inch position. It will be remembered that in this position every fiber was in contact with the aluminum alloy disk. The values for coefficient of friction are given for pressures which produced the maximum transmission efficiency. To derive the results tabulated below, the general formula for friction wheels was used:

$$H.P. = \frac{\frac{3.1416\,DN}{12} \times Pp}{33,000}$$

D = Mean diameter of friction wheel in inches.
N = No. revolutions per minute.
P = Pressure of contact over working face.
p = Value of coefficient of friction.

TABLE 31

Fiber No.	1	2	3	4	5	6	7	8	9	10
Pressure of contact, pound	375.0	425.0	425.0	425.0	475.0	350.0	300.0	425.0	375.0	375.0
Coefficient of friction	0.368	0.362	0.260	0.316	0.304	0.392	0.276	0.250	0.070	0.218

From the results given it is readily seen that the four highest values for the coefficient of friction are shown by fibers Nos. 1, 2, 4 and 6. In the case of fiber No. 1 the value submitted is almost identical with the manufacturer's given safe value of 0.364 for tarred fiber and zinc alloy and approaches closely the value of 0.390 for tarred fiber and aluminum alloy. For fibers 2 and 4 (the most efficient fiber) the value is not considered low. Fiber No. 6, being fairly soft material, presented the highest coefficient of friction. Fibers Nos. 3, 7 and 8, all very hard materials, appear to have suffered from the fact that hardening a material usually reduces the coefficient value. The low results in fiber No. 9 can be attributed to the oil at the surface. That fiber No. 10 possesses a low value is not surprising for the low coefficients of friction of leather against metal are well known.

DISCUSSION OF EXPERIMENTAL RESULTS

Two elements appear to influence the suitability of friction fibers with regard to transmission capacity and durability under the conditions of service; the pressure of contact and the value of the coefficient of friction between disk and friction wheel. Since the power capable of being transmitted varies with the pressure of contact it is believed that a fiber of high compression strength will give longer service. However, as the friction material is increased in hardness to withstand the greater pressures, the high value of the coefficient of friction decreases. Still another limiting factor, as set forth in this investigation, is the efficiency of transmission. The actual decrease in efficiency value with additional pressures beyond that intended to produce the maximum is now well known.

From one point of view it might be advisable that the fiber transmit the power with maximum efficiency at a low contact pressure, thereby saving the material from the effects of high compression. The heavy requirements of disk friction drives are known, however; the frequent applications of high contact pressures under all sorts of conditions even for short periods of time necessitate the material being sufficiently hard. That the transmission efficiency at those pressures will be less and that it will be increasingly difficult to secure a high value for the coefficient of friction is also fairly certain; but if a fiber can be secured wherein it would be possible to retain a good coefficient and to keep the decrease in transmission efficiency at a minimum, the elimination of present difficulties will have been accomplished in a great measure.

A serious fault of the present friction fillers is the insufficiently developed or, as appeared from one or two of the fibers tested, the absence of pressure withstanding quality. Even such fairly hard materials as the tarred fiber No. 1 or the "micarta" fiber No. 4 suffered from the highcontact pressure. Just what takes place when a fiber is subjected to high pressure is best described by W. D. Hamerstadt in a recent paper entitled, "History and Development of Friction Drive." He states that "when a pressure is applied by a metal against a fiber wheel, there results a certain flexing or compression of the fiber material. This results in the disturbance of the structure of the fiber in the relation of adjoining individual fibers in the mass. Should this pressure be great enough, it is readily conceivable that the bond or interlocking of certain fibers with the adjoining ones becomes broken. This results in a sheering action through the fiber as encountered in cast iron, steel, wood, stone, or cement, in a compression test. Having once broken the original bond or structure, it now becomes much easier for the fiber to compress with subsequent applications of pressure as the wheels pass in contact at high speeds, this working of one fiber on another soon results in generating very high heat. This heat is the damaging factor, as it very soon dries out the fiber—makes it brashy like charred or burned leather and the further destruction of the material is very much hastened." This action was ideally set forth in the applications of high pressures to fiber No 6. The constant kneading of one fiber over the other produced the high heat which caused the rapid disintegration of the material. Fibers No. 5 and No. 10, insufficiently hard, likewise suffered from the same action. In cases where the fiber is not brought in contact with the disk to prevent excessive slipping, the heating effect is decidedly more rapid. Here it does not suffer from the shearing action and working of one fiber upon another with the consequent drying of the material. The action is, on the contrary, that produced by sliding friction. The constant slipping produces a high heat and charring on the surface which gradually works its way inward and causes brittleness and disintegration.

In the light of all these generalizations, it is possible to classify our materials into two groups: (a) Hard fibers; (b) soft fibers. In the first classification can be placed fibers Nos. 3, 7, 8 and 9. A careful examination of Table 31 will show that the values of coefficient of friction for these fibers are all very low. Fiber No. 4 in the true sense is also a hard material but its coefficient of friction is fairly high and its general behavior in every way is such as to make its classification advisable under that of soft fibers. From Table 28 it will be readily seen that the horse-power transmitting capacity of all these fibers, especially in the low-speed positions, is lower than that for the soft fibers. As regards transmission efficiency, these fibers gave good results in the short tests, especially in the high speed positions; but the great amount of slip and the general disappointing behavior of Fiber No. 9 throughout and of Fiber No. 7 in the endurance test is now apparent. It is believed that any of these fibers, if submitted to actual operation, over long periods of time would give poor efficiency in the long run, produce great slipping and because of the low coefficient of friction transmit little horse-power.

Of the so-called soft fibers, Nos. 5, 6 and 10 must be eliminated from favorable consideration, the first two because of very poor pressure withstanding qualities and the last because of its general softness and the very low coefficient of friction. These three fibers have not only shown a low value for maximum transmission efficiency but also decided falling off in percentage with the increasing pressures throughout all four positions on the disk. Of the three remaining fibers, No. 2 has given the lowest efficiency results, standing out particularly as a poor transmission medium at high contact pressures. In spite of the tendency on the part of the radial pieces of cotton duck to twist out of a lateral position, this fiber has shown good wearing qualities. The tarred paper fiber No. 1 and the "Micarta" fiber No. 4 therefore stand out as the foremost materials in every respect. In the endurance test the fibers stood up equally well. Both fibers have shown equally good horse-power transmitting capacity. Fiber No. 4 showed the higher maximum transmission efficiency throughout and a decidedly lower decrease in efficiency as the pressures increased beyond those which gave the maximum value in each position. On the other hand, this fiber showed a greater percentage slip in the high speed positions and possessed a lower coefficient of friction.

From a consideration of all the results submitted it would seem advisable to concentrate further investigation as regards efficiency and durability under other conditions of operation and to make further developments and favorable changes in composition and design upon fibers Nos. 1, 2, and 4. Such a course, it is believed, is preferable to overcoming the many weaknesses presented by the remaining fibers."

SECTION XIV

Special Bevel Gears

The unprecedented demand for high grade bevel gears and ring gears which can be produced in large quantities economically for automobile drives has led to the development of certain distinctive types of bevels, which can very properly be classified as special gears. For the most part the tooth forms of these gears differ radically from that of the customarily employed octoid system with radial teeth and their production entails operations previously foreign to gear manufacture.

SPIRAL TYPE BEVEL GEARS

The Gleason Works have developed one of the most distinctive of these special gears (Fig. 147) in which the teeth are arranged, not in a spiral manner

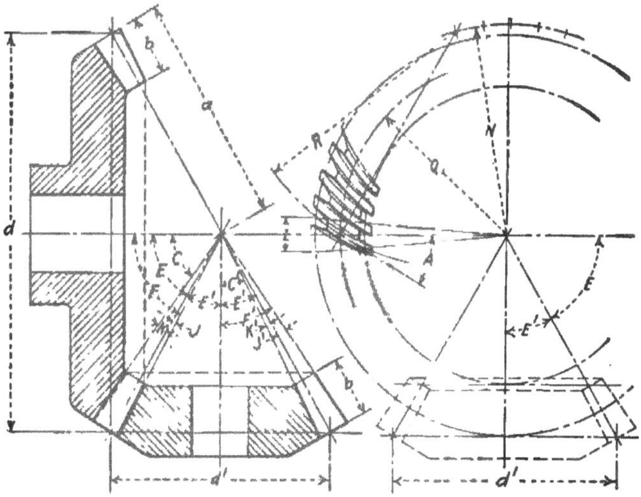

FIG. 247. DIAGRAM OF SPIRAL TYPE BEVEL GEARS.

as the designation of the gear would imply, but in the form of circular arcs which if prolonged would intersect at the cone center of the gear. That is, the various curved tooth elements converge and would all intersect at the cone center. Another peculiarity of the gear teeth is that, though they are generated to true octoid form by means of rotary cutters ground to rack form, the pressure angle (contact slope) is slightly different for either side of the teeth. This is due to the fact that the radii of the two tooth profiles, the concave and the convex, differ by the thickness of the rotary cutting tool, opposite sides of the cutter being employed to cut the two sides of the gear teeth, the outer edge of the cutter forming the concave sides of the gear teeth and the inner edge the convex sides.

Another characteristic of spiral type bevel gears is the peculiar thrust developed by the gears in action, diagrammatically depicted in Fig. 248. The transmitted load is naturally normal to radial elements of the gear, while the total tooth load is normal to the tooth. Considering the tooth pressure concentrated at the center point of the tooth, arrow *A* (Fig. 248), either

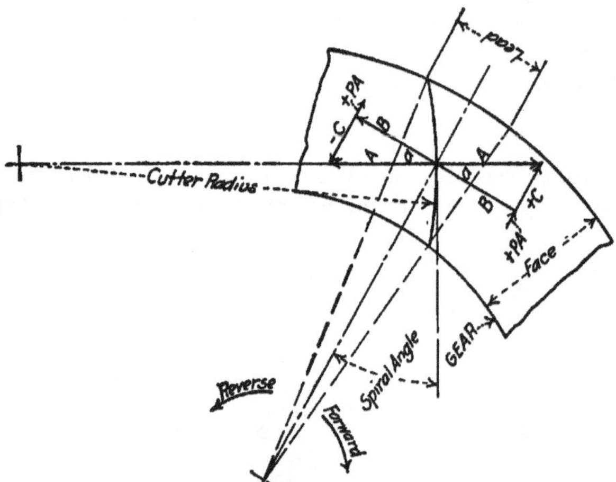

FIG. 248. DIAGRAM OF RIGHT-HAND SPIRAL ILLUSTRATING PRESSURES IN FORWARD AND REVERSE ROTATION.

above or below the center line of the tooth, depending upon the direction of rotation, represents the total tooth load; arrow *B*, the transmitted load; and arrow *C*, the balancing load, represents the thrust either in toward the apex center of the gear, or in the opposite direction, depending upon the direction of rotation. In addition to this spiral angle thrust is the ordinary pressure angle thrust common to all types of bevel gears. For rotation in one direction, this pressure angle thrust augments the spiral angle thrust, while for rotation in the reverse direction, the spiral angle thrust is toward the center of the gear and the pressure angle thrust acts against it, so the resulting thrust is the difference between the two forces.

The direction of rotation of a pair of spiral type bevel gears is customarily referred to that of the pinion. Viewing the pinion from the rear, rotation in a clockwise direction is designated as forward and in an anti-clockwise direction as reverse. In the case of the gear, the rotation is, of course, opposite and a positive spiral angle is taken as one in which the advancing profile planes of the teeth are convex in forward rotation (see Fig. 248).

The pressure angle thrust which either increases or reduces the spiral angle thrust really depends upon several conditions, such as the amount of power transmitted, pitch diameter and pitch angle of the pinion, revolutions of the gear combination and the spiral angle, but, for all practical purposes,

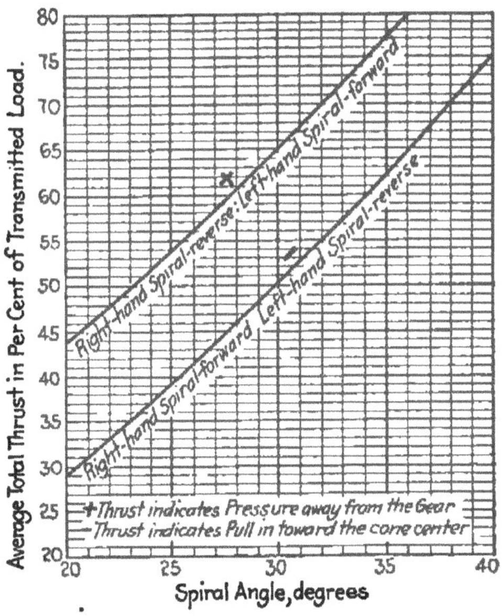

CHART 16. VALUE AND DIRECTION OF THRUST IN SPIRAL TYPE BEVEL GEARS.

may be reduced to a mean percentage of the load transmitted, or of the spiral thrust which is a definite proportion of the transmitted load for any given angle of spiral. Chart 16 depicts the average total load thrust for spiral angles of from 20 to 40 degrees in percentage of transmitted load, the plus values indicating pressures away from the gear and the minus values pressures toward the gear apex.

The increased load thrown on the teeth by reason of their spiral arrangement—the transmitted load divided by the cosine of the spiral angle—necessitates that the load be distributed over several teeth, so if the same general tooth form as is employed for straight tooth bevels is to be employed there is a definite relation between the lead, or spiral angle, and the circular pitch of the gears. The advisable relation between the spiral angle and the circular pitch is shown on Chart 17 and also the total tooth load in percentage of transmitted load for the various spiral angles between 20 and 40 degrees, the advisable lead being made proportional to the total tooth load plus a safety allowance of 10 per cent.

The pitch of the gear also affects the question of advisable spiral angle, as the finer the pitch the smaller can be the spiral angle for a given radius of tooth curve. It is desirable, furthermore, to employ cutters of the same diameter for the various pitches and sizes of gears which may be manufactured so the hypothesis may be made that the advisable spiral angle varies directly

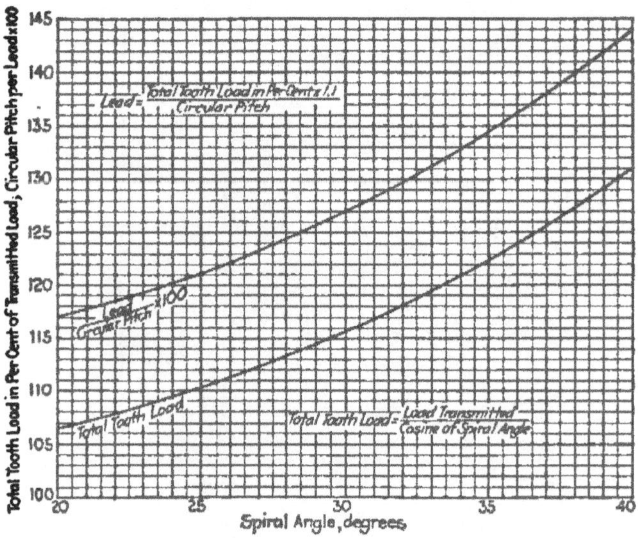

CHART 17. EFFECT OF SPIRAL ANGLE ON TOOTH PRESSURE AND MINIMUM ADVISABLE LEAD—SPIRAL TYPE BEVEL GEARS.

with the diametral pitch of the gear. Based on such relation, Chart 18 depicts good practice.

As the amount of lead for a given spiral angle and constant cutter radius depends upon the face of the gear, there is from the point of view of practical design an advisable minimum face for a given angle of spiral. A face any wider than such minimum permits and generally calls for a reduction in angle of spiral, if the cutter radius is not increased, for unless this is done too much importance is placed upon the perfection of workmanship in cutting the gears. The wider face secures no other advantage than an increase in power capacity of gears and this in the case of spiral type bevel gears is better regulated by the angle of spiral than it is by the face of the gear. However, the face of the gear should be sufficient to enable the advisable lead to be secured for the given angle of spiral with fixed radius of cutter.

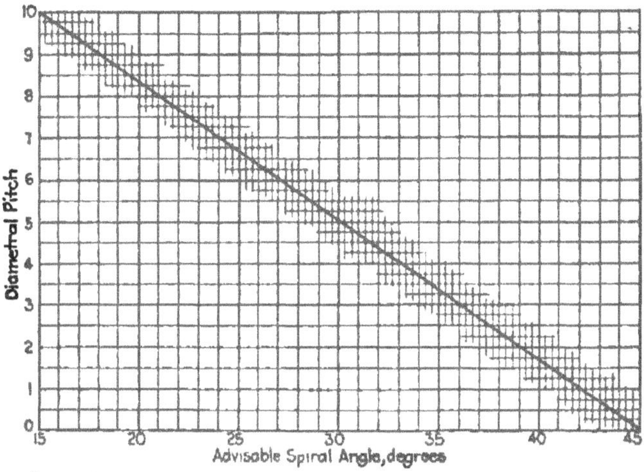

CHART 18. DIAMETRAL PITCH AND ADVISABLE SPIRAL ANGLE—SPIRAL TYPE BEVEL GEARS.

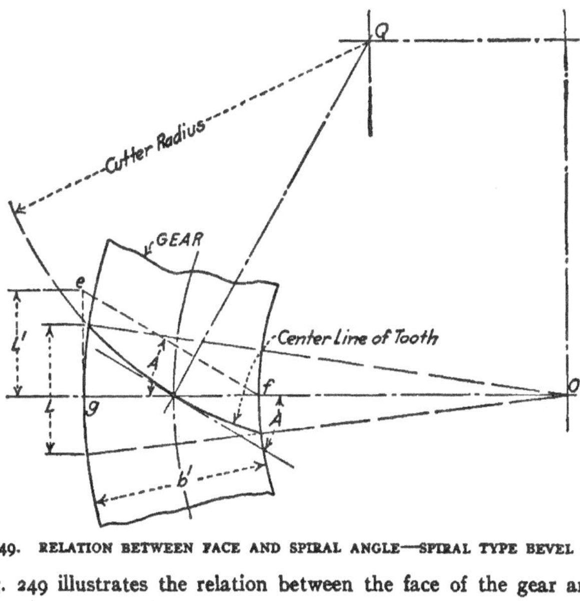

FIG. 249. RELATION BETWEEN FACE AND SPIRAL ANGLE—SPIRAL TYPE BEVEL GEARS.

Fig. 249 illustrates the relation between the face of the gear and the spiral angle. The angle *efg* is equal to the spiral angle and the dimension

L' is equal to the face of the gear multiplied by the tangent of the angle of spiral and is always slightly less than the actual lead. Consequently, if the face is made equal to the lead divided by the tangent of the spiral angle it will be slightly in excess of that required to secure the lead. This slight excess width can be ignored and the formula for minimum advisable face of gear becomes:

$$b = \frac{L}{\tan. A}$$

Where, L = lead, A = spiral angle, b = face.

Referring to Fig. 249,

$$L' = b' \tan. A < L \text{ (slightly less)}$$

$$b' = \frac{L'}{\tan. A} < \frac{L}{\tan. A}.$$

TOOTH PROPORTIONS

The octoid form of tooth with a mean pressure angle of 14½ degrees has been adopted as the standard for spiral type bevel gears. The blades of the cutter employed for forming the teeth are ground to rack form, so that the cutter and gear blank rolling together with a generating motion produce a true octoid tooth, but the curved arrangement of teeth necessitates a slight modification of the standard 14½-degree octoid tooth, the pressure angle on the convex side of the teeth being made slightly greater than 14½ degrees and on the concave side slightly less. The increase and decrease are equal, however, so that the sum of the two pressure angles is always 29 degrees.

As the spiral type of bevel gear permits and is usually employed in greater speed ratios than can be realized with straight tooth bevels, there is danger of undercutting the pinion teeth if the addendum and dedendum dimensions of the gears are made equal, so the standard adopted by the Gleason Works is to make the pitch depth of the pinion teeth seven-tenths of the total working depth of the tooth and the pitch depth of the gear, consequently, three-tenths of the working depth. This modification affects the face and cutting angles of the gears and the outside diameters of the gear blanks.

MACHINING OPERATIONS

In machining the teeth of spiral type bevel gears, two cutters are employed. The teeth are first blocked out with a roughing cutter and then a finishing cutter is employed to finish first one side of the teeth and then the other, the same finishing cutter being used to finish both sides of the teeth, but on slightly different cutter settings. These cutter settings must be accurately made and two methods of doing this have evolved—the "formula" method and the "layout" method.

The horizontal and vertical settings of the roughing cutter center for

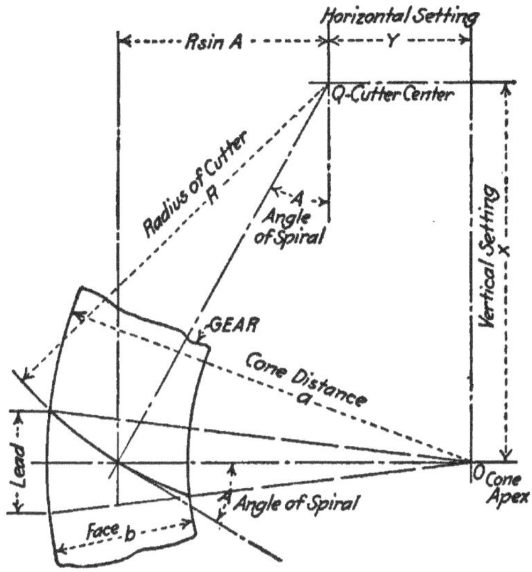

FIG. 250. SETTINGS FOR ROUGHING CUT—SPIRAL TYPE BEVEL GEARS.

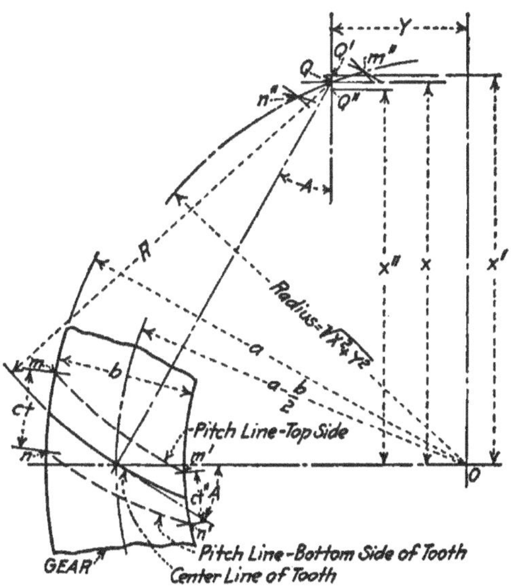

FIG. 251. SETTINGS FOR FINISHING CUTS—SPIRAL TYPE BEVEL GEARS.

right-hand spiral gears are shown in Fig. 250 and those for the finishing cuts in Fig. 251. The equations employed for computing the various dimensions according to the "formula" method follow and a brief explanation of their derivation.

NOTATION FOR SPIRAL TYPE BEVEL GEARS

DESIGN AND GENERAL	GEAR	PINION
Diametral pitch	p	p
Circular pitch	p'	p'
Number of teeth	n	n'
Spiral angle	A	A
Lead	L	L
Face (actual)	b	b
(minimum)	b'	b'
Depth of tooth	W'	W'
Addendum	s	s'
Dedendum	u	u'
Clearance	f	f
Circular thickness of tooth (outer pitch circumference)	α	α'
Center angle	E	E'
Pitch diameter	d	d'
Cone distance	e	e
Angle increment	J	J'
Angle decrement	K	K'
Face angle	F	F'
Cutting angle	C	C'
Diameter increment	V	V'
Outside diameter	D	D'
SHOP		
Diameter of cutter	D''	D''
Radius of cutter (mean)	R	R
Inner cutter angle	B	B
Outer cutter angle	T	T
Thickness of cutter at apex	t''	t''
Horizontal setting	Y	Y
Vertical setting (roughing cut)	X	X
Vertical setting (finishing cuts)		
Top side of tooth (concave)	X'	X'
Bottom side of tooth (convex)	X''	X''

FORMULAS FOR SPIRAL TYPE BEVEL GEARS

Test Formula:

$$b' = \frac{L}{Tan.\,A} \qquad \text{or, advisably } Tan.\,A = \frac{L}{b}$$ Formula A.

FORMULA

Design Calculations:

$$Tan.\,E = \frac{n}{n'} \qquad 1$$

$$E' = 90 - E \qquad 1'$$

$$d = \frac{n}{p} \qquad 2$$

$$d' = \frac{n'}{p} \qquad 2'$$

$$W' = \frac{2.157}{p} \text{ (14½-deg. standard tooth)} \qquad 3$$

$$s = \frac{\text{pitch depth} \times 2}{p} = \frac{0.6}{p} \qquad 4$$

$$s' = \frac{\text{pitch depth} \times 2}{p} = \frac{1.4}{p} \quad \text{.......... } 4'$$

$$u = W' - s \quad \text{.......... } 5$$

$$u' = W' - s' \quad \text{.......... } 5'$$

$$d = 0.5\,p' - \frac{0.2069}{p} \text{ (14½-deg. standard tooth)} \quad \text{.......... } 6$$

$$d' = 0.5\,p' + \frac{0.2069}{p} \text{ (14½-deg. standard tooth)} \quad \text{.......... } 6'$$

$$a = \frac{0.5\,d}{Sin.\,E} = \frac{0.5\,d'}{Sin.\,E'} \quad \text{.......... } 7$$

$$Tan.\,J = \frac{s}{a} \quad \text{.......... } 8$$

$$Tan.\,J = \frac{s'}{a} \quad \text{.......... } 8'$$

$$Tan.\,K = \frac{u}{a} \quad \text{.......... } 9$$

$$Tan.\,K' = \frac{u'}{a} \quad \text{.......... } 9'$$

$$F = E + J \quad \text{.......... } 10$$

$$F' = E' + J' \quad \text{.......... } 10'$$

$$C = E - K \quad \text{.......... } 11$$

$$C' = E' - K' \quad \text{.......... } 11'$$

$$V = s\,cos.\,E \quad \text{.......... } 12$$

$$V' = s'\,cos.\,E' \quad \text{.......... } 12'$$

$$D = d + 2V \quad \text{.......... } 13$$

$$D' = d' + 2V' \quad \text{.......... } 13'$$

NOTATIONS FOR SHOP CALCULATIONS

	GEAR	PINION
Outer cutting radius (concave side—mean value)	R'	R_1'
Inner cutting radius (convex side—mean value)	R''	R_1''
Dedendum at mid-tooth	u''	u_1''

All other notation similar to that employed for design calculations.

FORMULAS FOR SHOP USE

FORMULA

$$u'' = \frac{(a - 0.5b)u}{a} \quad \text{.......... I}$$

$$u'' = \frac{(a - 0.5b)u'}{a} \quad \text{.......... I'}$$

$$R' = R + 0.5(t'' + 0.5172\,u'') \quad \text{.......... II}$$

$$R_1' = R + 0.5(t'' + 0.5172\,u_1'') \quad \text{.......... II'}$$

$$R'' = R - 0.5(t'' + 0.5172\,u'') \quad \text{.......... III}$$

$$R_1'' = R - 0.5(t'' + 0.5172\,u_1'') \quad \text{.......... III'}$$

$$Y = a - (0.5\,b + R\,sin.A) \quad \text{.......... IV}$$

$$X = R\,cos.\,A \quad \text{.......... V}$$

$$X' = \frac{R'X}{R} \quad \text{.......... VI}$$

$$X_1' = \frac{R_1'X}{R} \quad \text{.......... VI'}$$

$$X'' = \frac{R''X}{R} \quad \text{.......... VII}$$

$$X_1'' = \frac{R_1''X}{R} \quad \text{.......... VIII,}$$

DERIVATION OF FOREGOING FORMULAS

All formulas pertaining to tooth proportions, pitch diameters, center angles, etc., which are not affected by the pitch depth of gear or pinion are derived in a manner similar to that for equivalent formulas for straight tooth bevel gears.

The addendum of the gear or pinion is found by dividing the proportional pitch depth by the diametral pitch. For gears in which the pitch depth equals 0.3 of the working depth of tooth, this is equal to 0.6 divided by the diametral pitch, and for pinions in which the pitch depth equals 0.7 of the working depth equals 1.4 divided by the diametral pitch.

The dedendum is most readily figured by subtracting the addendum from the working depth of the tooth, which latter is the same as in the case of ordinary straight tooth bevel gears.

The circular thickness of tooth on the pitch circumference of the gear is equal to half the circular pitch minus twice the product of the tangent of the pressure angle by the difference between the pitch depth and half the working depth of the tooth, divided by the diametral pitch. The circular thickness of the tooth on the pitch circumference of the pinion is equal to half the circular pitch plus this same amount.

The cone distance of spiral type bevel gears is the same as that for ordinary bevels, but the angles increment and decrement are affected by the pitch depth of the gear and pinion, their tangents being respectively the addendums and the dedendums, divided by the cone distance.

The face and cutting angles of the gears and pinions are simply affected by the change in angles increment and decrement, the formulas being similar to these for straight bevel gears; likewise, the diameter increment and the outside diameters of the gear and pinion.

The finishing cutter radii must necessarily be proportioned to the bottom width of the tooth spaces, necessitating taking into consideration the relative position of the pitch line: *i.e.*, pitch depth of gear and pinion. The outer cutting radius, that for the concave side of the tooth, is then equal to the mean cutter radius, plus one-half the apex thickness of the cutter, plus the dedendum (at center point of tooth) times the tangent of the outside cutter angle. The inner cutting radius is equal to the mean cutter radius, minus one-half the apex thickness of the cutter, minus the product of the mean dedendum and the tangent of the inner cutter angle.

The horizontal and vertical settings for both roughing and finishing cuts have already been explained, but it must always be borne in mind that the formulas presented for the finishing cut settings are based on the respective cutting radii normal to the pitch planes at the center point of the tooth profiles, not at either the outer or inner pitch circumferences. The finishing cutter radius (for either side of the tooth) varies from outer to inner end of tooth, but by locating the vertical setting from the cutting radius at mid-tooth the angle decrement automatically causes the required variation in radius on the pitch plane and over the entire tooth profile.

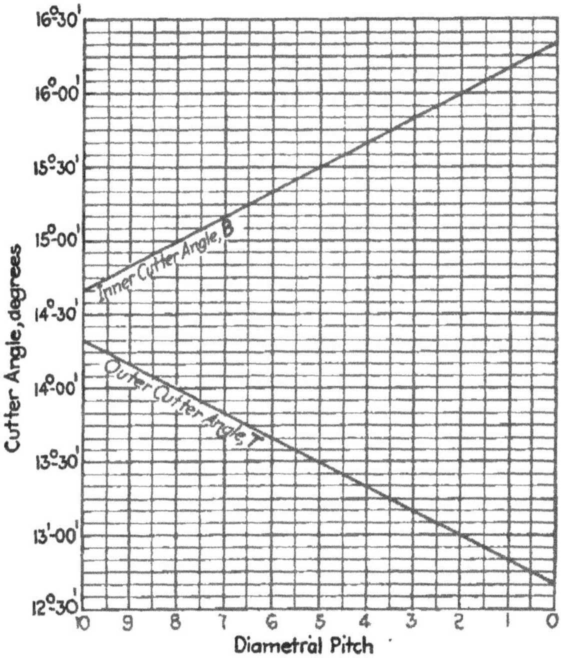

CHART 19.

The accepted standards for the finishing cutter angles, inner and outer, for different diametral pitches are given on Chart 19 and the details of the

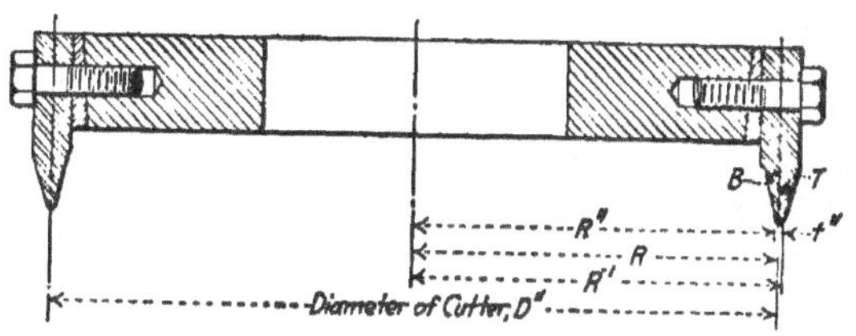

FIG. 252. DETAIL OF FINISHING CUTTER—SPIRAL TYPE BEVEL GEARS.

finishing cutter are shown in Fig. 252. The mean radius of the cutter, R, equals one-half the diameter of the cutter, D'', while the cutter angles B and T are obtained from Chart 19 and t'' is arbitrarily set customarily at $\frac{1}{16}$ inch.

THE "LAYOUT METHOD"

This method of obtaining the vertical settings for the finishing cuts is based on a somewhat different theory than that governing the *formula method*, and, as its advocates claim greater precision results from its use, a brief description of this method is of interest.

Referring to Fig. 253, a diagram of the gear is laid out on an enlarged scale in which the curve *lm* depicts the center line of the tooth—the curve being the arc of a circle having its center located by the horizontal and

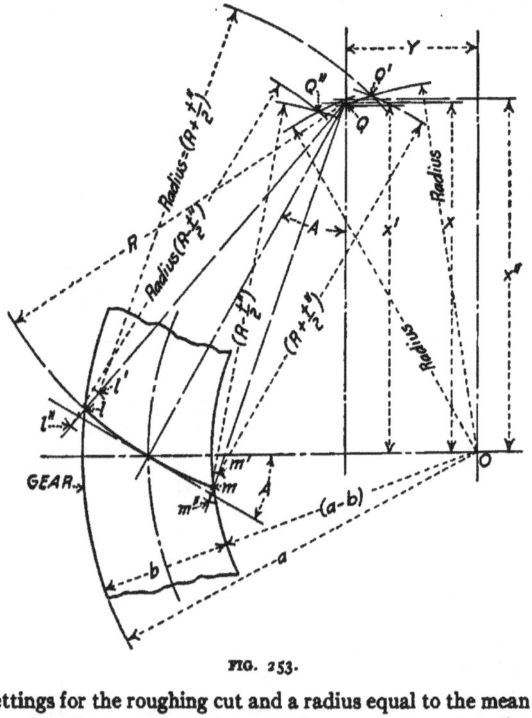

FIG. 253.

vertical settings for the roughing cut and a radius equal to the mean radius of the cutter. For the top side of the tooth (concave), the points *l'* and *m'* are struck off from points *l* and *m* respectively on the radii to these points—*i.e.*, on lines radiating from the cutter center to the points of intersection of the center line of the tooth with the inner and outer pitch circumferences—*ll'* and *mm'* equaling the dedendum times the tangent of the pressure angle at the respective points, the ratio *ll'*: *mm'* equaling *a*: (*a* - *b*). With the points *l'* and *m'* as centers and a radius equal to the mean radius of the cutter plus one-half the thickness of the cutter blade at its apex arcs are struck off intersecting at *Q'*.

For the bottom side of the tooth (convex), similar steps are taken, locating the point Q'', with a radius equal to the mean radius of the cutter minus one-half the thickness of the cutter blade at its apex from the points l'' and m''—the latter points being located as were the points l' and m'.

Finally, with O as a center, the points Q' and Q'' are swung over to the plane of vertical settings and the respective vertical settings, X' and X'', carefully scaled. The vertical settings thus secured differ from those obtained by the *formula method* as the three points Q, Q' and Q'' do not lie on the arc of a circle having point O as a center. In some cases—for instance, on the layout depicted on Fig. 253—the vertical setting for the convex side of the tooth may be greater than that for the concave side when the settings are obtained by the *layout method*, which would never occur in the *formula method*.

COMPARISON OF METHODS

The relative accuracy of the two methods is open to argument. The two sides of the teeth being finished with slightly different radii of cutter, there

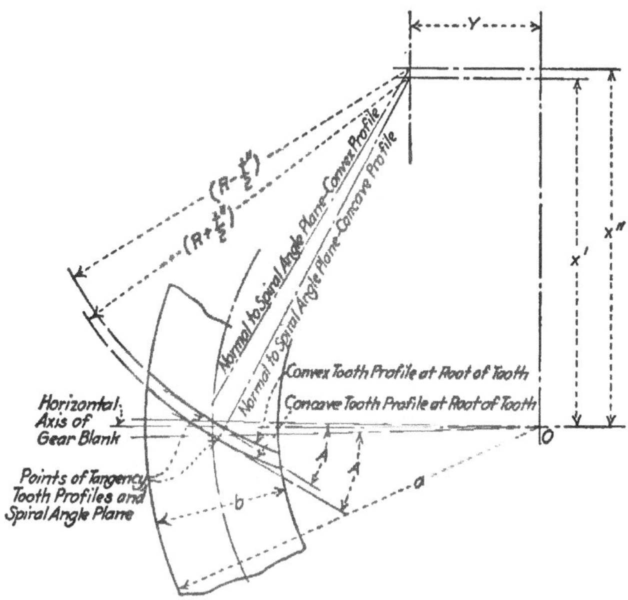

FIG. 254.

can actually be but one point of perfect tooth contact, but for all practical purposes contact takes place over quite an appreciable length of tooth, even before the gears have been run together and worn into perfect mesh, as the difference in the heights of the respective arcs—that of the concave side and that of the convex side of meshing teeth—is extremely slight. In gears finished on vertical settings arrived at by the *formula method*, the point of perfect contact is located on both convex and concave profiles at the same relative point—on the mid-pitch circumference where the curved profiles of the teeth are tangent to the spiral angle plane. In gears finished on vertical settings arrived at by the *layout method* the point of tangency between the spiral angle plane and the tooth profile is not located at the same relative point on both the concave and convex profiles, the points of tangency depending upon the relation of the horizontal setting to the vertical settings, the angle of spiral, etc. (see Fig. 254).

When the more remote point of tangency is on the convex profile, as depicted on Fig. 254, this failure of the spiral angle plane to be tangent to both sides of the tooth at the same relative point is actually a slight advantage as it locates the point of contact at a point where the power transmitted is slightly greater than at mid-tooth and where the tooth is slightly heavier. In cases where the contact point on the convex profile is slightly nearer the inner edge of the gear than mid-tooth no appreciable disadvantage exists on account of the flatness of the contact arcs. The slight eccentricity given to the load on the teeth by the failure of the spiral angle plane to be tangent at corresponding points on both sides of the tooth tends to accelerate the slight weave of parts which produces the perfect tooth contact of spiral type bevel gears which have been run together.

The *layout method* is open to the possibility of error in scaling and laying out the diagram and, in addition, consumes considerably more time than the simple calculations required for the *formula method*. This excess time could be put to far better advantage in running the gears together to perfect tooth contact—at least, the time saved by the *formula method* would compensate for the longer time required in truing up the tooth contact of gears generated on vertical settings derived from the formulas. Both methods of arriving at the vertical settings for the finishing cuts have their advocates so that the choice of methods is almost entirely a matter of personal preference on the part of the manufacturer or machinist.

WILLIAMS "MASTER FORM" BEVEL GEARING

In Section V mention was made of the radical departure from the so-called standard proportions of bevel gear teeth evolved in Pentz "parallel depth" bevel gears, in which the gear teeth are of constant depth. This innovation was brought about chiefly to reduce the cost of cutting bevel gears, the efficient bevel gear generating machine of today not having been perfected at that time. Eliminating the necessity of tapering the depth of the teeth unquestionably simplified gear cutting, but the complications incident to the variation in profile curvature of the teeth, from end to end of tooth, remained and necessitated the use of formed cutters of special cutting profile to finish the gears and a multiplicity of adjustments of the gear blank on the work arbor.

In the Williams "Master Form" System of Bevel Gearing, developed by Harvey D. Williams, the radical step is taken of avoiding the variations in tooth profile in the larger of a pair of gears—the gear, rather than the pinion—by making the tooth spaces simply straight gashes across the face of the gear. The "master form" feature of such gears is that the tooth profiles are straight, like those of the familiar involute rack teeth. The thickness of the teeth varies from end to end, owing to the difference in the inner and outer circumferences of the gears, but the tooth spaces are uniform as to depth and section.

The pinion meshing with this "master form" gear is a generated member. That is, its teeth have curved profiles which are most accurately and cheaply

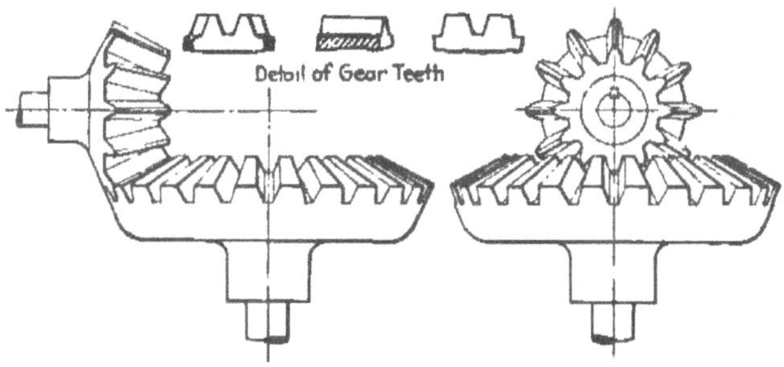

FIG. 255. WILLIAMS "MASTER FORM" BEVEL GEARING.

produced by the generating principle with cutting tools which are the conjugate of the simple rack tooth employed in gashing out the tooth spaces in the "master form" gear.

This radical departure from the usual design of bevel gearing (see Fig. 255) is particularly noteworthy in that it is a reversion to and adaptation of fundamental mechanical principles which appear to have been overlooked until quite recently in the evolution of gearing—spur gearing as well as bevel gearing. Toothed gears will roll together smoothly and efficiently provided their teeth and tooth spaces are conjugately related. This is the one essential requirement and the form and shape of the teeth is of quite secondary importance.

The development of the almost universal octoid system for bevel gearing, with its varying tooth profile, not only from end to end of tooth, but for every size of gear, is based on the requirement that each and every bevel gear be conjugately related to a crown gear with flat sided teeth. This is in line with the simple rack tooth base for the involute system of spur gearing, and though there may be some advantage to such a basis in spur gearing—in commercial gear production in large quantities this is now recognized as more or less of a fallacy—there is none in the case of bevel gearing.

In the Williams "Master Form" bevel gear, the teeth are almost as readily cut as those of an ordinary spur rack and more expeditiously and cheaply than they can be generated. Each gear cut with these "master form" teeth becomes the standard to which its mating pinion or pinions are proportioned. As these pinions can be generated as readily as those with octoid teeth no added difficulty of expense is entailed in the manufacture of the pinions.

PRODUCTION ECONOMIES

The cost of cutting Williams "Master Form" bevel gears depends largely upon the number of teeth on the gears and it has been estimated that about 80 per cent. of the combined number of teeth on the gear and pinions of automobile transmissions—for which mechanisms these distinctive gears are chiefly used—are on the gear, or larger wheel. The Williams system materially reduces the cost of cutting the teeth on the gear, resulting in a considerable reduction in the cost of manufacturing the complete transmission, as the cost of generating the Williams pinion is no greater than that of generating an octoid pinion of the same size.

The economy in manufacturing cost is probably the chief advantage of the Williams system of bevel gearing, but it also possesses other features which have considerable merit. The peculiar shape of the teeth makes the gears exceedingly smooth running and they are said to develop unusually high efficiency in operation. Another advantage of the gears—one which is of particular value in automobile drives—is the distinctive capacity of the gears to discount disalignment.

EFFECT OF SPRUNG SHAFTS

The strains and wrenches to which the transmission of an automobile is frequently unavoidably subjected have a tendency to spring the driving shaft out of alignment, seriously interfering with the satisfactory operation of the transmission, which depends largely upon the exactitude of shaft alignment. The teeth of the gear may remain in mesh with those of the pinion, but the sprung shaft throws a heavy unbalanced pressure on the gears, causing noisy operation and undue tooth wear. This is well nigh unavoidable whatever system of gearing is employed, but with Williams gearing the troubles produced by sprung shafts are greatly discounted.

Fig. 256 shows diagrammatically what might be expected to occur should the transmission shaft of an automobile be sprung out of alignment, from O to O'. The normal position of the gears with their centers at O is depicted in full lines and in sprung position with centers at O'. For the gears to remain in mesh, the inner corner of the tooth, A, remains in the same relative position, but if it were not for the restraining influence of the meshing teeth the diagonally opposite corner would move from B to B' and the strain to which the gearing is subjected in preventing such movement is measured by the distance between B and B'. In the case of the ordinary octoid form of

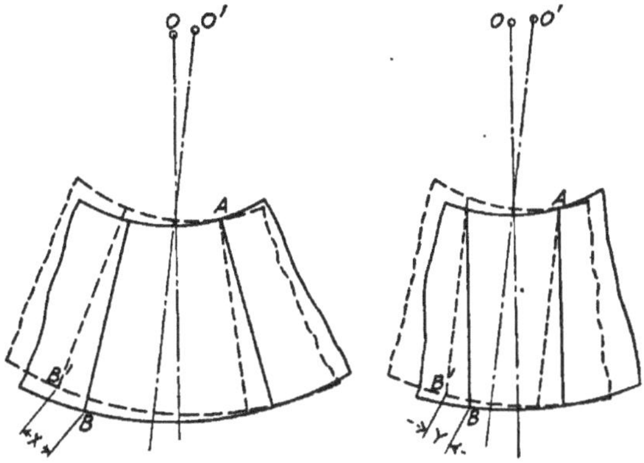

FIG. 256. EFFECTS OF SHAFT DISALIGNMENT.

tooth, the disalignment strain is measured by the distance X and in the case of the Williams "Master Form" of tooth by the distance Y, and Y is very appreciably less than X. The difference in intensity of strain, furthermore, is much more than the difference between the points B and B' for the respective strains and, if the gears are under considerable load, may produce fracture in one case and not in the other.

SECTION XV

Williams System of Internal Gearing*

The advantages of internal gearing may be summarized briefly as a compactness of design impossible of realization with externally meshing gears, an important safety feature in that the gear forms its own gear guard, tooth contact over a considerably longer arc than in either external or rack gearing, a reduced sliding action between meshing teeth, improved operating action, reduced wear and a considerably longer life of usefulness than can be possessed by externally engaging gears.

With such a formidable list of advantages favoring the internal gear, its chief limitations in the past have been the difficulty and cost of accurately cutting the internal gear teeth and the fact that true involute profile is rarely practical or feasible for the internal gear teeth, on account of the interference developed with the teeth of mating pinions. The difficulty and cost of cutting the internal gear teeth have been overcome, to a considerable extent, by the development of generating machines of approved type for cutting gear teeth, but the limitation set by interference—"fouling" of teeth—has only been reduced by marked modifications of the involute profile of the gear teeth. That is, the profiles of "standard" involute tooth—the form of tooth almost exclusively employed for internal gears—diverge to a marked degree from the true involute curve for the greater part of their depth. The necessary modification of the tooth profile causes the "standard" involute tooth for internal gears to lose to a large extent one advantage the involute system of gearing is supposed to possess—that of interchangeability.

LIMITATIONS OF INVOLUTE INTERNAL GEARING

A more serious limitation placed upon internal gearing by the involute system is that, despite the modification to the profiles of the gear teeth, there is a practical limit to the minimum number of teeth an involute gear may have—generally taken as 12—so that the smallest feasible pinion must have that number of teeth. For a 12-tooth pinion with teeth of involute form to possess the required strength, it is quite frequently necessary to employ a heavy pressure angle and adopt the stub form of tooth.

Increasing the pressure angle tends to increase also the arc of contact, by enlarging the angle of approach, but this gain is secured only through a certain sacrifice in the efficiency of power transmission. The heavier pressure angle increases the axial, or radial, component of the power transmitted and decreases the effective component.

The involute system of internal gearing and its limitations have been touched upon before discussing the Williams system of internal gearing to

* From a report on the system by Reginald Trautschold.

show that profiles of the internal gear teeth cannot be produced in commercial gearing of true involute form and the questionable advantage of interchangeability has to be sacrificed in any case, the limitation in the minimum number of teeth and the loss in transmission efficiency entailed in the use of the heavy pressure angle necessary for an involute pinion to possess adequate strength.

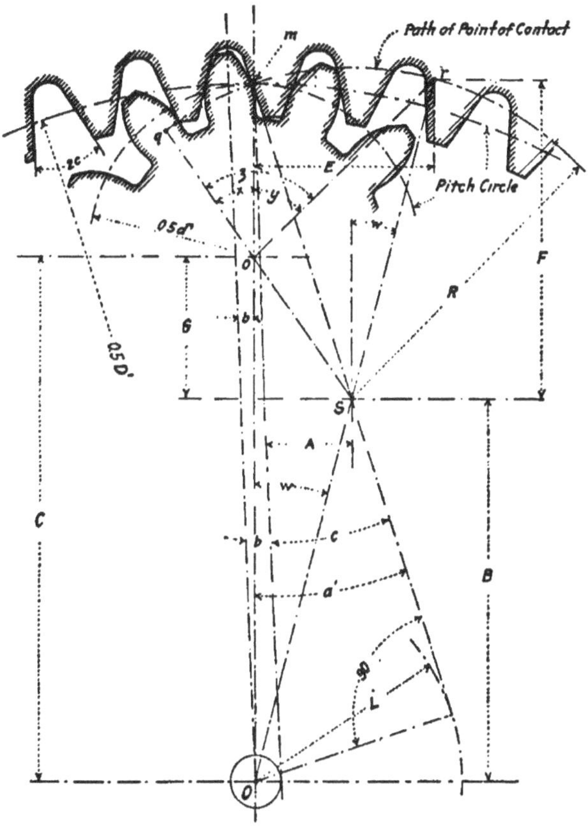

FIG. 257. WILLIAMS SYSTEM OF INTERNAL GEARING.

FORMULAS

$b = \frac{90}{N}$ $\quad a' = c + b$ $\quad L = 0.5D' \sin. a'$

$R = 0.25\,D' \cos. a'$ $\quad A = R \sin. a'$ $\quad B = 0.5D' - R \cos. a'$

$\tan. w = \frac{A}{B}$ $\quad \mathrm{Tan}.\, x = \frac{A}{C - B}$ $\quad E = 2R \sin. w$ $\quad F = \frac{E - A}{\tan. w}$

$\tan. y = \frac{E}{F - G}$ $\quad z = x + y$

WILLIAMS SYSTEM OF INTERNAL GEARING

In the Williams system of internal gearing, the forms of the gear teeth in both the internal gear and the mating pinion depart radically from those designed according to other systems but, in so doing secure certain inherent advantages that give to the gearing an improved operating action, increase the strength of the pinion teeth, reduce wear and greatly reduce the cost of producing accurate and efficient internal gearing of the spur or spiral type. The limitations placed upon internal gearing designed according to other systems are materially reduced, while all the advantages are retained and accentuated.

THE WILLIAMS SYSTEM

In the Williams system, the teeth of the internal gear are made with straight line profiles—the bounding profiles of the tooth spaces being similar to those of an involute rack—while the teeth of the mating pinion are provided with curved profiles of conjugate form, giving a combination that results in a distinctive path to the point of contact between the gear and pinion teeth throughout their arc of contact (see Fig. 257). An extension of each flat profile plane—the profile planes of the internal gear teeth—is tangent to a circle concentric with the circumference of the internal gear and the radial normal to a tangent from any point on the path of point of contact will, with the radial line to such point of contact, form a right angle triangle. The path traced by the apex of such triangle is over that section of the path of point of contact that can be made of use in practical gear combinations and is —within the limits set by attainable accuracy in commercial manufacture— the arc of a circle of a radius equal to one-half the length of the tangent from the instant axis, m, of the gears to the concentric circle of tangency, about its mid-point. That is, the usable section of the path of point of contact is, to all intents and purposes, the arc of a circle passing normally through the instant axis of the gears and tangent to the radial normal to the profile plane passing through the instant axis. The point on this path of point of contact at which contact can first occur is located by the radius of the circle of path of point of contact passing through the center of the pinion mating with the internal gear, and the point at which contact must cease is fixed by a radial line from the center of the internal gear passing through the center of the circle of path of point of contact. Referring to Fig. 257, in which the rotation of the gears is in a clockwise direction, contact between pinion and gear teeth—if the teeth were suitably proportioned—would commence at point q, on the path of point of contact, and end at the point r. The angle of approach, in such a case, would be measured by the angle x and the angle of recession by the angle y, or the total arc of pinion contact measured by the angle e.

In the combination illustrated, contact does not commence as soon, nor is it maintained so long. Contact begins at the point at which the inner circle of the internal gear crosses the usable path of point of contact and ends at the point at which the outer circle of the pinion crosses the same path. Despite the failure to utilize the maximum usable section of the path of point of contact, the actual arc of pinion contact is unusually long for the gearing combination shown. The arc of contact is, in fact, materially longer than that securable with gear and pinion teeth proportioned according to any other of the systems of internal gearing in common usage. To appreciate this important fact in full, a comparison of the paths of point of contact for internal gearing combinations of similar ratios, but designed according to other systems of tooth form, is necessary.

PATHS OF POINT OF CONTACT

The four diagrams shown in Fig. 258 illustrate in a graphic and illuminating manner the paths of point of contact in the principle standardized systems of internal gearing. In each instance, the direction of rotation in clockwise

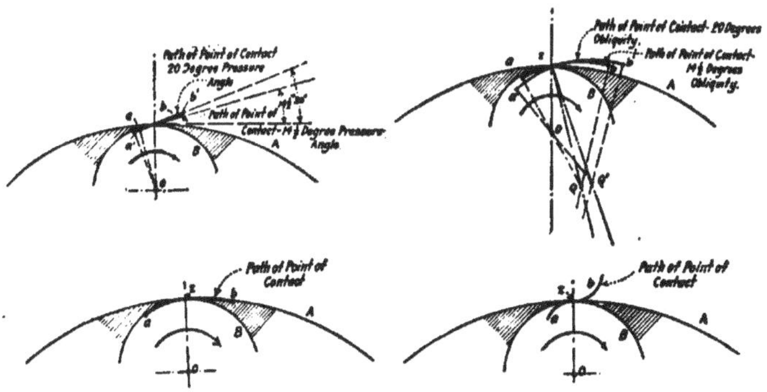

FIG. 258. COMPARISON OF PATHS OF POINT OF CONTACT.

and the shaded area between the pitch circles of the gears and pinions—*A* and *B* respectively—indicate areas within which no part of the path of point of contact can lie, for in such areas interference between gear and pinion teeth would result.

Diagram (I) depicts a combination of gears, in which the point of contact is concentrated and localized at the instant axis, *Z*, or on the pitch circles of the gears. If it were possible to devise a system of toothed gearing in which the path of point of contact was so localized on the pitch circles, it would result in concentration of wear at a point first on the pinion tooth, say on the section of the pitch circle from *a* to *Z* and then on the internal gear tooth, on its pitch circle from *Z* to *b* but, at the same time, a path of point of contact of a long arc of action. The long arc of action would be a decided advantage, but the concentration of wear a decided disadvantage. A radial path of point of contact passing through the instant axis, as every path of point of contact must do, would give no arc of action at all. However, these hypothetical paths of point of contact, the one following first the pitch circle of the pinion and then that of the internal gear and the radial path passing through the instant axis, represent the maximum and minimum arcs of tooth contact.

The path of point of contact shown in diagram (II) illustrates that of the epicycloidal system of gear teeth. The profiles of the teeth formed in this system are developed by a point on a generating circle rolling first on one side and then on the other of the pitch circles, so that path of point of contact is a curve normal to such profiles and similar to that depicted as *aZb*. The epicycloidal system of forming gear teeth has now been virtually abandoned except for special combinations of internal gearing where the gear and pinion have very nearly the same number of teeth and the form of involute teeth would make it impossible for such teeth to clear themselves, on account of the cost and difficulty of accurately proportioning the teeth. Furthermore, the diagram shows that about the instant axis of the gears the epicycloidal path of point of contact closely follows the pitch circles, concentrating the wear on the teeth—particularly in the case of the pinion. At the same time, the extremities of the path of point of contact tend to approach a radial direction and so curtail the length of the arc of contact, or action. A limited arc of contact and concentration of wear at and near the pitch circles, as well as the cost and difficulty of accurately forming the teeth, well justify the general abandonment of the epicycloidal system of internal gearing.

In diagram (III) are depicted the paths of point of contact for 14½-degree and 20-degree teeth in the involute system of gears, *aZb* and *a'Zb'* respectively. The paths are flat planes passing through the instant axis and inclined to the common tangent plane at the respective pressure angles. The commencement of tooth contact on the flank of the pinion, and therefore the limiting factor in the useful depth of the gear teeth face, is fixed by the pinion radius normal to the approach section of the path of point of contact, while the limit to the recession section of the path of point of contact is—theoretically, at least—unlimited. However, there is a practical limit set to the length of the recession section of the path of point of contact by the length of the feasible pinion tooth addendum, for if the flank of the internal gear teeth is too long the tops of the pinion teeth are cut away by the conjugate profiles of the mating gear.

The quite general adoption of 20-degree involute teeth for internal gears has been brought about by an attempt to secure a longer arc of approach than can be secured with 14½-degree involute teeth. In spite of such practice, the angle of approach remains quite limited and an effective angle of contact for involute gears is to be secured only through adopting the stub type of tooth with heavy pressure angle for the internal gear and decreasing the addendum of the pinion teeth, upon the face of which the wear is more concentrated than it is upon the flank of the internal gear teeth.

The fourth of the diagrams illustrates the paths of point of contact for 14½-degree and 20-degree obliquity teeth at the instant axis formed according to the Williams system, *aZb* and *a'Zb'* respectively. The much greater approach section to the path of point of contact is very apparent and well typifies the reason for the improved operating action of the Williams gear. A Williams gear tooth of 14½-degree obliquity has a considerably longer arc of approach than has a 20-degree involute tooth and the usable section of the recession path of point of contact for the 14½-degree Williams tooth is also much longer than can be made of use for a 20-degree involute tooth. In the case of 20-degree Williams tooth, the increase in length of arc of action is as pronounced as in the case of involute teeth.

It will also be noted that the wear on the teeth is quite evidently much more uniformly distributed in the Williams system of gearing than it is in the involute and, furthermore, that such concentration of wear as does occur is toward the bottom of the flanks of both pinion and gear teeth, where the greatest amount of metal is and where, owing to the distinctive form of the teeth, wear can occur with less sacrifice of tooth strength than in any other form of tooth. Wear at such points will also have the tendency to ease off the shock of initial contact and so produce quieter and smoother operating gears.

STRENGTH OF PINION TEETH

In the matter of strength, the pinions—always the weaker of a pair of mating gears—proportioned according to the Williams system are considerably superior to those of equivalent involute form and this peculiarity is clearly demonstrated by a comparison of the forms of the respective internal gear teeth. The teeth of the Williams pinion have curved profiles of exact conjugate form for the particular Williams internal gear with which it is to operate, while the accurate curves of the profiles of the involute pinion teeth are generated by the same or similar involute rack tool employed for the generation of the mating internal gear.

The tooth forms illustrated in Fig. 259 are accurate representations of—(a) involute internal gear teeth; (b) an accurately mating pinion tooth; (c) equivalent Williams internal gear teeth; (d) a mating Williams pinion tooth; and (e) the two forms of pinion teeth superimposed to emphasize the difference in their respective tooth profiles. The curved profiles of the involute internal gear teeth, both above and below the pitch circle, it will be noted, are concave and so curve away from the contact planes tangent to the tooth profiles on the pitch line. The straight profiles of the Williams internal gear teeth, on the other hand, lie in the contact planes of the gear teeth. The extra root thickness of the involute internal gear tooth, though adding to its strength, results in a corresponding decrease in the thickness of the top of the mating pinion tooth, which though perhaps not weakening the tooth does tend to limit the depth of its addendum. The extra top thickness of the gear tooth, however, has a more serious effect, for it tends to weaken the pinion tooth by the necessary undercutting required for clearance. The composite diagram (e) forcibly illustrates the strength superiority of the Williams pinion tooth, for though the pinion teeth selected for comparison have been chosen as only slightly undercut—the involute tooth is a 20-degree 80 per cent. stub tooth of a 12-tooth pinion—the difference in their root

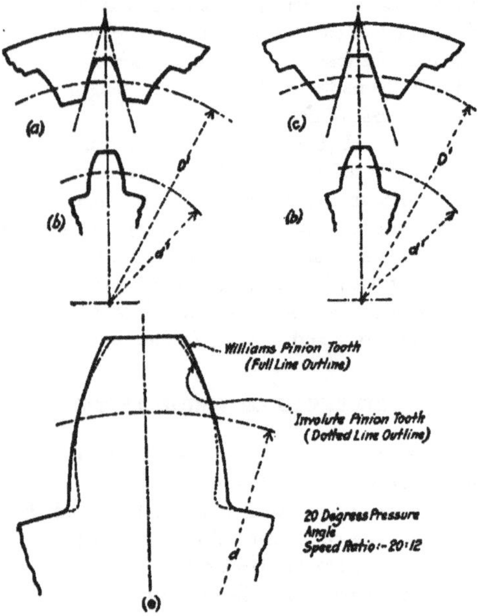

FIG. 259. COMPARISON OF TOOTH FORMS.

thicknesses is quite sufficient to indicate an appreciable difference in the strength of the two types of pinion teeth. The strength of the respective teeth are proportional to the square of their root thicknesses.

The proportions of Williams internal gear teeth, furthermore, are such as to permit a modification in customary design whereby the strength of the gearing—measured by the strength of the pinion teeth—is materially increased, by making the pinion teeth as strong as the gear teeth. This is accomplished, as illustrated by the eight tooth (6 to 1) combination shown in Fig. 260, by increasing the thickness of the pinion teeth and correspondingly decreasing the thickness of the gear teeth, so the root thicknesses of the respective teeth are equal. The pitch of the gears is not altered, simply the thickness of the pinion teeth increased and the space between adjacent gear teeth widened by reducing the thickness of the gear teeth.

This method of securing a high-speed ratio with a pinion of few and unusually strong teeth cannot be resorted to as effectively in the case of gears proportioned according to the involute system. In gearing so proportioned, the undercutting of the pinion with few teeth, though they be amply thick on the pitch circle, would be so excessive as to necessitate unduly reducing the thickness of the gear teeth to equalize the strength of the pinion and gear teeth.

WEAR

The superiority of the Williams system of internal gearing in the matter of longer arc of contact, or action, is well demonstrated by a graphic comparison of similar gear combinations of Williams and involute form. Fig. 261 diagrammatically illustrates the operating relation between the Williams internal gear and its mating pinion shown in Fig. 257, and also the operating relations for involute gears of the same ratio and proportions.

The section of the path of point of contact actually utilized in this combination of Williams gears is from the point *s* to the point *t*, or the section included between the intersections of the path of point of contact and the inner circle of the internal gear and the outer circle of the mating pinion respectively. The approach of the pinion teeth is measured by the angle *e* and their recession by the angle *f*, or the length of the pinion arc of contact, or action, is measured by the angle *g*. These angles, it will be noted, are materially less than the usable angles of approach and recession as limited by the available arc of contact, fixed by the extreme points *q* and *r* of the path of point of contact, yet they are considerably greater than the equivalent angles for an involute gear combination of the same speed ratio and pressure angle—*i.e.*, when the addendum and dedendum dimensions of the teeth are the same. The oblique line tangent to the Williams path of point of contact at the instant axis of the gears, the line *s't'*, depicts the path of point of contact for such involute gear combination. Contact between opposing involute profiles cannot commence in advance of the point *s'*, the intersection of the path of point contact with its radial normal from the pinion center, and cannot continue beyond the point *t'*, where the outer circle of the pinion crosses the path of point of contact.

In the involute gear combination, the angle of pinion approach, *a*, is equal to the pressure angle and is quite noticeably less than the angle of approach, *e*, for the Williams gear combination. The angle of recession, *k*, is also less than the corresponding angle for the Williams gear combination, so that the total pinion angle of contact, *l*, or action, is, in the involute gear combination, only about 85 per cent. of its value in the case of the Williams gears. Furthermore, the delayed commencement of contact in the involute combination limits the tooth profile area over which contact occurs to approximately 75 per cent. of that so utilized in the more efficient construction.

With an arc of contact, or action, 17½ per cent. greater and 50 per cent. more tooth profile area over which contact between the mating teeth takes place, the Williams gear combination has a wear resisting capacity some 75

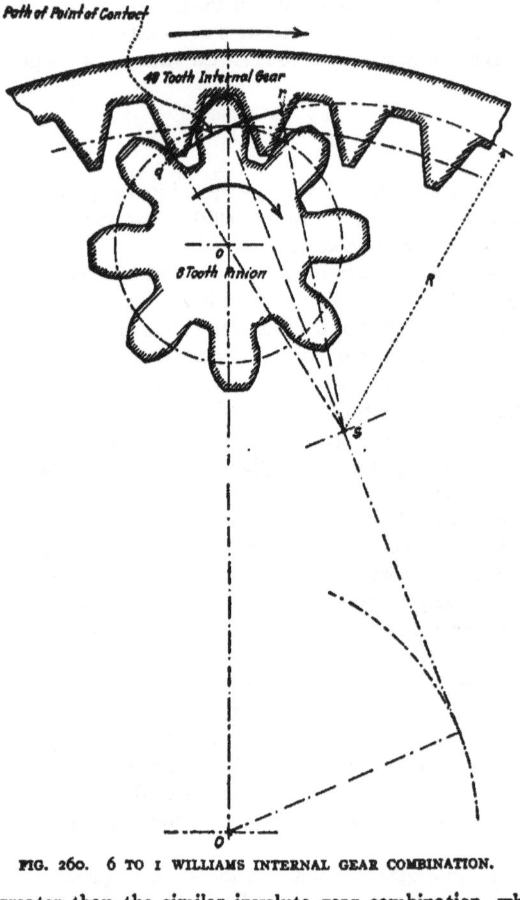

FIG. 260. 6 TO 1 WILLIAMS INTERNAL GEAR COMBINATION.

per cent. greater than the similar involute gear combination, when transmitting the same power at the same speed of rotation—provided, of course, that the respective gears are constructed of materials of similar wear resistance. These comparative values are, naturally, applicable only to the particular gears under consideration and will vary to some extent for other gear combinations, but they are practical.

To secure a utilizaton of the full depth of the involute tooth profiles for contact, in a combination such as that depicted in Figs. 1 and 5, would necessitate increasing the pressure angle to something over 20 degrees—the obliquity of pressure at the instant axis for the gears as illustrated being 14½ degrees—and though such modification would increase the angle of pinion approach to about that for the 14½-degree Williams gears, the angle of

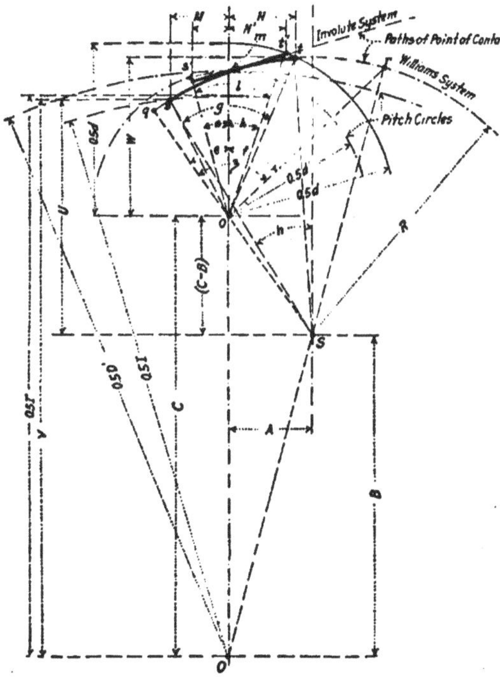

FIG. 261. COMPARISON OF ARCS OF CONTACT.

recession would be quite materially reduced, unless the flank of the gear teeth was considerably increased. Such modifications would be too excessive to be feasible, but if they could be made and were also adopted for the Williams gear combination, the superiority of the latter form of construction in the matter of angle of contact would still be marked, and if the flank of the pinion teeth was also increased—which could be profitably done in the Williams system, but not in the involute—the superiority of the Williams construction in lengthened arc of contact and also increased wear resistance would be substantially the same as in the gear combinations that have been discussed.

REDUCTION IN THE NUMBER OF PINION TEETH

The teeth of the pinion in a Williams combination are proportioned to operate with a particular gear, or, in other words, the pitch and obliquity of the internal gear teeth positively control the profile curvature of the mating pinion teeth. It is obvious, therefore, that pinions with fewer teeth than are

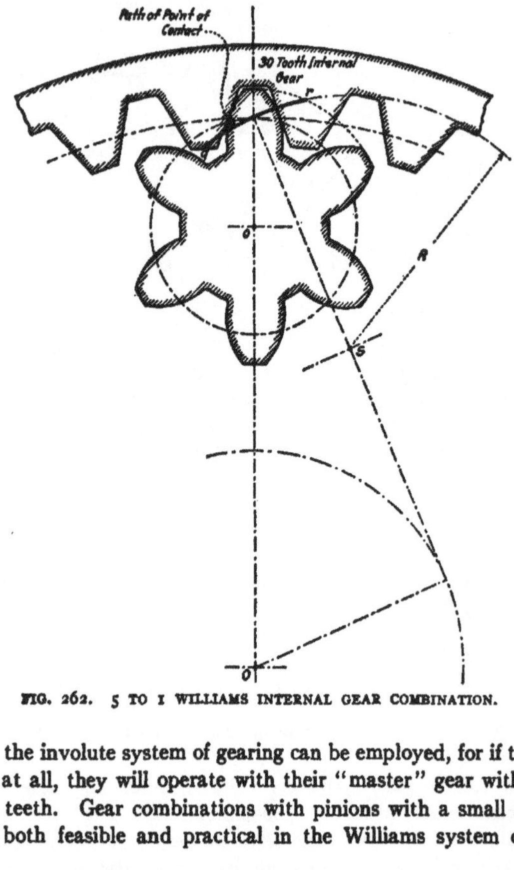

FIG. 262. 5 TO 1 WILLIAMS INTERNAL GEAR COMBINATION.

feasible in the involute system of gearing can be employed, for if they can be generated at all, they will operate with their "master" gear without interference of teeth. Gear combinations with pinions with a small number of teeth are both feasible and practical in the Williams system of internal gearing.

Fig. 262 illustrates a combination of Williams gears—a 5 to 1 ratio—in which the pinion has only six teeth, yet they are well proportioned for strength and free from objectionable undercutting. The obliquity of the gear teeth is marked, it is true, but the reduction in the efficiency of power transmisson is far less than it would be in the case of involute gears with a correspondingly heavy pressure angle, for during the recession period of the engaging teeth—the portion of tooth contact during which the transmission of power from the pinion to the gear is considerably more effective than during the approach to the position of so-called full mesh—the obliquity of the direction of pressure reduces steadily.

The gear combination shown in Fig. 260 illustrates a method of proportioning the pinion and gear teeth by which a pinion with only eight teeth was employed in a 6 to 1 reduction without sacrifice of strength or adopting a tooth of unusual obliquity.

This distinctive feature of ability to generate and use pinions with a small number of teeth is quite obviously an important advantage of the Williams system of internal gearing. Greater speed ratios are obtainable without unduly increasing the diameter of the internal gear, or a coarser pitch may be employed than is feasible with involute gears, when available space for the accommodation of the gearing is limited.

The ability to employ gears with fewer teeth of heavier pitch than can be used in the involute system—practice, always to be recommended—is particularly advantageous in the Williams system on account of its superiority in length of arc of tooth contact and improved operating action.

MACHINING TEETH

The present increasing demand for internal gears of the involute form has been brought about largely by the perfection of efficient and accurate gear cutting or generating machines, machines which automatically modify the involute profile of the gear teeth to decrease in large measure the limitations set by the interference of teeth of true involute profile. In the Williams system of internal gearing, modifications of tooth profiles are unnecessary, for the pinion is developed to operate with a particular mating internal gear and only with such or similar gear. In this manner, interference in gear combinations for which commercial demand exists is virtually eliminated and no provisions have to be made in machining the gears.

The straight tooth profiles of the Williams internal gear make the accurate cutting of such gear teeth a comparatively simple and inexpensive operation. A plain "V" shaped, reciprocating cutting tool will gash out the tooth spaces rapidly and finish accurately both of the bounding tooth profiles, necessitating only the simple adjustment act of indexing the gear blank from tooth space to tooth space. Any ordinary shaper with a suitable indexing mechanism can be employed. If the internal gear is of the ring type, the tooth spaces can be milled with even greater rapidity, the cutter—as in the case of the simple reciprocating tool—being easily ground with the utmost precision as to the accuracy of the cutting edge and its obliquity.

The machining of the pinion teeth is somewhat more complicated, as they either have to be generated or a cutting tool employed, conforming in profile to the outline of the tooth space by being accurately fitted to a generated pinion tooth space template. The generation of the pinion teeth or of the pinion teeth space template requires the use of a simple generating machine in which the simple cutting tool, conforming in shape to the straight profile internal gear tooth, swings about the rotating pinion blank on a radius equal to one-half the pitch diameter of the mating internal gear, so that the ratio of the rotating speeds of the cutter and pinion blank is the same as that of the speed ratio of the gear combination in which the pinion is to be employed. As the pinion of the ordinary internal gear combination has comparatively few teeth their generation is a simple task compared to the task of generating the teeth of the mating gear, but when a large number of pinions are required of a specific size to operate with internal gears of given size and proportions, as in the manufacture of gears for a standardized product produced in quantities, a formed cutter, fitted to a generated pinion tooth space template, will be found to expedite materially the cutting of the pinion teeth.

COSTS OF MACHINING WILLIAMS GEARS

The cost of cutting Williams internal gears and gear combinations is quite obviously very much less than that for machining gears of other tooth form, and is more and more marked as the difference in the number of teeth in the internal gears and their mating pinions increases. If the average speed ratio of internal gear combinations is taken as 4 to 1, the relative time consumed in cutting a gear combination proportioned according to the Williams system, compared to that required to generate similar gears proportioned according to the involute system, is substantially as follows: When generating all the Williams pinion teeth, 50 to 60 per cent., and when cutting the Williams pinion teeth with a formed cutter, 30 to 40 per cent. In addition to this very material saving in the time required to machine the gear teeth, less skilled operators may be employed for the simple operations entailed in cutting Williams gears than can be safely trusted to operate the more intricate machines required for generating the involute type of tooth, so that the total cost of manufacturing Williams internal gears and mating pinions is very substantially less than that of generating similar gears designed according to the involute system of gearing.

DISTINCTIVE ADVANTAGES

The advantages of the internal type of gearing are now quite generally appreciated and conceded, and in every instance the superiority of the Williams system of internal gearing is conspicuous. Briefly summarized, the more evident of the advantages that are possessed to an accentuated degree by the gear combinations proportioned according to this system are:

1. Greater length of arc of tooth contact, due to the distinctive path of point of contact.

2. Reduction in wear, due to the prolonged contact and the greater proportion of the tooth profile areas utilized for contact.

3. Increased strength of pinion teeth, on account of the reduction in undercutting.

4. Improved operating action, due to the slight relief in shock of tooth contact occasioned by the concentration of wear on the flanks of the pinion and internal gear teeth.

5. Reduction in the required number of pinion teeth.

6. Reduction in the diameter of the internal gear for a given speed ratio and load.

7. Greater speed ratios without increase in diameter of internal gear.

8. Possibility of employing a coarser pitch without increasing the diameter of the internal gear.

9. Increased strength of gear combinations.

10. Simplicity of design.

11. Accuracy and ease of gear tooth reproduction.

12. Greatly reduced cost of manufacture.

SECTION XVI

ROLLED GEARING*

The evolution in commercial gear production which entails the simple process of forging teeth on the gear blanks heated to a semiplastic condition, by causing them to roll under heavy pressure in positive synchronized contact with hardened die rolls, is unquestionably the outstanding achievement in modern methods, for not only is a much lower cost of production realized, but the gears produced are stronger, tougher and more accurately formed than the best types of gears produced by machining processes of generation. The process involved is essentially one of molding generation and the gears so forged operate with high degree of pure rolling action, insuring their high mechanical efficiency in service. To appreciate the validity of these claims necessitates a clear understanding of approved gear cutting methods.

Originally the machining of toothed gearing involved the operation of cutting each individual tooth space with the aid of a formed cutter, so that the section of the tooth space was conjugately related to that of the teeth formed on the gear blanks. The next step in the development of modern gear production practice was the evolution of the so-called generation of gear teeth, whereby an attempt was made to duplicate in the production process the action of engaging gears, by causing the cutter and the work to advance as if in mesh while the tooth spaces were machined. The gear shaper, in which the cutter takes the form of a pinion which is made to cross the face of the gear blank rapidly—removing metal on the forward stroke—while the cutter and blank roll together much as if they were pinion and gear, and the gear hobber, in which the teeth of the hob are patterned after the teeth of an involute rack and the gear blank and hob are caused to move together much as a mounted gear would be actuated by an advancing rack with which it was in mesh, illustrate the approach made toward a simplification of process by attempting to employ the same simple rolling action, which is the objective of gearing, in the process of gear fabrication.

WEAKNESSES OF GEAR GENERATION

Although these methods of gear generation exemplify very decided advancement in the art of producing high grade gearing, both economically and expeditiously, they possess certain weaknesses which in the light of more recent developments limit the precision with which the gear teeth may be formed and also entail considerable scrapping of metal. These weaknesses

* Reginald Trautschold, Consulting Engineer, The Anderson Rolled Gear Co.

are inherent respectively to the generation method of machining and to machining operations in general.

The generation method entails two quite unrelated operations, the revolution of the gear blank as if actuated by a pinion conforming in tooth dimensions to the dimensions of the cutting tool and the travel of the cutting tool across the face of the gear blank. These two operations cannot be performed simultaneously without some sacrifice in the accuracy of tooth formation. Since neither the shaping nor the hobbing process can give a continuous amount of metal, else there could be no advancement of the gear blank, a series of more or less pronounced ridges and flats mar the surfaces of "generated" gear teeth.

Another weakness common to all machined, or cut, gearing is the unavoidable removal of metal to form the tooth spaces of the gears. The metal so removed, though of high grade, has no value except a quite nominal one as scrap.

DEVELOPMENT IN GEAR ROLLING

The method of overcoming these drawbacks, or weaknesses, to the generation method of gear cutting, is by rolling the teeth on gear blanks heated to an effective forging temperature by means of a hardened die roll. This process is not a new discovery, but its practical development for commercial gear production is distinctly so. As early as 1872, John Comley attempted to roll teeth on hot gear blanks by a simple knurling process and since that date there have been a number of similar attempts, but none of these experimenters succeeded. In each case, knurling was employed—that is, either the shaft carrying the forming die or that supporting the heated gear blank was positively driven, but not both—with the result that though teeth were produced they were malformed and poorly spaced. The failures were due to unavoidable slippage between the heated gear blanks and the die rolls. Nevertheless the attempts were highly valuable in the light of later development as they demonstrated the entire feasibility of the rolling process. They proved that by rendering gear blanks sufficiently plastic by heat to permit the forging of teeth on their peripheries, a simple process of rolling the heated gear blank with a hardened steel die roll constituted a practical means of gear tooth formation.

It remained for H. N. Anderson, however, to conceive and develop a practical method of accomplishing this result and he did so by the simple expedient of driving positively, through substantial timing gears, both the shaft carrying the heated gear blank and that supporting the die roll. In this manner the slippage between die roll and blank which caused the failures of the earlier attempts to roll gears by a simple knurling process was entirely eliminated and the teeth rolled on the blank accurately spaced and well shaped.

The first gear rolling machine to be constructed on this principle was completed in 1911 for the intended purpose of employing the gear tooth rolling method simply to supplant the tooth gashing operation in cut gear production then in general use. However, the teeth rolled by this pioneer machine were found to be so well and accurately proportioned and formed that any subsequent machining operation on the teeth, such as grinding, promised to be unnecessary. By virtue of the basic molding generation process entailed, it was at once apparent that a gear rolling machine of precision could be built which would produce gearing of extreme tooth accuracy in one operation and that such a machine would make it possible to avoid error or approximation in the form of the teeth produced.

MASTER FORM SYSTEM OF GEARING

Further development entailed—in addition to the refinement of the rolling machine necessary to convert it into a practical tool for the quantity production of gears—the adoption of a simple system of gearing which would meet the requirements of low cost, precision and ease in die-roll design and construction. Naturally, many tooth forms can be employed for the die-roll or master rolling gear, but it is quite essential to adopt a standard tooth section which can be reproduced with the utmost precision and at a minimum expense, if accuracy in the formation of the rolled teeth and low costs of production are to be realized. That is, the die-roll tooth form should be susceptible of accurate and economical reproduction.

The surface which is reproduced with greatest precision and at the same time at minimum expense by mechanical means is the plane. Consequently, the adoption of a straight-sided tooth, the profile surfaces of which are perfectly flat, is the logical selection for die-roll teeth. Such tooth form is also distinctive of the basic involute rack and of the octoid crown gear, so may very properly be termed a "master form." It is eminently suitable, for the die-roll teeth, meeting the requirements for their design and construction, and the teeth rolled by die rolls so proportioned resemble in appearance the familiar involute gear tooth.

The rolled teeth differ slightly in profile curvature from the true involute, it is true, but as the gears of different size rolled by a common die roll have teeth which are conjugately related as to section this is quite immaterial. In fact, the effective length of tooth profile—the limits of the path of tooth contact—is somewhat greater than in the case of the involute system of gearing as adapted in commercial gear production, for a certain relief is provided at top and bottom of teeth by the straight die-roll tooth profile, without sacrificing the duration of tooth contact.

The reason for this relief in the case of spur, helical and herringbone gears—all of which types can be rolled with equal facility—is readily understood, as the die-roll teeth resemble in section the teeth of an involute rack. That is, the thicknesses of the teeth at top and bottom are somewhat greater than if the teeth were of true involute form. In the cases of gears of the bevel form, the same relief is secured by making the die rolls in the form of very flat bevel gears, instead of true crown gears. The straight profile teeth are retained, as if the die rolls were of the true crown gear form, but the center angle of the die roll is made somewhat less than 90 degrees. The effect of this is the same as in the case of die rolls for spur gears, for a similar relationship exists between the straight line profile teeth of the flat bevel die roll and the teeth of a true crown gear as between the straight line profile teeth of the spur die roll and the teeth of an involute rack. In the case of the bevel gears, however, a true crown gear form of die roll can be used by making certain modifications in tooth proportions, as will later be explained, but this is not usually attempted.

FIG. 263. FRONT AND SIDE VIEW OF BEVEL RING GEAR ROLLING MACHINE.

PROCESS OF GEAR ROLLING

The rolling, or forging, machine is shown in Figs. 263 and 264. It consists essentially of two independent shafts—one supporting the gear blank and the other the die roll—driven at synchronized speeds through substantial timing gears, and can best be described by a brief account of the actual process of gear rolling. A gear blank heated to its most suitable forging temperature—ordinarily in the neighborhood of 2,000 or 2,100 degrees Fahrenheit—is mounted on the work shaft and clamped securely in place as shown on the right-hand arbor, Fig. 265. The water-cooled die, carried on the other functioning shaft, on starting the machine is advanced by a smooth positive loam action and gradually forced—under a pressure of 10 to 20 tons—

into engagement with the semi-plastic gear blank, progressively displacing the metal and gradually building up the molded gear teeth. The work and die are kept in this rotational contact until the teeth are fully formed,

FIG. 264. REAR AND SIDE VIEW OF BEVEL RING GEAR ROLLING MACHINE.

the die roll meanwhile advancing to full mesh. The heated gear blank has in the meantime cooled to below its critical temperature, so that the formation of forging scale has also ceased. The die roll is then withdrawn and the rolled gear removed to cool.

FIG. 265. DIE BLANK AND TIMING GEARS—BEVEL RING GEAR ROLLING MACHINE.

As the rotary speeds of the die roll and gear blank are positively synchronized at the speeds of their engaging pitch surfaces through the heavy timing gears, there is no driving action between the die roll and gear blank and the teeth of the die roll are constrained to enter the blank on radial lines and in the same relative position on each successive revolution. The advancement of the die roll, with the accompanying displacement of gear blank metal, is slight per revolution, so that teeth on the gear blank are molded gradually and without strain. Throughout the process of rolling, the temperature of the die roll does not rise above that which is bearable to the hand, being kept cool by a stream of water directed against its face at the point farthest from that of its contact with the heated blank.

When gears of extreme accuracy of tooth structure are required, the work and die-roll shafts are made to reverse their direction of rotation frequently during the greater part of the process of building up the teeth.

FIG. 266. FIG. 267.

FIG. 266. BEVEL RING GEAR AND PINION BLANKS.
FIG. 267. ROLLED BEVEL RING GEAR AND PINION BLANKS.

This reversal is performed at a speed of about 150 r. p. m. and it has the effect of balancing the displacement of metal on either side of the teeth, so forming a perfectly symmetrical tooth structure. The reversal in rotation, automatically performed, has heretofore been accomplished through the agency of powerful friction clutches, but in the latest types of machines special D. C. reversable motors have been substituted with a considerable saving in machine cost and in the over all dimensions of the rolling machine.

ELIMINATION OF FORGING SCALE

During the rolling operation, the stream of cooling water directed against the die roll serves to wash free any forging scale which may tend to cling to the die-roll teeth and is also instrumental in ridding the gear blank of the scale as rapidly as it is formed. The cool, wetted die-roll teeth coming in contact with the hot gear blank accentuates its rate of shrinkage, loosening the forging scale as it forms. The speed of blank rotation then throws the scale free of the machine, leaving the surfaces of the rolled teeth entirely free of clinging scale. As rolling process is continued until the gear blank cools below its critical temperature, the forged teeth are highly polished, free from any imbedded scale or other blemish when the formed blanks are removed from the rolling machine.

FINISHING ROLLED GEARS

In fact, the teeth of the gears are finished with the rolling operation and nothing remains to be done other than to bore and face the hubs and to dress the shrouding formed by the molding of the gear teeth. For these finishing operations, the rolled gear blanks are chucked on the accurately rolled pitch surfaces of the gears—to assure perfect centering—and the machining performed on suitable automatic machines of standard type.

FIG. 268 FIG. 270.

FIG. 268. FINISHED BEVEL RING GEAR AND PINION.
FIG. 270. HERRINGBONE BEVEL GEAR AND PINION.

The shrouding tying the formed teeth and gear body, or web, into an integral unit is due to the use of die teeth somewhat shorter than the face width of the gear blanks and may be entirely cut away, producing gears of the usual form, or may be retained in whole or in part to add strength to the gear. In the case of a pair of meshing gears, a full shroud or a part shroud may be retained on the member which is customarily the weaker of the two, making it the equal or superior of the other in strength. Or, shrouds on both members can be so proportioned as to develop the maximum or the most effective strength of the gear combination.

INHERENT STRENGTH SUPERIORITY

The heavy pressure under which the semi-plastic blank metal is worked into teeth during the rolling process is also instrumental in increasing the strength of rolled gearing by bringing about a marked rearrangment and modification of the metal structure. The plastic metal during the gradual molding of the teeth is subjected to a thorough and powerful kneading in process which breaks up the more or less heterogeneous crystalline structure of the blank into one of much finer and uniform crystals arranged in a trussed formation of almost fibrous characteristic. This distinctive rearrangement of metal structure serves to tie the teeth to the body of the gear blank in a way which adds greatly to the strength of the teeth and serves also to equalize any warpage effects which might otherwise develop in subsequent heat treatment. The density of the tooth structure is increased, developing, as well, teeth of superior toughness.

Exacting laboratory tests, which have been duplicated or bettered under working conditions, have shown that the average superiority of rolled gear teeth—comparing them with high grade cut teeth from similar metal blanks—is a gain of 25 per cent. in strength and one of about 20 per cent. in hardness. In operation, this superiority means greatly increased wear resisting qualities and a capacity to withstand successfully heavy and sudden overloads.

PRODUCTION ECONOMIES

As the process of gear rolling is essentially one for large quantity production, the resulting economies are probably of even greater importance than such vital considerations as accuracy of tooth formation, resistance to wear, hardness, toughness and strength of gear teeth, and in this connection the rolling process shows a number of marked advantages. Important savings are realized by the process in material, labor and equipment costs and also by the much smaller space required for the accommodation of the plant needed for a given production output of gears.

The teeth being formed on the gear blanks by a molding process, rather than by the removal of any metal, as in all gear cutting processes, the gear blanks can be so proportioned that only a minimum amount of metal is trimmed away in finishing the gear hubs and the ends of the gear teeth. This enables a saving to be made of from 20 to 40 per cent. in the weight of the rough blank.

The crew for operating a rolling machine and its supply furnace for heating the gear blanks consists of only two men, an operator and a helper, who under intelligent direction from a competent foreman may be recruited from the class of intelligent laborers. Such a crew can easily attain and maintain an average hourly output of 90 rolled gears per machine. Compared to the labor expense in simply cutting finished teeth by the most economical process of machining, at a rate of 90 gears per hour, the saving by the rolling method amounts to between 90 and 97 per cent. Finishing up the gears somewhat cuts down the superiority of the rolling method, but the saving is still very marked, being from 25 to 80 per cent., or even more, depending upon the size and type of gearing produced.

The big output of a gear rolling machine—a 10-inch gear with 2-inch face consuming only a trifle over 14 seconds in the rolling operation—results in considerably fewer of them being required for a given rate of production than the number of gear cutting machines needed for the same gear production. Consequently, a substantial saving is effected in necessary equipment investment and a considerable saving in floor space. As for the investment saving, the case of a shop having a daily production capacity of 700 ring bevel gears, 700 bevel pinions, 700 miter bevel gears and 1,400 spur gears will prove typical.

The gear rolling machines and heating furnaces for such an output would cost 75 per cent. less than the equivalent equipment in generating machines and machine tools. In the finishing up operations—the boring and facing

FIG. 269. CROSS-SECTION OF ROLLED GEAR TOOTH SHOWING STRUCTURE.

of hubs, etc.—the additional equipment cost favors the machining process to some extent, but in the over all equipment cost, that for all machines entailed, the rolling process is the more economical by some 55 or 60 per cent. This comparison of equipment costs is based upon established costs of gear rolling machines employing friction clutch reversing, or oscillating, mechanism. The latest type of rolling machines employing direct connected reversing motors in place of the friction clutches will be very much less costly—a saving of 30 or 40 per cent. being realized by the substitution—making possible a substantially greater saving in necessary equipment investment.

In addition to the foregoing distinctive economies of the rolling process, a substantial saving is realized in the space of floor area required for the gear rolling equipment, but even disregarding this latter item, which can quite evidently be very important, the other savings realized through rolling instead of machining gears are so substantial as to be startling. Quite aside from the fact that hot rolled gearing can be produced which is far superior to the best quality of generated gearing in accuracy of tooth formation, finish and strength of teeth and in wearing qualities, production of gears by this method shows a saving of from 20 to 40 per cent. in material, anywhere from 25 to 80 per cent. in labor expense and from 55 to 60 per cent. in the fixed charges represented by investment outlay.

DESIGN OF ROLLED GEARING

Great as are these economic, physical and mechanical superiorities of rolled gearing, still other advantages distinguish the process which enhance its value in quantity gear production and substantially increase the scope of commercial gearing. The simple straight line profile teeth of the die-rolls introduce simplification in gear design and make commercially feasible certain types of gears heretofore beyond the scope of practical gear production.

The question of gear design in rolled gearing resolves itself into a relatively simple matter of die-roll construction and proportions, based upon a few simple formulas and rules, applicable to spur and bevel gearing generally, whether the gear teeth are of the straight, helical or herringbone form. These formulas are appended and a brief explanation of the three general forms of die rolls—two for bevel gears and one for spur gears—will make plain not only the steps in the design of die rolls but the enlarged scope or field in gear mechanisms opened up by the rolling process.

BEVEL GEAR DIE ROLLS

As previously stated, there are two varieties of die rolls for producing bevel gears, the flat bevel die roll and the crown die roll. The former is nearly invariably employed for rolling bevel gears of so-called standard pitches. The reason for this is that the radius of the crown die roll is governed by the cone distance of the gear to be rolled, when the gear blank is at its critical temperature, and this distance is rarely an even multiple of the circular pitch. Consequently, it becomes necessary to decrease the center angle of the die roll until its cone distance (effective) equals that of the rolled gear blank at its critical temperature, thus providing for a full complement of whole teeth for the die roll. The circumference of a circle having a radius equal to the cone distance of the gear to be rolled divided by the circular pitch of the gear gives the number of teeth for a crown die roll. In the case of a bevel die roll, the quotient is the number of die-roll teeth plus some fraction. To ascertain the pitch diameter of the flat bevel die roll, the fraction is dropped and the product of the whole number multiplied by the coefficient of expansion is divided by the diametral pitch. The center angle of the flat bevel die roll is then readily obtained, the necessary offset to the die roll shaft etc. (See formulas for bevel gear die rolls.)

The use of the flat bevel die roll is necessary in order to produce a rolled gear of established pitch and consequently of fixed diameter, but it is quite conceivable that many occasions might arise where gear proportions slightly smaller or larger might prove very much more desirable than proportions established definitely by a standard of pitch. That is, gearing controlled by considerations of available space or by diameter offers a wider scope than a system in which the diameters of the gears are fixed by the pitch. In any production process, this would entail some modification in

FIG. 271. MISCELLANEOUS GEARS MADE BY THE ROLLING PROCESS.

pitch, but in the gear rolling system this is accomplished with the utmost precision by the use of a crown die roll having an effective diameter equal to double the cone distance of the gearing required when at its critical temperature. In the case of any gear cutting process, on the other hand, it would entail the use of special and costly cutting tools which could not be ground or reproduced with the same precision. The greater strength of rolled gearing, furthermore, affords somewhat greater leeway in the matter of pitch than may be taken with cut gearing.

The redressing, or grinding, of bevel gear die rolls—an operation which may be necessary after rolling from 500 to 1,000 gears—is a simple proposition, as all proportions of die-roll teeth are kept constant, but in the case of die rolls for spur gears the situation is somewhat different. The spur die roll is in the form of a gear with straight line profile teeth, which if redressed will reduce to some extent the diameter of the die roll and consequently the circular pitch of its teeth. Although this is true, it is possible to redress such die rolls until there is quite a measurable reduction in their pitch diameters without destroying the accuracy and smooth running qualities of the gears they can roll.

This peculiarity is made possible by the positive driving of the two functioning shafts through heavy timing gears. As these timing gears control the angular advance of the die rolls and of the gear blanks, the effect of a

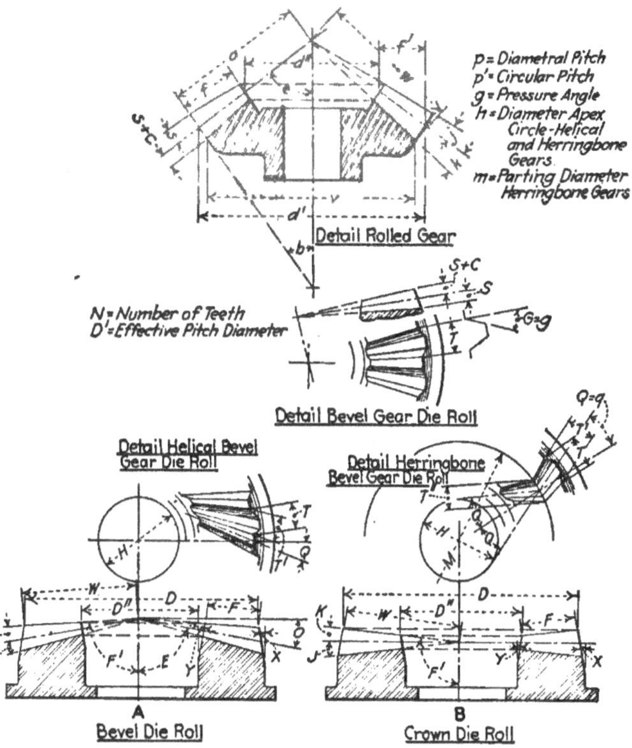

FIG. 272.

variation in the respective diameters of die and blank—within reasonable limits—is the introduction of a slight creep between the die and blank as they are brought into and run in synchronized contact, the width of the tooth spaces of the die-roll and not the thickness of the die-roll teeth governing the thickness of the teeth rolled on the plastic gear blanks. Consequently, as the radial spacing of the formed teeth is controlled and kept constant by the timing gears, it is essential that the die tooth spaces be maintained constant. If this is done, a slight reduction in the thickness of the die-roll teeth, due to the reduction in die roll diameter through tooth redressing, will not appreciably modify the profile curvature of the rolled teeth and will not affect the accuracy of tooth spacing on the rolled gear. By maintaining constant the tooth spaces of spur gear die rolls, they can be redressed several times before they should be discarded, making the life of such dies measurable by a production of several thousand gears.

HELICAL AND HERRINGBONE DIE ROLLS

The simple straight line profile tooth adopted for die rolls being equally suitable for axial and helical, or spiral, arrangment of teeth, die rolls for helical gears, both of the spur and bevel variety, can be made and kept in

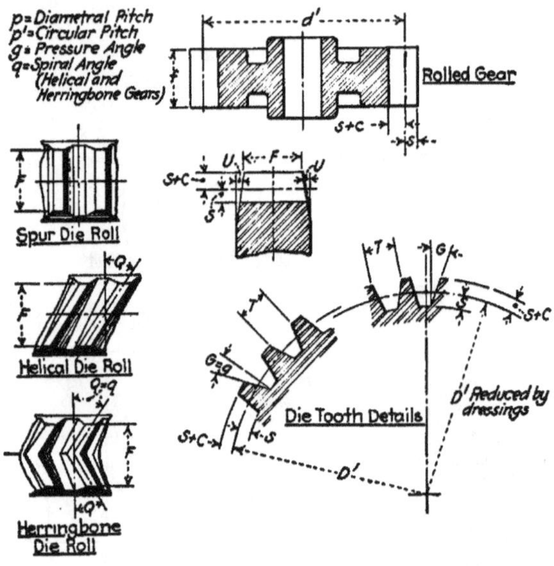

FIG. 273.

condition with almost the same ease as similar die rolls with straight axial teeth. This enables helical gearing to be rolled as cheaply as that with ordinary gear teeth. By making the die rolls in concentric sections, separable for tooth dressing, but otherwise keyed into an integral unit, herringbone varieties of either spur or bevel gearing with rigid binding tooth prows can be rolled just as cheaply. Herringbone bevel gears, which exemplify the scope of rolled gearing, cannot be produced by any practical process of gear cutting. They will doubtless prove of even greater commercial value than the herringbone gear type of spur gear, as they will eliminate the objectionable side thrust in most installations of bevel gearing.

DIE ROLL FORMULAS AND DESIGN

NOTATION BEVEL GEARING

DIE ROLL

Z = Coefficient of expansion
D' = Effective pitch diameter
N = Number of teeth
E = Center angle
F = Face
F' = Face angle
K = Addendum angle
J = Dedendum angle
D = Outer diameter
D'' = Inner angle
S = Dedendum
C = Clearance
G = Pressure angle
X = Bevel angle (outer)
Y = Bevel angle (inner)
T = Width of tooth space
T' = Normal width of tooth space
W = Face cone distance
Q = Spiral angle
H = Diameter of apex circle
M = Parting diameter (herringbone gears)
T'' = Normal width of tooth space (parting circle)
O = Offset angle

ROLLED GEAR

p = Diametral pitch
p' = Circular pitch
d' = Pitch diameter (outer)
d'' = Pitch diameter (inner)
f = Face
f' = Profected face
s = Addendum
c = Clearance
e = Center angle
b = Back angle
j = Addendum angle
k = Dedendum angle
g = Pressure angle
a = Pitch cone distance
v = Root diameter
h = Diameter apex angle
w = Root cone distance
q = Spiral angle
m = Parting diameter

FORMULAS FOR BEVEL GEAR DIE ROLLS

$$Z = (1 + \text{temperature range})\ 0.00000672$$

BEVEL DIE ROLLS

$$N + \text{fraction} = \frac{6.2832a}{p'}$$

(N = largest whole number)

$$D' = \frac{ZN}{p}$$

$$Sin.\ E = \frac{0.5N}{ap}$$

$$O = 90^\circ - E$$

$$F' = E + K$$

$(K = k)$

$$v = d' - 2(s + c)\ sin.\ b$$

$$w = \frac{0.5\,v}{Sin.\ (e - k)}$$

$$W = Zw + 0.0625$$

$$D = 2Z\ cos.\ (90^\circ - F')$$

$$D'' = D - 2F\ cos.\ (90^\circ - F')$$

CROWN DIE ROLLS

$$N = \frac{6.2832a}{p'}$$

$$D' = 6.2832aZ$$

$$D = D' + 0.125$$

$$E = 90\text{ deg.}$$

$$O = 0\text{ deg.}$$

$$F' = 90^\circ + K$$

$(K = k)$

$$W = \frac{0.5D}{Cos.\ K}$$

$$D'' = D - 2F\ cos.\ K$$

ALL BEVEL GEAR DIE ROLLS

$$S + C = \frac{D(s + c)}{d'} \qquad S = \frac{Ds}{d'} \qquad F = Zf + 0.125$$

$$T = 2W\ sin.\ \frac{180}{N}$$

X and Y are arbitrary within limits

HELICAL BEVEL GEAR DIE ROLLS

$$h = d'\ sin.\ q \qquad H = Zh \qquad T' = T\ cos.\ Q$$

HERRINGBONE BEVEL GEAR DIE ROLLS

$$M = Z\sqrt{(d')^2 - (d' + d'')f'}$$

$H = Zh$ $\qquad$ $T' = T \cos. Q$ $\qquad$ $T'' = \dfrac{T'M}{D}$

NOTATION FOR SPUR GEAR DIE ROLLS

DIE ROLL

Z = Coefficient of expansion
D' = Effective pitch diameter
N = Number of teeth
F = Face
S = Dedendum
C = Clearance
T = Width of tooth space
U = Bevel angle
Q = Spiral angle
T'' = Normal width of tooth space

ROLLED GEAR

p = Diametral pitch
p' = Circular pitch
d' = Pitch diameter
f = Face
s = Addendum
c = Clearance
g = Pressure angle
i = Advance of gear
q = Spiral angle

FORMULAS FOR SPUR GEAR DIE ROLLS

$$Z = (1 + \text{temperature range})\ 0.00000672$$

$N = 24p + 1$ $\qquad$ $D' = \dfrac{ZN}{p}$ $\qquad$ $F = Zf + 0.125$

$S = 0.3183p'Z$ $\qquad$ $S + C = 0.3683p'Z$

$i = 1.1p'$ (Helical and herringbone gears)

$Tan.\ G = \dfrac{i}{f}$ (Helical gears) $\qquad$ $Tan.\ G = \dfrac{2i}{f}$ (Herringbone gears)

U is arbitrary within limits

INDEX

F

G

H

I

www.ingramcontent.com/pod-product-compliance
Lightning Source LLC
Chambersburg PA
CBHW021548210326
41599CB00010B/361